ADVANCES IN ELECTROMAGNETIC FIELDS IN LIVING SYSTEMS

HEALTH EFFECTS OF CELL PHONE RADIATION

Volume 5

Advances in Electromagnetic Fields in Living Systems

Health Effects of Cell Phone Radiation

Volume 5

Edited by

James C. Lin

University of Illinois at Chicago,
Illinois, USA

James C. Lin
University of Illinois
Chicago, IL 60607-7053
USA

ISBN 978-0-387-92733-6 e-ISBN 978-0-387-92736-7
DOI 10.1007/978-0-387-92736-7
Springer Dordrecht Heidelberg London New York

Library of Congress Control Number: 2009929205

Printed on acid-free paper

Springer is part of Springer Science+Business Media (www.springer.com)

Preface to Volume 5

While the objective of the fifth volume in the series on *Advances in Electromagnetic Fields in Living Systems* remains the same as previous volumes, the editorial approach for this volume is to focus on a portion of the nonionizing electromagnetic radiation spectrum and a particular set of applications, namely, radio frequency (RF) and microwave and their use in cellular mobile communication devices and systems. This emphasis has prompted the insertion of the subtitle for this volume: *Health Effects of Cell Phone Radiation.* It is recommended that readers who desire a more fundamental understanding of RF electromagnetic interaction with biological systems examine chapters on related subjects in previous volumes of this series.

The popularity and rapid deployment of wireless communication technology has led to increasing numbers of new devices and systems that emit RF electromagnetic energy. It has resulted in large numbers of individuals at the workplace or in the general public being exposed to RF fields. In most cases, the RF sources are in close proximity to the human body. The increased exposures at the workplace or in daily life have prompted the need for further research to evaluate RF safety and health implications. It is estimated that more than 3.5 billion people have access to cellular mobile telephones – nearly half of the world population, at present. Indeed, at the current rate of growth more of the world's population will have access to mobile phone services than to electricity. However, exposure to RF electromagnetic fields is not limited to mobile or wireless communication; widespread applications of RF and microwave energy are found in RF article identification and surveillance, inductive heating devices and appliances, adaptive vehicular cruise control, advanced magnetic resonance imaging, on-body biomedical sensing and interrogation, novel active and passive security and detection technology, and proposed digital living network applications. Given the technological, regulatory, and marketing challenges, the timing of the introduction or deployment of many new applications is somewhat uncertain. It should be noted that the experience of the cellular mobile telephone industry indicates that once new technology is deployed, the adoption rate can easily explode. While it takes advanced technology to develop a product, the availability of low-price, high-quality, and high-performance components from around the world brings down the cost of a new product through large-scale production. Without a doubt, the total level of human exposure will rise because of the superposition of new and existing sources. There is a real need for reliable scientific answers on health effects associated with exposures resulting from widespread use of RF electromagnetic fields in new and existing devices and systems.

The biological effects and health implications of RF and microwave radiation associated with cellular mobile telephones and related wireless systems and devices have become a focus of international scientific interest and world-wide public concern, and show no sign of relenting soon. Although our knowledge regarding the biological effects of RF and microwave radiation has increased considerably, the scientific evidence on health effects of RF and microwave radiation associated with these wireless devices is still tentative. The uncertainties persist, in part, because of the limited number and scope of studies that have been conducted to date. Aside from the lack of a scientific consensus on experimental studies that provide clear evidence either refuting or supporting a health effect, there is also uncertainty in epidemiological studies on the cancer induction or promotion potential of RF radiation from cell or mobile phones. One concern has been that an established effect from wireless radiation, even small, could have a considerable impact in terms of public health. Chapters in this volume provide an updated account of recent research results on the potential health risks and discuss the biological effects of RF and microwave radiation from cellular mobile and wireless personal communication devices and systems.

The line-up of articles for this volume is organized along the hierarchical chain of cells, animals, and humans, and concludes with a chapter on characterization of the physical interactions and consideration of guidelines to ensure safe human exposure to RF and microwave fields employed for wireless communications. Specifically, this volume begins with a chapter summarizing the cellular effects of RF fields induced by the use of cell phones and their base stations. Studies on the effects of RF fields on cells in vitro are classified into two main categories: (1) genotoxic and (2) nongenotoxic effects. The genotoxic effects include DNA strand breaks, micronucleus formation, mutation, and chromosomal aberration, that is, changes involving damage to DNA. The nongenotoxic effects described include changes in cellular function, such as cell proliferation, cellular signal transduction, and gene expression (mRNA and protein).

In common usages, the source of RF radiation from cell phones is located in close proximity to the human head or body. Thus, a particular area of interest is tumorigenesis in the brain – tumors that start in the brain including the malignant astrocytoma and glioblastoma multiforme. The second chapter provides an assessment of recent research results on the carcinogenic potential of RF radiation from cellular mobile and personal communication devices. Specifically, the topics included are experimental studies involving brain and other cancer induction and promotion, and long-term survival of laboratory mice and rats exposed to various types of cell phone RF fields.

The observational and laboratory studies conducted on humans form the subjects of the four chapters that follow. The designs and results of published epidemiological studies on users exposed to cell phone-emitted RF radiation and risk of cancer are described in Chap. 3, which also discusses the overall body of evidence regarding a potential association. It starts with a description of published studies of risk of intracranial tumors (glioma, meningioma, acoustic neuroma), and then proceeds to studies of other types of neoplasm.

The next chapter reviews the current research designed to examine some of the possible interactions of cell phone electromagnetic fields with human cognitive behavior. Cognition is a complex topic that involves neurophysiology and its effects on behavior such as responsiveness or decrease in choice reaction time of human volunteers. This chapter considers the question whether there is a plausible link between physiological effects and cognitive changes. The published literature suggests that the research is of variable quality and the results are inconsistent, although there is a trend toward improved quality with better experimental design and careful execution.

This is followed by a chapter that summarizes hypersensitivity reported to be caused by exposure to RF electromagnetic fields emitted from cellular telephones and cell phone base stations. A particularly vexing challenge in studying this phenomenon is that the symptoms reported by electromagnetically hypersensitive individuals, such as headache and fatigue, are common and nonspecific: they may have many causes. The published laboratory research such as provocation experiments, to date, on electromagnetic hypersensitivity and subjective symptoms from exposures to cell phone fields are very limited. Nevertheless, the evidence now available suggests that while the reported hypersensitivity and subjective symptoms may be real, the question as to whether they are associated with cell phone use must await more comprehensive studies.

The situation concerning occupationally exposed populations is the last of the four chapters on human subjects. The protection of workers exposed to RF energy radiating sources has begun to attract global attention. Assessment of the risk and protection afforded workers from exposure to different sources may soon become an issue. This paper highlights occupational exposures in connection with wireless communication – handheld phones and base station antennas. It is worthy of note that occupational exposures, where there is a possibility of exceeding international guidelines, occur only in work environments close to mobile phone base stations.

An important task in assessing the health and safety of RF exposure from wireless communication devices and systems is the determination of electromagnetic fields and absorbed energy in biological tissues. This volume concludes with a final chapter devoted to a comprehensive summary of the dosimetric investigations and the well known biological effect resulting from either partial-body or whole-body exposures, SAR-induced temperature rises in humans. The descriptions include SAR distributions and peak temperature elevations, their derivation and computation, and implications for guidelines designed to limit human exposure in the wireless communication frequency band. It is hoped that they will serve as a common ground for a better understanding of human exposure to the cellular mobile telephone radiations.

In closing, I wish to pay special tribute to the authors for their tremendous contributions and to the anonymous reviewers for their exceptional advice, which has been an enormous aid in finalizing chapters in this volume.

James C. Lin
Chicago

Contents

Contributors

Paolo Bernardi
Department of Electronic Engineering, Università di Roma "La Sapienza", Roma, Italy

Marta Cavagnaro
Department of Electronic Engineering, Università di Roma "La Sapienza", Roma, Italy

Norbert Leitgeb
Institute of Health Care Engineering, Graz University of Technology, Graz, Austria

James C. Lin
Department of Electrical and Computer Engineering, and Department of Bioengineering, University of Illinois at Chicago, Chicago, IL, USA

Kjell Hansson Mild
Department of Radiation Physics, Umeå University, Umeå, Sweden

Junji Miyakoshi
Department of Radiological Life Sciences, Graduate School of Health Sciences, Hirosaki University, Hirosaki, Japan

Stefano Pisa
Department of Electronic Engineering, Università di Roma "La Sapienza", Via Eudossiana 18, 00184 Roma, Italy

Emanuel Piuzzi
Department of Electronic Engineering, Università di Roma "La Sapienza", Roma, Italy

Alan W. Preece
Bristol Oncology Centre, Bristol, UK

Minouk J. Schoemaker
Section of Epidemiology, Institute of Cancer Research, Sutton, Surrey, UK

Anthony J. Swerdlow
Section of Epidemiology, Institute of Cancer Research, Sutton, Surrey, UK

Jonna Wilén
Department of Radiation Physics, Umeå University, Umeå, Sweden

Chapter 1

Cellular Biology Aspects of Mobile Phone Radiation

Junji Miyakoshi

ABSTRACT

This chapter provides a summary of the cellular effects of radiofrequency (RF) fields generated by the increased use of cell phones and their base stations. In vitro studies of the effects of RF fields can mainly be classified into studies of genotoxic and nongenotoxic effects. Genotoxic effects include DNA strand breaks, micronucleus formation, mutation, and chromosomal aberration; i.e., changes involving damage to DNA. Nongenotoxic effects refer to changes in cellular function, including cell proliferation, cellular signal transduction, and gene expression (mRNA and protein). In general, currently available reports suggest that (1) RF energy does not cleave intracellular DNA directly, since most genotoxicity studies have shown negative effects. Cells may be damaged at extremely high SARs, mainly due to the thermal effect of RF fields; (2) some interesting cellular responses associated with stress proteins; i.e., heat-shock protein production and phosphorylation are induced by RF field. However, the results are inconsistent, perhaps due to differences in cell lines, RF exposure conditions, and exposure devices – the reproduction of results in different laboratories would be important; and (3) Microarray analysis has not provided definite evidence of an effect of RF exposure on cellular functions, including

J. Miyakoshi Department of Radiological Life Sciences, Graduate School of Health Sciences, Hirosaki University, 66-1 Hon-cho, Hirosaki, 036-8564, Japan, e-mail: miyakosh@cc.hirosaki-u.ac.jp

J.C. Lin (ed.), *Advances in Electromagnetic Fields in Living Systems, Health Effects of Cell Phone Radiation, Volume 5*,
DOI 10.1007/978-0-387-92736-7_1,

apoptosis, the immune system, and ROS production. Thus the current published evidence does not allow a definite conclusion regarding the effects at a cellular level. Studies on RF effects are ongoing worldwide. The rapid development of biotechnology has increased the potential for detection of microresponses in cells and genes, and future studies of RF effects should be performed using improved biotechnological methods.

1. INTRODUCTION

Cellular studies of the effects of high frequency electromagnetic fields have been conducted more often than epidemiological and animal studies. In particular, research on hyperthermia has been performed to elucidate the effects of these fields on human cancer therapy. This chapter provides a summary of the cellular effects of radiofrequency (RF) fields generated by the increased use of cell phones and their base stations. In vitro studies of the effects of RF fields can mainly be classified into studies of (1) genotoxic and (2) nongenotoxic effects. Genotoxic effects include DNA strand breaks, micronucleus formation, mutation, and chromosomal aberration; i.e., changes involving damage to DNA. Nongenotoxic effects refer to changes in cellular function, including cell proliferation, cellular signal transduction, and gene expression (mRNA and protein) (Table 1). In vitro studies for RF fields examining on these criteria are summarized below. In addition, several reviews of in vitro studies of RF fields have been published (Meltz, 2003; Vijayalaxmi and Obe, 2004; Verschaeva, 2005) and referral to these reports is recommended.

2. GENOTOXIC EFFECTS

Direct and indirect effects of external factors on intracellular DNA are studied and such effects are referred to as "genotoxic effects." Given contemporary emphasis on genes, searching for effects of RF exposure on genotoxicity is an active area of research. Genes are the coding, the program, for the life of cells, and if an environmental

Table 1. Cellular and molecular phenomena for the evaluation of RF effects

Genotoxic effects	Non-genotoxic effects
• Chromosomal aberration	• Proliferation
• Chromatid aberration	• Cell cycle (distribution)
• DNA strand break	• DNA synthesis
• Micronucleus formation	• Gene expression
• Mutation	• Signal transduction
• Others	• Ion channel
	• Transformation
	• Apoptosis
	• Transcriptomics
	• Immune system
	• Others

stimulus were to affect the code, it would have obvious adverse implications. The genotoxic effects routinely assessed are (1) chromosomal aberration, (2) DNA strand breaks, (3) micronucleus formation, and (4) mutation.

Typically the signals used in recent experiments are based on one of the many signal patterns used in mobile telephony. Given the rapidly developing technology and expanding market penetration of mobile telephony, many different signal modulation schemes, such as Code Division Multiple Access (CDMA), Global System for Mobile Communication (GSM), Time Division Multiple Access (TDMA), etc. are being used. Biologists typically base their independent variables on these schemes. They also try to cover a range of specific absorption rate (SAR) values, with the upper limit being exposures producing over hyperthermia.

2.1. Chromosomal Aberration

Chromosomal aberration is induced directly by DNA damage and also other factors in condensed chromosomes in mitotic phase. For example, it is well known that irradiation of ionizing radiation to cells breaks DNA strands and induces chromosomal aberration. In cultured cells, chromosomal aberration occurs spontaneously and extremely infrequently.

To observe chromosomal aberration, cell division is arrested using colcemid when chromosomes are condensed in metaphase. Next, cells are suspended in hypotonic solution and centrifuged, then, soaked in a fixing solution. Then, cells are plated on a slide glass and stained using Giemsa stain and observed using microscope. Various types of chromosomal aberration are observed, i.e., severe ones such as chromosomal break, ring, dicentric chromosome, large fragment, rearrangement, loss and amplification, and slight ones like gap. However, it should be noted that, when an extremely important gene is included in an abnormal site, the severity of chromosomal aberration is not consistent with the severity of the type of chromosomal aberration (Yaguchi et al., 2000). Typical chromosomal aberrations are shown in Fig. 1.

Chromosomes consist of two chromatids and chromatid-type aberration and sister chromatid exchange (SCE) are observed. In a broad sense, this chromatid-type aberration is included in chromosomal aberration. It is known that SCE occurs infrequently in normal cultured cells. Especially, it is well known that SCE and chromatid-type aberration are very frequently observed in mitomycin-C (MMC)-treated cells. However, it has not been confirmed whether or not SCE alone gives great damage on cells and the severity depends on genes that are involved with SCE (Yaguchi et al., 1999).

Both positive and negative results have been obtained regarding the effect of RF fields on chromosomal aberration. Relatively old studies (up to the 1990s) showed positive results; in contrast, recent studies have been negative.

Some studies have shown that exposure of cells to an RF field produces increased chromosomal aberration. Chinese hamster V79 cells were exposed to 7,700 MHz for 10–60 min at 0.5–30 W/cm^2, and the frequency of chromosomal aberration was investigated. Compared to sham controls, the frequency of chromosomal aberration

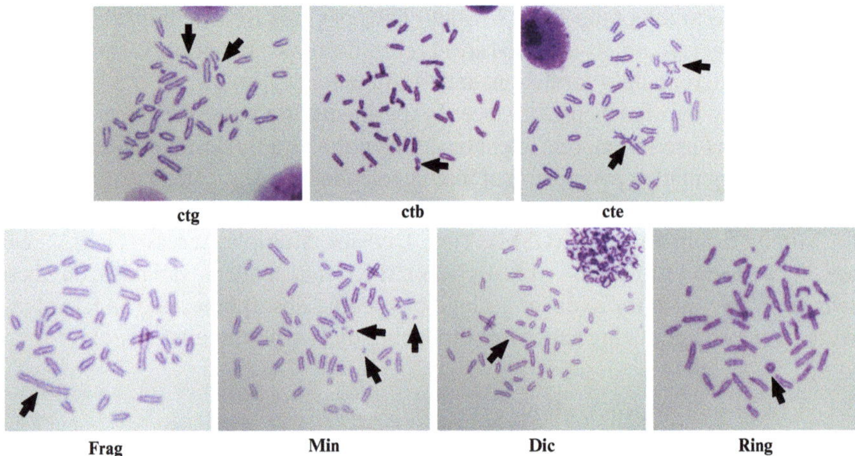

Figure 1. Typical chromosomal and chromatid aberrations in mouse m5S cells. *ctg* Chromatid gap, *ctb* chromatid break, *cte* chromatid exchange, *Frag* fragment, *Min* minute, *Dic* dicentric, *Ring* ring.

was increased with exposure time (Garaj-Vrhovac et al., 1991), which suggested that this exposure condition had considerable toxicity on cells.

Human blood-derived lymphocytes were exposed to 830 MHz (continuous wave) for 72 h at SARs in the range of 0–8.8 W/kg; chromosome 17 aneuploidy and frequency of abnormal DNA replication were determined (Mashevich et al., 2003). Under some circumstances, the temperature in the exposure device could be increased during irradiation; thus, effects of increased temperature also were investigated. No effects were observed at 38.5°C or less. In this study, the temperature was less than 38°C at 8.8 W/kg; therefore, any biological effect observed here was not the result of increased temperature. At SARs of 2.6–8.8 W/kg, chromosome 17 aneuploidy and frequency of asynchronous replication were increased.

Human blood-derived lymphocytes were exposed to 2,450 MHz, with a 50 Hz (1/3 loaded) pulse, for 30 or 120 min at the very high SAR of 75 W/kg, and the frequency of chromosomal aberration was counted (Maes et al., 1993). In the group exposed for 120 min, the frequency of chromosomal aberration was increased; however, no changes were found in SCE and the number of cell divisions. Human blood-derived lymphocytes were exposed to 7,700-MHz electromagnetic radiation for 10–60 min at 0.5–30 mW/cm^2. The frequency of chromosomal aberration was increased in the groups exposed at 10 mW/cm^2 for 30 min and at 30 mW/cm^2 for 10 min or longer (Garaj-Vrhovac et al., 1992).

Several studies indicate that exposure of cells to RF fields did not cause chromosomal aberration or SCE. Vijayalaxmi and colleagues exposed human peripheral blood-derived lymphocytes to 835.62 MHz, continuous wave (frequency division multiple access, FDMA) at either 4.4 or 5.0 W/kg of SAR for 24 h. They investigated mitotic index, chromosomal aberration, percentage of binucleate cells, and formation of micronuclei (MN). In both RF fields, no difference was found from the sham-exposed group (Vijayalaxmi et al., 2001a). They also conducted another study with 847.74 MHz, continuous wave (CDMA) at a SAR of either 4.9 or 5.5 W/kg for 24 h. They measured mitotic index and the frequency of chromosomal aberration, finding no differences from the sham group (Vijayalaxmi et al., 2001b).

Human blood-derived lymphocytes were exposed to 900 MHz (continuous wave, pseudorandom signal, and dummy burst signal) for 2 h at 0–10 W/kg of SAR and the frequency of chromosomal aberration was investigated. In addition, interactive effects with X-rays (1 Gy) and mitomycin C (MMC, at 0.1 μg/ml) were also examined. With either continuous wave or pseudo-random signals, RF exposure alone induced no changes in the frequency of chromosomal aberration, and interactive effects with X-rays were not found. As for interactive effects with MMC, in the 2 W/kg group, half of cells showed significant differences but the other half did not (Maes et al., 2001). Chinese hamster ovary (CHO)-K1 cells were exposed to pulsed 2,450 MHz for 2 h at a SAR of 33.8 W/kg; cells were RF exposed and treated concurrently with MMC or adriamycin to investigate effects of combined exposure. RF exposure alone and with MMC had no effect. With RF exposure combined with adriamycin, the number of changed cells per 100 cells was than that of the control at 37°C. However, it was found that this increase was not induced by RF field but by increased temperature which was suggested by results of the temperature-controlled group. There were no differences in mitotic index between the RF-exposed and temperature-controlled groups (Kerbacher et al., 1990). Human blood-derived lymphocytes were exposed to 2,450-MHz RF field (continuous wave) for 90 min at a SAR of 12.46 W/kg and the frequency of chromosomal aberration and mitotic index (MI) were investigated. No changes were found in the frequency of chromosomal aberration or of MN formation, and MI was not affected (Vijayalaxmi et al., 1997).

Human blood-derived lymphocytes were exposed to 935.2 MHz (GSM modulation) for 2 h at SARs of 0.3–0.4 W/kg. No differences were found in the frequency of chromosomal aberration (Maes et al., 1997). They also conducted another study in which human lymphocytes were exposed to 954 MHz (GSM modulation) for 2 h at 1.5 W/kg of SAR; cells were treated with MMC after exposure. SCE and the number of cell divisions were counted. RF treatment alone had no effect on SCE. After RF exposure, cells were treated with MMC. SCE was increased slightly; However, the number of cell division was not changed (Maes et al., 1996).

In a recent study, human blood-derived lymphocytes were exposed to 900 MHz (GSM modulation) for 2 h at 0.3 or 1.0 W/kg of SAR; MI and frequency of SCE were determined. In both 0.3 and 1 W/kg groups, none of the dependent variables were changed (Zeni et al., 2005). Peripheral blood samples collected from healthy human volunteers were exposed in vitro to 2,450 MHz (2.13 W/kg) or 8,200 MHz (20.71 W/kg) pulsed-wave RF-field radiation for 2 h. Cultured lymphocytes were examined to determine the extent of cytogenetic damage assessed from the incidence of chromosomal aberrations and micronuclei. Under the conditions used to perform the experiments, the levels of damage in RF-radiation-exposed and sham-exposed lymphocytes were not significantly different (Vijayalaxmi, 2006).

In addition, Komatsubara et al. reported the effects of CW or PW fields at 2,450 MHz (SAR = 5–100 W/kg) in mouse m5S cells. The RF exposure for 2 h, in both RF fields (CW and PW) at average SAR of 100 W/kg, does not induce chromosomal aberrations (Fig. 2) (Komatsubara et al., 2005).

Some studies have reported that RF-field exposure in vitro can result in increased chromosomal aberration. However, most of the studies have shown negative results. It generally is believed that RF fields do not induce chromosomal aberration at SARs which are so low as to be incapable of producing tissue heating.

2.2. DNA Strand Breaks (Comet Assay)

DNA strand breaks are an index to show whether or not DNA strand is directly broken by cell genotoxicity. DNA strand breaks are usually examined using comet assay (Miyakoshi et al., 2000a, 2002). The following is a brief explanation of comet assay:

1. Cells are treated with exposure to the electromagnetic fields or external stimuli including chemical agents and collected.
2. Cell suspension is mixed with agarose and fixed on a slide glass.
3. The slide glass is soaked in Lysis solution and cells are unfixed.
4. After electrophoresis under basic or neutral conditions, a slide glass is soaked in 70–80% ethanol to fix cells.
5. Air-dried slide glass is stained using SYBR Green I.
6. Using fluorescent microscope, stained DNA is observed and photographed. Images are analyzed using analysis program for comet assay.

Figure 3 shows a typical photograph of comet after exposure to X-rays. Under basic condition, single-strand breaks are analyzed and double-strand breaks are analyzed under neutral condition. Usually, three image analysis values, i.e., tail length, tail percent, and tail moment are compared with those of untreated controls to evaluate effects on DNA strand breaks (Miyakoshi et al., 2000a, 2002).

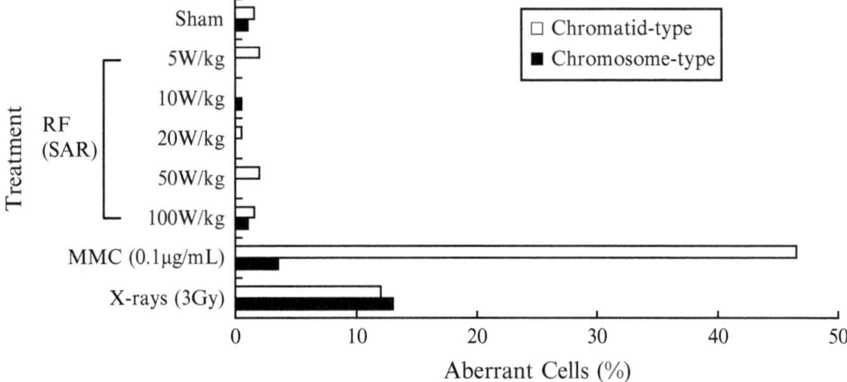

Figure 2. Chromosomal- and chromatid-type aberrations frequency in mouse m5S cells exposed to RF, MMC, and X-rays (redrawing data from Komatsubara et al., 2005).

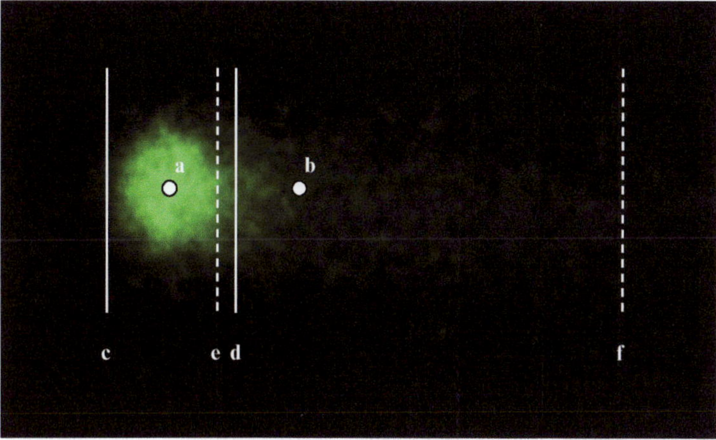

a: Head center
b: Tail center
c: Head top
d: Head end
e: Tail top
f: Tail end
Tail length = f – e
Tail percent = Tail (f-e) content / Comet (f-c) content ×100
Tail moment = Tail percent × (b-a) / 100

Figure 3. Picture of a sample result from a comet assay. *a*, Head center; *b*, Tail center, *c*, Head top, *d*, Head end, *e*, Tail top, and *f*, Tail end. Tail length = $f-e$, Tail percent = Tail $(f-e)$ content/Comet $(f-c)$ content ×100, Tail moment = Tail percent × $(b-a)$/100.

Several studies have reported positive results that show breakage of DNA strands by RF exposure.

In the 1980s, studies using plasmid (PUC8.c2) were conducted (Sagripanti and Swicord, 1986). Plasmid DNA solution was exposed to an RF field (2,550 MHz; 2 and 8.5 W/kg, or 21 and 85 W/kg; 20 min; 20°C), and effects of exposure were evaluated with plasmid form conversion. SAR-dependent decreases in Form I, plus SAR-dependent increases in Forms II and III, were observed. The maximum temperature increase during RF exposure was 0.8°C. When the effect of temperature on plasmid form distribution was assessed form conversion was not found, even when temperature was increased by 8°C. Therefore, it was suggested that this RF effect on plasmids, i.e., effect on DNA strands, was caused directly by RF exposure in the absence of appreciable temperature rise.

Zhang et al. (2002) used the comet assay to evaluate human lymphocytes treated with one of the three kinds of exposure (1) 2,450-MHz RF field (5 mW/cm^2) for 2 h, (2) 24-h MMC treatment only, and (3) 24 h MMC treatment after 2 h of RF exposure. RF fields alone had no effect. However, cells exposed to RF field followed by treatment with MMC (at a high concentration of 0.025–0.1 μg/ml) had increased Comet length, as compared with MMC alone. MN formation also was assessed. A similar increasing trend was found.

Phillips et al. (1998) used MOLT-4 lymphoblast cells to examine effects of (1) iDEN signal (813.5625 MHz; SAR=2.4 and 24 mW/kg) and (2) TDMA signal (836.55 MHz; SAR=2.6 and 26 mW/kg) on single-strand DNA breaks. They used the alkaline comet assay. DNA damage was decreased in two groups: 2.4 μW/g iDEN signal for 21 h, and 2.6 μW/g TDMA signal for 21 h. In the two respective groups with SARs that were 10-fold greater, DNA damage was increased with iDEN signal and decreased with TDMA signal. The authors interpret this pattern of mixed results as indicating that RF fields can have an effect on DNA directly and DNA repair mechanisms.

In recent studies, RF exposure (1,800 MHz; SAR=1.2 or 2 W/kg; different modulations; during 4, 16, and 24 h; intermittent 5 min ON/10 min OFF or continuous wave) induced DNA single- and double-strand breaks. Effects occurred after 16-h exposure in both cell types (human fibroblasts and rat granulose cells) and after different mobile-phone modulations (pulse and talk). The intermittent exposure showed a stronger effect in the comet assay than continuous exposure (REFLEX, Risk Evaluation of Potential Environmental Hazards From Low Frequency Electromagnetic Field Exposure Using Sensitive in vitro Methods, Final Report (2004)).

The chronological effects (in cells cultured after 2-, 4-, 24- and 2+4-h exposures) of 835.62-MHz frequency-modulated, continuous wave or to 847.74-MHz CDMA signals were investigated using alkaline Comet assay in U87MG and C3H10T1/2 cells in the logarithmic growth phase and C3H10T1/2 cells at confluent state. In these conditions, no increase in temperature was observed; the mean among all exposed conditions was 37 ± 0.3°C. The positive control was gamma ray irradiation. In all RF-exposed groups, no differences from the sham group were found (Malyapa et al., 1997a).

Similarly, chronological effects of 2,450-MHz exposure of cells in the logarithmic growth phase at 0.7 or 1.9 W/kg were investigated using alkaline Comet assay. In all RF-treatment groups, no differences were found from the sham group (Malyapa et al., 1997b). Effects of exposure for 2 h to a pulse-modulated 1,900-MHz RF field at SARs between 0 and 10 W/kg were assessed. DNA damage of human lymphocytes was investigated using Comet assay and MN formation. Under these conditions, no effects of the RF exposure were found (McNamee et al., 2002a).

Also using the Comet assay, Miyakoshi and colleagues have investigated the effects of RF exposure on cultured cells human brain tumor-derived MO54 cells. The temperature during exposure at 2,450 MHz reached 38.9°C, at the exposure of 100 W/kg. No effect from RF exposure was found from 2-h exposures at SARs of 25, 78, and 100 W/kg (Input power: 13 W) (Miyakoshi et al., 2002).

In one study, four kinds of RF fields were used (1) 837 MHz (phonetic modulation, analog RF field) with TDMA, (2) 837 MHz with CDMA, (3) 837 MHz (nonphonetic modulation), and (4) 1,909.8 MHz [phonetic modulation with global system of mobile communication-type personal communication systems (GSM-PCS)]. Exposures were for either 3 or 24 h at SARs of 1–10 W/kg. Results of Comet assay analysis showed no effect of exposure (Tice et al., 2002). In this study, the frequency of MN formation was increased in one group that was exposed at 10 W/kg of SAR for 24 h.

Lagroye et al. (2004a) used C3H10T1/2 cells to assess effects of exposure to 2,450 MHz on DNA damage, DNA-protein cross-linking, and DNA–DNA cross-linking. DNA damage was evaluated using Comet assay, and DNA-protein cross-linking was detected using proteinase K. As a positive control for cross-linking, cisplatinum was also used. For the exposure condition at 1.9 W/kg for 2 h, no effects were found in the above three examinations.

To assess effects with an in vivo exposure, the same group exposed Sprague–Dawley rats to an RF field (2,450 MHz at 1.2 W/kg) for 2 h. After 4 h, brain cells were isolated and evaluated using two kinds of Comet assay (Singh and Olive methods), with and without proteinase K. No exposure effects were found by either comet method, with or without proteinase K (Lagroye et al., 2004b).

In recent studies, no significant differences in DNA strand breaks have been observed between test groups exposed to W-CDMA or CW radiation and sham-exposed negative controls, as evaluated by an alkaline comet assay performed immediately after the exposure period in human glioblastoma A172 cells and normal human IMR 90 fibroblasts from fetal lung (Sakuma et al., 2006). These results confirm that low-level exposure does not exert genotoxicity up to an SAR of 800 mW/kg.

The study used several standard in vitro tests for DNA damage in G_0 human lymphocytes exposed in vitro to a combination of X-rays and RF fields, and comprehensively examined whether 24-h continuous exposure to a 935-MHz GSM basic signal delivering SAR of 1 or 2 W/kg is genotoxic per se or whether it can influence the genotoxicity of X-ray irradiation, which is a well-established clastogenic agent. Within the experimental parameters of the study, no effect of the RF-field signal was observed in all instances (Stronati et al., 2006).

Furthermore, the genotoxic effect observed in the report of REFLEX was not reproduced. ES1 cells were exposed to RF field (1,800 MHz; SAR=2 W/kg, continuous wave with intermittent exposure) for different time periods and then performed an alkaline comet assay (pH >13) and a MN test. Clear negative results were obtained in both tests in independent repeated experiments. The reasons for the differences between the results reported by the REFLEX project and their experiments remain unclear (Speit et al., 2006).

Despite the existence of some positive papers, the weight of the evidence supports the general consensus that RF exposures do not break DNA bonds. Scientists appropriately are conservative and will persist in supporting well-established fundamental findings until contrary evidence becomes overwhelming. From the perspective of biophysics, there is simply not enough energy in RF to break DNA bonds.

2.3. Micronucleus Formation

Many studies use micronucleus (MN) formation in the mitotic phase to evaluate chromosomal aberration and DNA damage in cells. Micronucleus (MN) is frequently used as an index for genotoxicity evaluation of cells as well as the above comet assay. MN is a phenomenon that DNA in a cell was damaged and a part of DNA was isolated from the original nucleus as a small nucleus (Koyama et al., 2003, 2004). The following is a brief explanation for analysis method of MN formation:

1. After cells are treated with exposure to the electromagnetic fields or chemical agents, cells are cultured in medium with cytochalasin B for 18–36 h (usually 1.5-fold of doubling time) to arrest the cell division cycle at binucleate cell status immediately after cell division.
2. Cells are collected and centrifuged, then, plated on a slide glass and fixed with ethanol.
3. Cells are stained using PI and stained-nuclei are observed with fluorescent microscope.
4. At least 1,000 binucleate cell images are examined per experiment and the number of cells including 1, 2, or 3 or more MNs is counted.Figure 4 shows a cell with MN formation. MN formation occurs extremely infrequently and spontaneously. Therefore, MN formation is evaluated both in untreated and treated groups with appropriate statistical method (Koyama et al., 2003, 2004).

Only a few studies have reported an increased MN frequency following RF exposure, and the majority of studies show no increase in MN frequency.

Garaj-Vrhovac et al. (1996) reported a time-dependent increase in MN frequency following exposure of human white blood cells to 415-MHz RF fields. At 2,450 and 7,700 MHz, the results showed induction of MN, compared to control cultures, at a power density of 30 mW/cm^2 and after exposures of 30 and 60 min (Zotti-Martelli et al., 2000). The study indicated that RF fields are able to cause cytogenetic damage in human lymphocytes at high power density and long exposure times. In contrast, D'Ambrosio et al. (1995) reported an increase in MN frequency due to temperature elevation induced by RF exposure.

C3H10T1/2 cells in either the resting or the proliferative phases were exposed to 835.62-MHz (FDMA) and 847.74-MHz (CDMA) RF fields for 3–24 h at 3.2 or 5.1 W/kg of SAR (FDMA) and at 3.2 or 4.8 W/kg of SAR (CDMA). When the frequency of MN formation was measured, no differences between exposed and control frequencies were found in any condition (Bisht et al., 2002). Human blood-derived lymphocytes were exposed to 1,900 MHz (continuous wave) for 2 h at 0–10 W/kg of SAR, and MN formation was measured (McNamee et al., 2002b). The exposure device kept the temperature at 37 ± 0.5°C, even at SAR of 10 W/kg. Increased frequency of MN formation was not observed.

Using MN formation as the dependent variable, Miyakoshi and colleagues have investigated the interactive effects of RF fields with chemicals using a wide

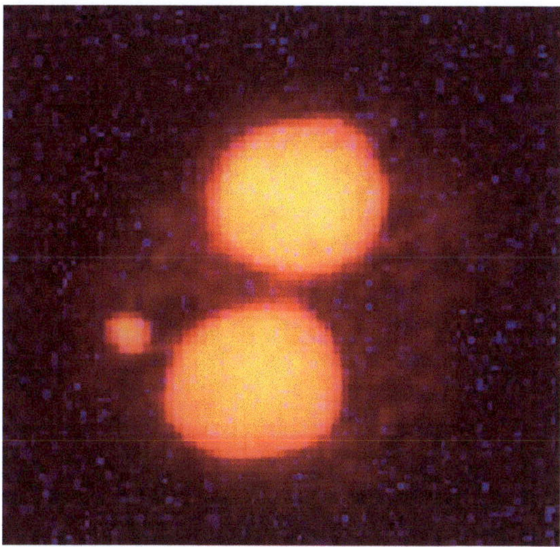

Figure 4. Photo of a typical micronucleus in Chinese hamster ovary K1 cell.

range of SARs, including the extremely high value of 100 W/kg (Koyama et al., 2003). CHO-K1 cells were exposed to 2,450 MHz for 18 h at 13–100 W/kg and effects of combined exposure with bleomycin also were investigated. No differences were found between the RF-exposed and sham-exposed groups with SARs of up to 50 W/kg. MN formation was increased with higher SAR, both in the RF-alone and RF+bleomycin groups. Also in the 39°C-treated group used as a temperature control, MN were increased, compared with sham exposure.

Koyama et al. exposed CHO-K1 cells to a 2,450-MHz electromagnetic field for 2 h, using SARs of 5–200 W/kg; effects of combined exposure with bleomycin also were investigated (Koyama et al., 2004). No differences were found between the RF-alone up to 50 W/kg of SAR (Fig. 5). However, at 100 W/kg and higher, MN formation was increased. As for combined RF+bleomycin exposure, an increase was observed only at 200 W/kg. Also in the 39°C- and higher-treated groups as temperature control, MN were increased. However, in combined exposure with heating and bleomycin, increased MN formation was observed only in the group exposed to 42°C. From these results, it was suggested that RF fields at SAR intensities less than those associated with the normal environments encountered daily have no effect on MN formation. However, MN formation was observed at extremely high SARs (more than 50 W/kg) with associated increased temperature.

Many other studies (Vijayalaxmi et al. 2001a, b; Bisht et al., 2002; McNamee et al., 2003; Zeni et al., 2003) have reached negative conclusions regarding an increase in MN frequency following RF exposure.

In general, cells exposed to RF fields do not show an increased incidence of MN formation. However, at relatively high SAR, such as 50 W/kg and greater, increased MN formation has been reported.

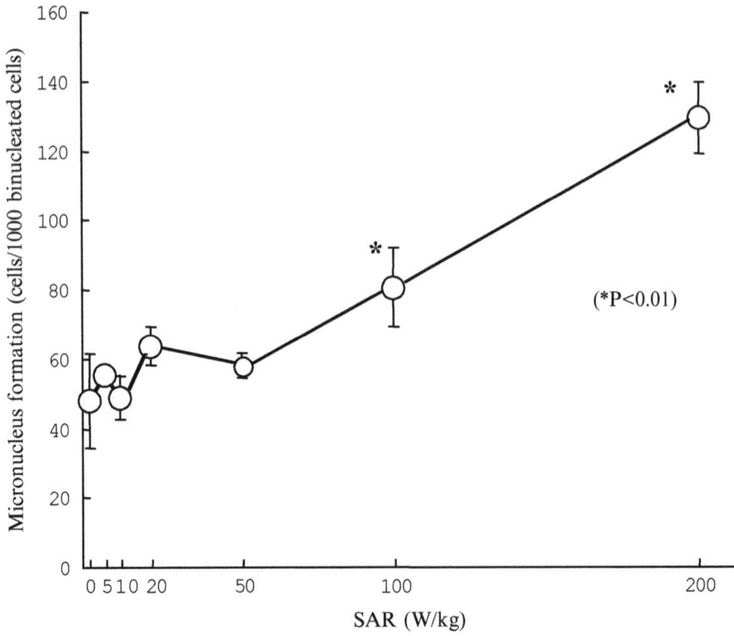

Figure 5. Micronucleus formation frequency in Chinese hamster ovary K1 cells after exposure to 2,450 MHz RF field (redrawing data from Koyama et al., 2004).

2.4. Mutation

Genotoxicity in cells that cannot be defined with the above chromosomal aberration, DNA strand breaks or MN formation, is mutation, such as, DNA base sequence changes. For example, human cells have 30,000 or more genes, therefore, it is impossible to examine mutation of all genes. Then, the following examination method to detect mutation has often been used. When a certain gene caused mutation, enzyme that is produced by this gene shows abnormal status, then, by addition of a specific agent, only cells producing abnormal enzyme are survived and the mutant is detected (Miyakoshi et al., 1997, 1999). The following is a mutation analysis procedure using HPRT gene locus on the X chromosome:

1. First, to exclude cells with spontaneous HPRT mutations, cells are cultured in medium containing hypoxanthine-aminopterin-thymidine (HAT medium).
2. After transferred to normal medium, cells are exposed to the electromagnetic field.
3. After exposure, cells are cultured for 6–10-fold of the doubling time as mutation expression time.
4. After that, cells are cultured in the medium containing 6-thioguanine (6-TG) where only cells with HPRT gene mutation can survive until the colony formation is gained.
5. Colonies are stained and counted to calculate mutation frequency.

Figure 6 shows the detection procedure for HPRT mutants. Mutation consists of various types of mutation including base change, deletion, frame shift, etc. It is also known that gene sites exist where mutation occurs extremely often (called as hot spot) (Miyakoshi et al., 1997, 1999).

At present, RF fields have not been found to induce mutations in cells. Meltz et al. (1990) investigated mutation of the thymidine kinase gene locus induced by RF fields alone and with concomitant chemical treatment. Mouse leukemia cells (L5178Y) were exposed for 4 h to pulsed 2,450-MHz RF field at SARs of up to 30 W/kg. No differences in the frequency of mutation were found between RF-exposed and sham-exposed groups. Furthermore, compared with MMC alone treated groups (positive control), no differences were found in (1) the RF field and MMC groups, and (2) the temperature and MMC groups. Similarly, no increase on frequency of induced mutation was observed under a different exposure condition (up to 40 W/kg of SAR; for 4 h), and no modifying effect of RF field on the response to Proflavin (positive control) was found. Their results suggested that RF fields, even at extremely high SAR, have no effect on induced mutation.

Eberle et al. (1997) also reported that RF fields (440 MHz, 900 MHz, 1,800 MHz) exposure did not change the hypoxanthine-guanine phosphoribosyl transferase (HPRT)-mutation frequency in human white blood cells.

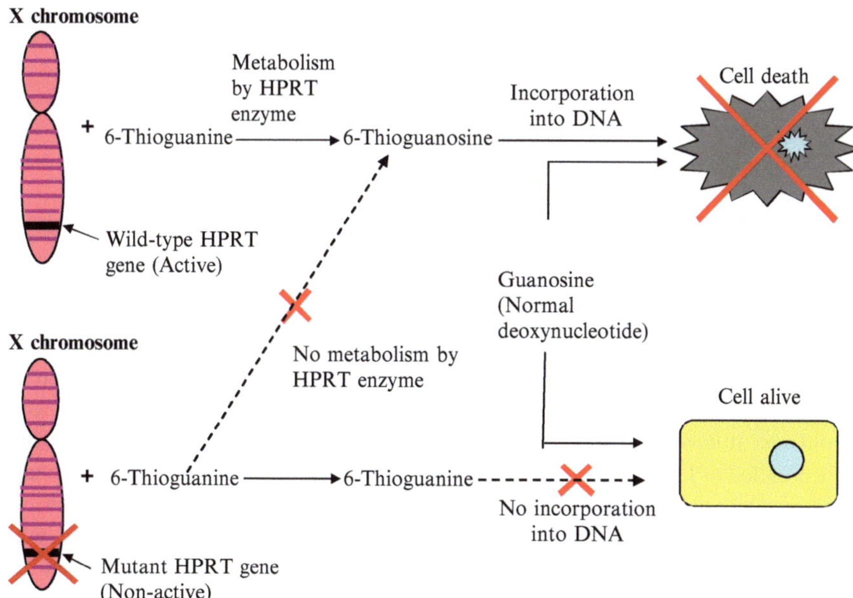

Figure 6. An outline of the method detecting mutation at the HPRT gene on X-chromosomes.

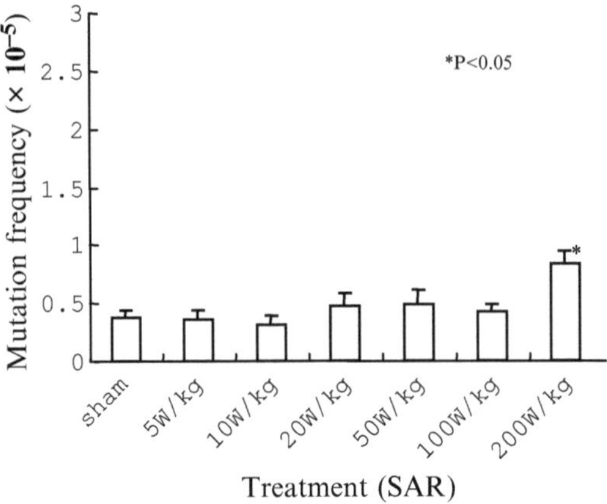

Figure 7. Mutation frequency at the HPRT gene in Chinese hamster ovary K1 cells by exposure to 2,450 MHz RF field (redrawing data from Koyama et al., 2007).

Most recently, our laboratory examined the bacterial and Chinese hamster gene mutations (Koyama et al., 2007). We have examined the effects of 2,450-MHz RF fields on bacterial mutations and the HPRT gene mutations. Using the Ames test, bacteria were exposed to RF field for 30 min at SARs from 5 to 200 W/kg. In all the strains examined, there was no significant difference in the frequency of revertant colonies between sham-exposed and RF-exposed groups. In examination of mutations of the HPRT gene, CHO-K1 cells were exposed to RF field for 2 h at SARs from 5 to 200 W/kg. RF field alone did not induce HPRT mutations up to 100 W/kg (Fig. 7).

3. NON-GENOTOXIC EFFECTS

3.1. Cell Proliferation

The major criteria for assessment of cellular effects of external factors are cell growth and survival, which are dependent on the extent of the effect. Severe damage inhibits or suppresses cell growth and leads to cell death. It is extremely rare for cell growth to be accelerated by external factors, other than specific cell growth factors.

Cultured cells usually grow in dishes exponentially such as 2-fold, 4-fold, 8-fold, etc. Immediately after plated on dishes, cells grow slowing for several hours (this is called as lag phase), but after that, enter into the exponentially growing phase and grow with cell-specific doubling time until reaching a confluent state. When cell growth rate was changed by internal or external factors, it can be indicated with changes in doubling time in the exponentially growing phase (Miyakoshi et al., 1994).

Cell cycle distribution and DNA synthesis may undergo induced changes due to the effect on cell proliferation.

Cell cycle is usually divided into four phases, i.e., mitotic phase (M-phase) from the beginning to the completion of cell division, gap 1 phase (G_1 phase) after the completion of cell division and before the beginning of DNA synthesis, DNA synthesizing phase (S-phase) from the beginning to the completion of DNA synthesis, and gap 2 phase (G_2 phase) after the completion of DNA synthesis and before beginning of cell division. Cell cycle is specific to cell lines, however, no significant differences are found in M-phase and S-phase generally between cell lines. Therefore, it is considered that differences in the whole cell cycle depend on G_1 phase and G_2 phase. When cells are damaged by stimuli of an external factor, a repair of the damage required time and time of normal cell cycle phases often changes. Usually, cell cycle progression is delayed (for example, prolonged G_1 phase), then, cell cycle distribution is changed. At present, cell cycle distribution is usually determined using FACS with collected and fixed cells whose DNA was stained with Propidium iodide (PI) (Nakahara et al., 2002).

Also in DNA synthesis, the synthesis rate is almost constant in normal cell growth. DNA synthesis rate is usually determined with the volume of radioisotope tritium (^3H)-labeled thymidine uptake per unit time. Also external stimuli or internal factors occasionally change DNA synthesis rate with cell characteristics. Changes in DNA synthesis rate are deeply involved to various synthesis-related enzymes and repair procedures for damage and its mechanism is not so simple.

Maes et al. (1997) found no effect. Garaj-Vrhovac et al. (1991) found a reduction in cell growth with RF exposure (7,700 MHz, CW, 0.5–30 mW/cm^2) dependent on the power density. Maes et al. (1993) reported no effect of RF field (2,450 MHz, 75 W/kg) on number of cell divisions. Tian et al. (2002) reported that at SARs >20 W/kg, cell survival rates were reduced. In the recent report, Takashima et al. (2006) compared the effects of continuous and intermittent exposure (2,450 MHz) at high SARs on cell growth, survival, and cell cycle distribution. When cells were exposed to a continuous RF field at SARs from 0.05 to 100 W/kg for 2 h, cellular growth rate, survival, and cell cycle distribution were not affected.

3.2. Gene Expression

Gene expression is a process that DNA base sequence (exactly exon) is interpreted and mRNA specific to this gene is produced and then protein is produced with polypeptide chain from the mRNA. As the endpoint for evaluation of gene expression, the existence/absence of gene expression is examined using mRNA or protein production that is derived from a specific gene. Usually the latter approach is used. The followings are analysis procedure for protein expression (Miyakoshi et al., 2000c, 2005; Hirose et al., 2003; Tian et al., 2002).

In recent biochemical experiment, various “kits” have been developed and researchers easily carry out experiments also in gene expression. The following is the procedure of Western Blot, which is generally used in protein expression analysis:

1. Cells are exposed to the electromagnetic field and heat and treated with chemical agents.

2. Immediately after or after appropriate expression time (it depends on treatment, protein and cell lines to be analyzed), cells are collected using trypsin treatment and scratching method.
3. Collected cells can be kept until the next process at −80°C.
4. Cells are treated with surface-active agent to isolate protein and protein density and volume are accurately determined.
5. A certain volume of protein is applied to gel for Western Blot to isolate protein by molecular weight with electrophoresis.
6. Isolated protein in gel is transferred (transcribed) into the membrane.
7. Protein-transcribed membrane and antibody specific to gene to be analyzed are mixed and shaken slowly to bind protein with antibody.
8. Using coloring reagent, a part of antibody is colored and protein production volume is estimated with the volume of colored part.
9. Protein expression volume is estimated using Image analysis software.

Figure 8 shows a sample of Western Blot. An antibody is specific to cell lines and target protein, and occasionally forms complex or phosphorylated, therefore, it is extremely important to select an appropriate antibody. (Miyakoshi et al., 2000c, 2005; Hirose et al., 2003; Tian et al., 2002

3.2.1. Heat-Shock Proteins

At present, searching for effects of RF exposure on gene expression is an active research area. The most attractive issue at the cell level is effects of heat-shock protein (HSP), whose expression is induced by various stresses, on gene expression. As described above, when RF energy is absorbed, heating can occur at the higher SARs. Therefore, it should be analyzed carefully whether HSP expression (1) is induced by heating associated with RF fields or (2) is a response initiated as nonthermal but RF-associated response. Such experiments must be conducted using RF exposure systems that accurately control and record temperature.

A Finnish research group (Leszczynski et al., 2002) exposed human endothelial-derived EA.hy926 cells to 900 MHz (GSM) at 2 W/kg of SAR for 1 h. Then protein expression volume and its phosphorylation were investigated using two-dimensional cataphoresis with the radioactive isotope ^{32}P. In addition, expression volumes of HSP27 and phosphorylated HSP27 were determined using the Western blot method. Transiently increased ^{32}P protein and phosphorylated HSP27 were found in the

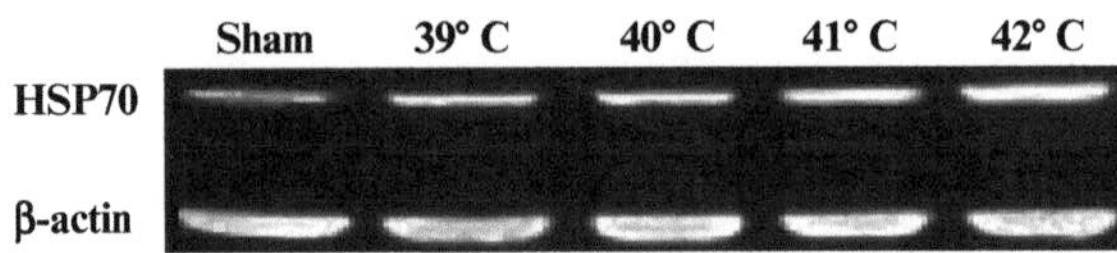

Figure 8. Expression of HSP70 in A172 cells after exposure to hyperthermia at 39°C, 40°C, 41°C, and 42°C for 3 h using Western blot analysis.

RF exposed group as compared to a sham-exposed control group. The expression volumes of HSP27 and p38 mitogen-activated protein kinase (MAPK) were increased transiently by exposure to RF field. These results suggested the possibility that RF exposure had some effects on signaling, especially on the stress-responding mechanisms of HSP27 and p38 mitogen-activated protein kinase.

Miyakoshi and colleagues also have investigated effects of RF field on HSP expression. Using an exposure dish with three sections, human brain tumor-derived MO54 cells were exposed to 2,450 MHz at SARs of 5, 20, 50, or 100 W/kg, and cell survival rates and HSP70 expression were determined using the Western blot method. At 5 W/kg, no effect on HSP70 expression was observed. However, at 20 W/kg and higher, HSP70 expression was increased in a manner dependent on SAR and on exposure time (Tian et al., 2002).

Similarly, MO54 cells were exposed to 1,950 MHz at 1–10 W/kg, and expression of HSP27, HSP70, and phosphorylated HSP27 (serine 78) was determined. Compared with the sham group, no differences in expression of HSP27 and HSP70 were found. However, expression of phosphorylated HSP27 was decreased by 1- and 2-h exposures (Fig. 9) (Miyakoshi et al., 2005).

The expression of HSP70 increased in the time- and dose-dependent manner at >50 W/kg SAR for 1–3 h. A similar effect was also observed in corresponding heat controls. There was no significant change in HSP27 expression caused by RF field at 5–200 W/kg or by comparable heating for 1–3 h. However, HSP27-phosphonylation increased transiently at 100 and 200 W/kg RF field (Wang et al., 2006). Our results suggest that exposure to a 2,450-MHz RF field has little or no apparent effect on HSP70 and HSP27 expression, but it may induce a transient increase in HSP27 phosphorylation in A172 cells at very high SAR (>100 W/kg).

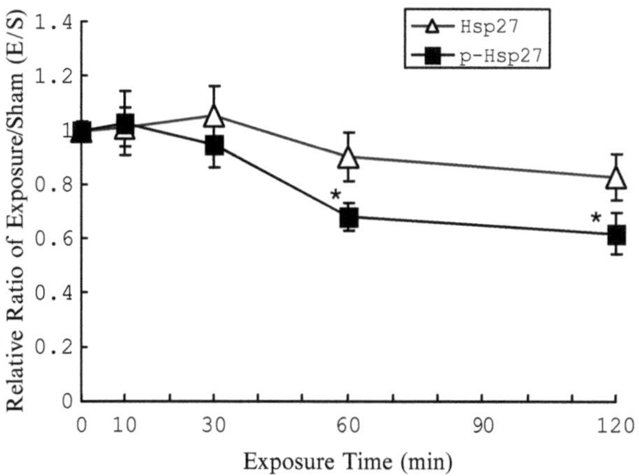

Figure 9. Changes in Hsp27 and p-Hsp27 expression after exposure to RF field at SAR of 10 W/kg for 10–120 min in MO54 cells. The data point represents the mean value and the experiment was performed six times (*$P<0.05$) (redrawing data from Miyakoshi et al., 2005).

No significant differences in the expression levels of phosphorylated HSP27 at serine 82 were observed between the test groups exposed to W-CDMA or CW signal (80 and 800 mW/kg for 2–48 h) and the sham-exposed negative controls, as evaluated immediately after the exposure periods by bead-based multiplex assays on human A172 and IMR 90 cells. Moreover, no noticeable differences in the gene expression of HSPs were observed between the test groups and the negative controls by DNA Chip analysis (Hirose et al., 2007).

A well-characterized exposure system, SXC 1800, built by the IT'IS foundation was used at 1,800 MHz, with a 217-Hz modulation. The expression of three HSPs (HSP70, HSC70, HSP27) using immunohistochemistry and induction of apoptosis by flow cytometry by exposure to RF field was investigated on human primary keratiocytes and fibroblasts (Sanchez et al., 2007). The results showed no effect of a 48 h GSM-1800 exposure at 2 W/kg on either keratiocytes or fibroblasts, in contrast to ultraviolet B (UVB)-radiation or heat-shock treatments, which injured cells.

Developing embryonic stem (ES) cells deficient in p53 were exposed to a GSM-217 RF field of 1,710 MHz at a SAR of 2 W/kg. The amount of mRNA of HSP70 was amplified and transient, slight increases were found simultaneously in c-jun, c-myc, and p21 genes. However, such changes were not found in wild-type ES cells with the normal p53 gene, suggesting that effects of RF fields on HSP70 are dependent on the genetic background of cells, including the p53 gene (Czyz et al., 2004).

3.2.2. Oncogenes

Oncogenes are genes that undergo changes in expression, structure, or function due to certain effects and cause normal cells to become cancerous. Since the identification of the first oncogene, src, numerous such genes have been discovered. Most oncogenes are involved in cell proliferation, including growth factors, signal transduction-related factors, and transcription factors. Regarding the relationship of oncogenes with the RF effect, early response genes, i.e., c-myc, c-fos, and c-jun, have been examined.

Rat pheochromocytoma (PC-12) cells treated with nerve growth factor (NGF) were exposed to 836.55 MHz (TDMA) for 20–60 min at 0.09–9 W/kg, and expression levels of c-jun and c-fos were determined using Northern blot analysis (Ivaschuk et al., 1997). The mRNA level for c-fos was not changed. However, expression of c-jun in cells that were exposed for 20 min at 9 mW/cm^2 was lower than that of the sham group. Additionally, in cells that were exposed for 40–60 min, the expression of c-jun did not differ from sham-exposure, perhaps implying recovery. These results suggest that RF exposure has a transitory inhibitory effect on c-jun expression.

In (1) the logarithmic growth phase, (2) the phase transiting to the plateau phase, and (3) the plateau phase, mouse-derived C3H10T1/2 cells were exposed to two kinds of RF field: 835.62 MHz (MCW) or 847.74 MHz (DMA) for 4 days at SAR of 0.6 W/kg. In all RNA that was isolated from cells, mRNAs of c-fos, c-jun, and c-myc were synthesized using the RT-PCR method and verified using gel electrophoresis.

No differences from the sham-exposed group were found. In addition, there was no difference in DNA binding capacity of the AP1, AP2, and NF-κB transcription factors. However, in the FMCW-exposed group in both (1) the phase transiting to plateau level and (2) the plateau phase, mRNA of c-fos was increased about 2-fold. A similar increase (approximately 1.4-fold) in mRNA of c-fos also was observed following CDMA RF exposure (Goswami et al., 1999).

Exponentially growing human lymphoblastoma cells (TK6) were exposed to 1,900 MHz pulse-modulated RF fields at average SARs of 1 and 10 W/kg (Chauhan et al., 2006). Their study found no evidence that the 1,900 MHz RF-field exposure caused a general stress response (expression of FOS, JUN, MYC, HSP27, and HSP70) in TK6 cells under the experimental conditions.

3.2.3. Others

The RF effect on expression of genes other than HSPs and oncogenes has been examined in several studies. The effect on Egr-1 gene expression of a modulated RF field of 900 MHz generated by a wire patch cell (WPC) antenna exposure system was studied as a function of time in SH-SY5Y neuroblastoma cells. Short-term exposure induced a transient increase in the Egr-1 mRNA level paralleled with the activation of the MAPK subtypes ERK1/2 and SAPK/JNK (Buttiglione et al., 2007). The results provide evidence that exposure to 900-MHz modulated RF field affects both Egr-1 gene expression and cell regulatory functions involving apoptosis inhibitors such as Bcl-2 and survivin, thus providing important insights into a potentially broad mechanism for controlling in vitro cell viability.

Intermittent exposure of human Mono Mac 6 (MM6) cells to RF EMF pulses for a total of 90 min, with a pulse width of 0.79 ± 0.01 ns and a pulse repetition rate of 250 pps (Natarajan et al., 2006), revealed no difference in NFκB-dependent gene expression profiles at 8- or 24-h postexposure, indicating that activated NFκB does not lead to differential expression of κB-dependent target genes.

At present, the results for effects of RF fields on gene expression, including HSPs and oncogenes, have been inconsistent. Many studies are ongoing and the results of recent whole human genome studies using microarray analysis are likely to be of importance, as described in Sect. 3.3.

3.3. Transcriptomics (Microarray Analysis)

The complete sequence of the human genome has been determined and analytical methods for screening of human gene expression have been developed, including microarray analysis using DNA chips.

Microarray analysis allows exhaustive assessment of the expression levels of mRNAs in a given cell. An example of a microarray experiment is described below:

1. Total RNA is extracted from cultured cells of the control and experimental groups in the normal manner.

2. After cDNA is synthesized by reverse transcription and transcribed in vitro, amplification of antisense RNA including aminoaryl-modified uracil is performed.
3. Control and experimental RNAs are labeled with different fluorochromes (Cy3 and Cy5).
4. Fluorochrome-labeled RNA is hybridized to a DNA chip on which a target gene is fixed at each spot.
5. The fluorescent intensity of each fluorochrome (Cy3 and Cy5) at each gene spot is obtained as an image using a fluorescent scanner.
6. Using the attached software, the fluorescent intensity of the fluorochrome at each gene spot is quantified.
7. The software is used to quantify the difference in gene expression level in the experimental group from that in the control group.These procedures are illustrated in Fig. 10.

This method has been used to study RF effects and several articles have been published. However, current microarrays do not always detect responsive genes accurately and have a high probability of detection of false positives. Candidate genes require confirmation by RT-PCR, and detection efficiency is not always good for small changes in expression. The following text refers to several recent articles.

Among 1,200 candidate genes, 24 upregulated genes and 10 downregulated genes were identified after 24-h intermittent exposure at an average SAR of 2 W/kg, which are associated with multiple cellular functions (cytoskeleton, signal transduction pathway, metabolism, etc.) after functional classification (Zhao et al., 2007). The results indicated that the gene expression of rat neuron could be altered by exposure to RF fields (1,800 MHz, 217-Hz Modulation, 2 W/kg) under our experimental conditions.

Gurisik et al. (2006) exposed two human cell lines (one of neuronal (SK-N-SH) and the other of monocytoid (U937) origin) to a 900-MHz RF signal, pulsed at 217 Hz, producing a SAR of 0.2 W/kg. They found no significant difference between sham-exposed vs. RF-exposed cells in gene microarray assessment, HSP expression, cell cycle distribution, and apoptosis.

In vitro experiments with C3H 10T1/2 mouse cells were performed to determine whether FDMA- or CDMA-modulated RF fields induce changes in gene expression (Whitehead et al., 2006). For both CDMA and FDMA radiations (5 W/kg), the number of probe sets with an expression change greater than 1.3-fold was less than or equal to the expected number of false positives. Thus the 24-h exposures to FDMA or CDMA RF field at 5 W/kg had no statistically significant effect on gene expression.

Qutob et al. (2006) examined the ability of exposure to a 1,900 MHz pulse-modulated RF field for 4 h at SARs of 0.1, 1.0 and 10.0 W/kg to affect global gene expression in U87MG glioblastoma cells. They found no evidence that nonthermal RF fields can affect gene expression in cultured U87MG cells relative to the nonirradiated control groups.

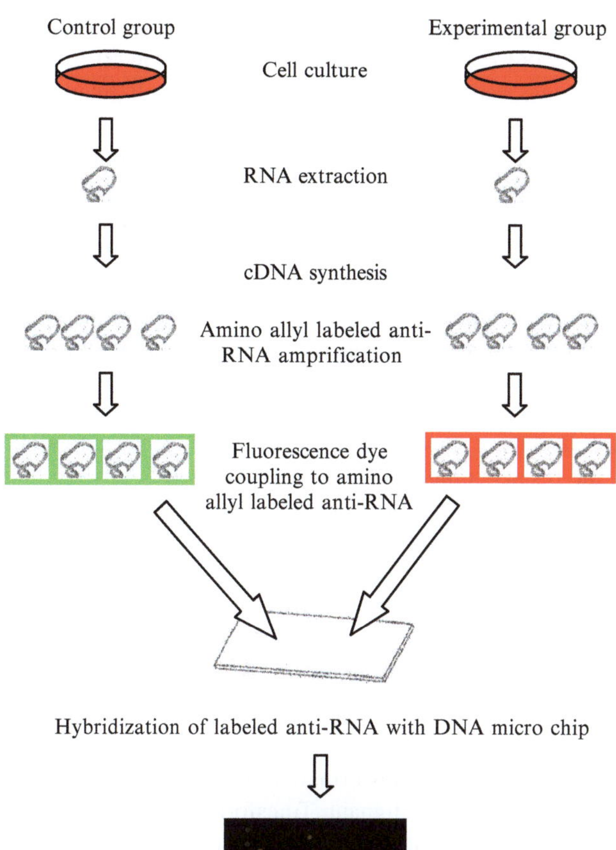

Hybridization of labeled anti-RNA with DNA micro chip

Scanning of fluorescence spots

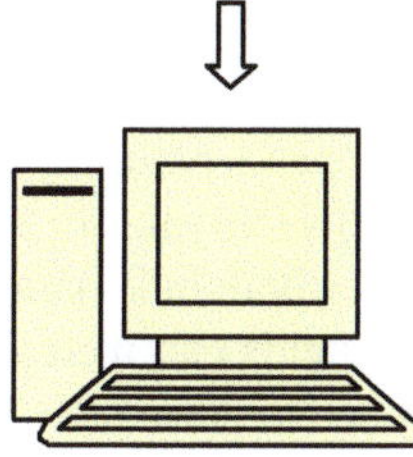

Measurement of the intensity of each spot
Analysis of alterations in mRNA expression

Figure 10. Scheme of a method for microarray assay.

The group of Leszczynski et al. reported the following data. Nylund and Leszczynski (2006) examined in vitro cell response to RF field (900-MHz GSM signal) using two variants of human endothelial cell line: EA.hy926 and EA. hy926v1. Obtained results show that gene and protein expression were altered, in both examined cell lines, in response to 1-h RF exposure at an average SAR of 2.8 W/kg. However, the same genes and proteins were differently affected by the exposure in each of the cell lines. Therefore, it is likely that different types of cells and from different species might respond differently to RF field or might have different sensitivity to this weak stimulus.

Six human cell types, immortalized cell lines, and primary cells were exposed to 900- and 1,800-MHz RF fields. RNA was isolated from exposed and sham-exposed cells and labeled for transcriptome analysis on whole-genome cDNA arrays (Remondini et al., 2006). NB69 neuroblastoma cells, T lymphocytes, and CHME5 microglial cells did not show significant changes in gene expression. For other three cell lines, the results come from 900 MHz-exposed EA.hy926 human endothelial cells (22 up-regulations, ten down-regulations), 900 MHz-exposed U937 human lymphoma cells (32 up-regulations, two down-regulations), and 1800 MHz-exposed HL-60 leukemia cells (11 up-regulations, one down-regulations). Analysis of the affected gene families does not point toward a stress response.

In addition, there are negative reports concerning microarray analysis of RF effects (Hirose et al., 2006; Zeng et al. 2006).

3.4. Cell Transformation

Transformation means cell transformation, i.e., characteristics of normal cells are changed and cells become malignant. Transformed cells are called as transformant, whose characteristics are changed morphologically in a normal culture condition. For example, normal cells stop cell division just when cells contact with each other and a monolayer of cells is formed in a dish bottom. On contrary, transformed cells lose this control system of normal cells and cell division continues even when cells contact with each other. In cells with high-grade malignancy, cells are accumulated and rise and form a colony (called as focus generally). Transformation is classified into Types I, II, and III (slight to serious grade) and Type-II and -III cells are defined as neoplastic transformant in general (Miyakoshi et al., 2000b). Figure 11 shows a photograph of typical transformants. However, in culturing cells successively, it was confirmed that a transformant appears naturally in extremely low rate even when no external treatment is given to cells (Miyakoshi et al., 2000b).

In 1967, Russian scientists (Stodolnik-Bara ska, 1967) published in Nature the first report that exposure to RF fields can increase frequency of cell transformation. Human lymphocytes were exposed to pulsed RF fields at 7 and 20 mW/cm^2 for 4 h per day for 3–5 days. The frequency of cellular transformation was increased in a manner apparently dependent on exposure duration. In addition, the percent of cells in the mitotic phase was increased, also in a manner related to increasing exposure duration.

Some positive effects of RF-field exposure combined with external factors were reported by a single research team during the 1980s. Balcer-Kubiczek and

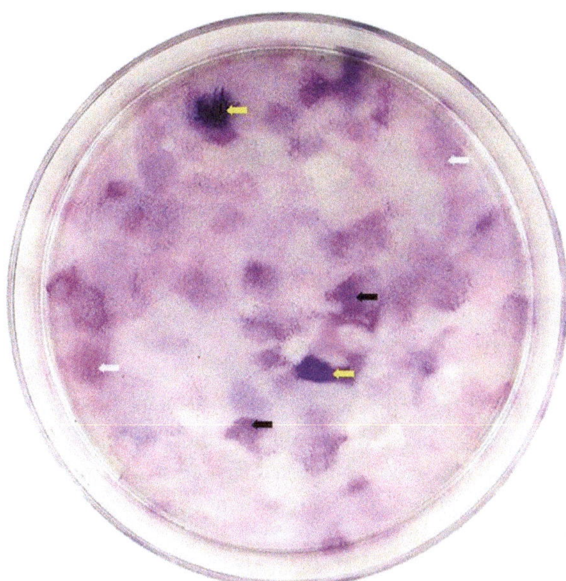

Yellow arrows: Type III
Black arrows: Type II
White arrows: Type I

Figure 11. Typical transformants in C3H10T½ cells. *Yellow arrows*: Type III, *Black arrows*: Type II, *White arrows*: Type I.

Harrison (1985) conducted a study in which C3H10T1/2 cell were exposed to RF (2,450 MHz, 4.4 W/kg) for 24 h with either benzopyrene or X-rays. Dose-dependence of benzopyrene and X-rays were evaluated with and without TPA treatment after the exposure. After correction for survival rate of each treatment, the frequency of transformation was increased by combined exposure to X-rays, RF fields, and TPA. To the authors, this suggested that RF field had copromotional effect on oncogenesis. Similarly, C3H10T1/2 cells were exposed to pulsed 2,450-MHz at 4.4 ± 0.8 W/kg for 24 h. Results were evaluated using the focus test method. No effect of RF field alone was observed. However, combined exposure to RF field, TPA (0.1 μg/ml), and X-rays (1.5 Gy) increased the frequency of transformation (Balcer-Kubiczek and Harrison, 1989). Finally, Balcer-Kubiczek and Harrison (1991) conducted another study in which C3H10T1/2 cells were exposed to 2,450 MHz, modulated at 120 Hz, for 24 h at 0.1, 1, or 4.4 W/kg. RF exposure was combined with X-rays; and after exposure, the additional effects of both the presence and the absence of TPA were evaluated. The frequency of transformation was not induced by RF field alone. However, combined exposure with TPA treatment increased the frequency of transformation.

From these results, Balcer-Kubiczek and Harrison concluded that exposure to RF field alone had no effect on the frequency of transformation. However, some studies indicated the possibility that the frequency of transformation was increased by combined exposure to RF field with initiators for the transformation-inducing process and or with other external factors promoting transformation.

However, the recent studies, which have used focus formation as the dependent variable; about effects of exposure to RF fields alone on transformation have provided negative results. For example, C3H10T1/2 cells were exposed to 835.62-MHz frequency-modulated continuous wave (FMCW) or to 847.74-MHz CDMA at 0.6 W/kg for 7 days, and results were tested using the focus-formation method. No transformation induced by exposure to RF field was found (Roti Roti et al., 2001). When RF field was combined with 12-O-tetradecanoylphorbol-13-acetate (TPA) a carcinogenic promoter, or with X-rays (4.2 Gy), transformation was not enhanced by a 42-day exposure.

Similarly, C3H10T1/2 cells exposed to an RF field (836.55 MHz, TDMA) at SARs of 0.15, 1.5, and 15 mW/kg combined with TPA treatment at concentrations of 10, 30, and 50 ng/ml showed no enhanced transformation, using the focus-formation test under any of the exposure and treatment conditions (Cain et al., 1997).

In our recent study (Wang et al., 2005), we exposed to RF field alone at a wide range of specific SARs of 5–200 W/kg for 2 h and examined the transformation frequency between the controls and RF field with or without TPA (0.5 ng/ml), a tumor promoter. RF exposure alone and RF with TPA did not elevate the transformation frequency (Fig. 12). However, the transformation frequency for RF field at SAR of more than 100 W/kg with methylcholanthrene (MC) or MC plus TPA was increased compared with MC alone or MC plus TPA. On the other hand, the corresponding heat groups (heat alone, heat + MC, and heat + MC + TPA) did not increase

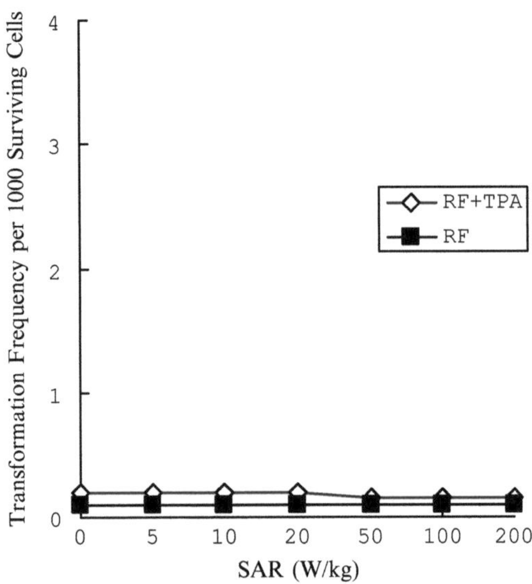

Figure 12. Transformation frequency of C3H10T½ cells after exposure to 2,450-MHz RF field alone and in combination with TPA (redrawing data from Wang et al., 2005).

transformation compared with each control level in C3H10T1/2 cells. This result suggests that 2,450-MHz RF field could not contribute to the initiation stage of tumor formation, but it may contribute to the promotion stage at the extremely high SAR (≥100 W/kg).

Consequently, it is difficult to conclude that exposure to RF alone raises the frequency of transformation; however, the possibility of a modifying effect cannot be ruled out for RF field in combination with chemical inducers and promoters that raise the frequency of transformation.

3.5. Apoptosis

Apoptosis, which is also referred to as "programmed cell death," is a mechanism for self-protection of damaged cells through self-induced cell death. Apoptosis is a form of cell death that is actively induced by the cell itself to maintain normal individual status. Cell death resulting from extrinsic damage and an undesirable cellular environment is referred to as necrosis, which is distinct from apoptosis. Signal transduction in induction of apoptosis has been elucidated using molecular biology techniques and can be used as an indicator to assess the effect of external factors on apoptosis. The signal transduction pathway of apoptosis is outlined in Fig. 13.

In our study (Hirose et al., 2006), no significant differences in the percentage of apoptotic cells were observed between the test groups exposed to RF signals (2,142.5 MHz, CW or W-CDMA, SAR up to 800 mW/kg) and the sham-exposed negative controls, as evaluated by the Annexin V affinity assay. No significant differences in expression levels of phosphorylated p53 at serine 15 or total p53 were also observed between the test groups and the negative controls by the bead-based multiplex assay. Moreover, microarray hybridization and real-time RT-PCR analysis showed no noticeable differences in gene expression of the subsequent downstream targets of p53 signaling involved in apoptosis between the test groups and the negative

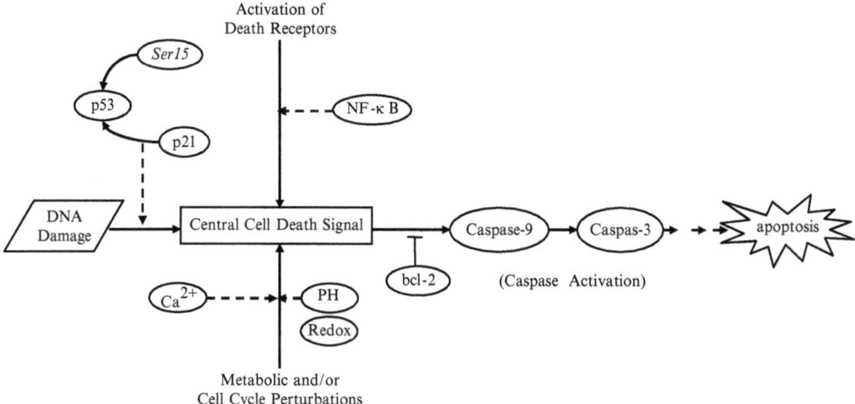

Figure 13. An outline of the signal transduction pathway of apoptosis.

controls. Our results confirm that exposure to low-level RF signals up to 800 mW/kg does not induce p53-dependent apoptosis.

Lantow et al. (2006) have used human Mono Mac 6 cells to investigate the influence of RF field in vitro on cell cycle alterations and BrdU uptake, as well as the induction of apoptosis and necrosis using flow cytometry after exposure to a 1,800 MHz, 2 W/kg SAR, GSM-DTX signal for 12 h. No statistically significant differences in the induction of apoptosis or necrosis, cell cycle kinetics, or BrdU uptake were detected after RF exposure compared to sham or incubator controls.

Joubert et al. (2007) also reported that no statistically significant difference in the apoptosis rate was observed between controls and 24-h GSM-exposed neurons, either 0- or 24-h postexposure at the average SAR of 0.25 W/kg.

Merola et al. (2006) reported that combined exposures to RF field and to the differentiative agent retinoic acid or to the apoptotic inducer camptothecin were carried out to test possible interference between RF field and chemical agents. Overall their data suggest that 900-MHz RF exposure up to 72 h does not induce significant alterations in the three principal cell activities (proliferation, differentiation, and apoptosis) in a neuroblastoma cell line.

Consequently, it is extremely unlikely that RF exposure induces apoptosis.

3.6. Immune System

The immune system protects hosts from infection and cancer. When an external organism invades the body, immune cells start to attack the organism for self-protection. Specifically, these cells produce many antibodies that inhibit the external threat and killer T cells then eliminate the invader. Immune cells have an important role in this process and the effect of RF fields on these cells has been examined.

Tuschl et al. (2006) reported the effect of GSM-modulated RF field on human immune cells. Exposure was performed at GSM Basic 1,950 MHz, a SAR of 1 mW/g in an intermittent mode (5 min "ON," 10 min "OFF"), and a maximum ΔT of 0.06°C for the duration of 8 h. No statistically significant effects of exposure were found and there is no indication that emissions from mobile phones are associated with adverse effects on the human immune system (IL-1, -2, and -4; INF-γ; and INF-α).

Thorlin et al. (2006) examined the effect of 900-MHz RF in cultured astroglial and microglial brain cells. Primary cultures enriched in astroglial cells were exposed to 900-MHz RF field in a temperature-controlled exposure system at SARs of 3 W/kg GSM modulated wave for 4, 8, and 24 h or 27 W/kg CW for 24 h, and the release into the extracellular medium of the two proinflammatory cytokines interleukin 6 (IL-6) and tumor necrosis factor-alpha (TNF-α) was analyzed. This study does not provide evidence for any effect of the RF fields used on damage-related factors in glial cells in culture.

A Polish group reported positive effects of RF field on immune cell activity. In an earlier study, G_0 phase peripheral blood mononuclear cells (PBMC) that are exposed to low-level (SAR = 0.18 W/kg) pulse-modulated 1,300-MHz RF fields and subsequently cultured demonstrate changed immune activity (Dabrowski et al., 2003). In addition, the microcultures of PBMC exposed to RF field (900 MHz, GSM, 27 V/m, SAR=0.024 W/kg) demonstrated significantly higher response to mitogens and higher immunogenic activity of monocytes (LM index) than control cultures (Stankiewicz et al., 2006).

Further studies are required to determine the effect of RF fields on the immune system.

3.7. Reactive Oxygen Species

Aging, exercise, UV, and many other forms of stress are known to increase reactive oxygen species (ROS) production, with subsequent ROS reactions with intracellular DNA and lipoprotein leading to altered cellular function. ROS include oxygen ions, free radicals, and both inorganic and organic peroxides. They are generally very small and highly reactive species. The harmful cellular effects of ROS include (1) damage of DNA, (2) oxidation of polydesaturated fatty acids in lipids, and (3) oxidation of amino acids in proteins. Therefore, damage to cells is increased with an increase in ROS levels. Only a few studies have examined the effects of RF fields on ROS production.

Lantow et al. (2006) examined the effect of RF field on ROS production. Heat and PMA treatment induced a significant increase in superoxide radical anions and in ROS production in the Mono Mac 6 cells when compared to sham and/or incubator conditions. No significant differences in free radical production were detected after RF fields (GSM 1,800 MHz, SAR=0.5–2.0 W/kg) exposure or in the respective controls, and no additional effects on superoxide radical anion production were detected after coexposure to RF field+TPA or RF field+lipopolysaccharide (LPS).

Zeni et al. (2007) investigated the induction of ROS in murine L929 fibrosarcoma cells exposed to RF field at 900 MHz, with or without coexposure to 3-chloro-4-(dichloromethyl)-5-hydroxy-2(5H)-furanone (MX), a potent environmental carcinogen produced during chlorination of drinking water. The study provided no indication that 900-MHz RF-field exposure, either alone or in combination with MX, induced formation of ROS under any of the experimental conditions investigated.

At present, no study has reported that ROS production is increased by RF exposure.

3.8. Summary of In Vitro Study

The effects of RF exposure on cells in currently available reports can be summarized as follows. (1) RF energy does not cleave intracellular DNA directly, since most genotoxicity studies have shown negative effects. Cells are damaged at an extremely high SAR, but this is thought to be due mainly to the thermal effect of RF fields. (2) An interesting cellular response induced by RF field is associated with stress proteins; i.e., HSP production and phosphorylation. However, the results of studies of these effects are inconsistent, perhaps due to differences in cell lines, RF exposure conditions and exposure devices, and reproduction of results in different laboratories is of importance. (3) Microarray analysis has not provided definite evidence of an effect of RF exposure on cellular functions, including apoptosis, the immune system, and ROS production.

Studies on RF effects are ongoing worldwide, but the current published evidence does not allow a definite conclusion regarding the effects at a cellular level. The rapid development of biotechnology has increased the potential for detection of microresponses in cells and genes, and future studies of RF effects should be performed using improved biotechnological methods.

REFERENCES

Balcer-Kubiczek, E.K., and Harrison, G.H., 1985, Evidence for microwave carcinogenesis in vitro. *Carcinogenesis* 6: 859–864.

Balcer-Kubiczek, E.K., and Harrison, G.H., 1989, Induction of neoplastic transformation in C3H/10T1/2 cells by 2.45-GHz microwaves and phorbol ester. *Radiat Res* 117: 531–537.

Balcer-Kubiczek, E.K., and Harrison, G.H., 1991, Neoplastic transformation of C3H/10T1/2 cells following exposure to 120-Hz modulated 2.45-GHz microwaves and phorbol ester tumor promoter. *Radiat Res* 126: 65–72.

Bisht, K.S., Moros, E.G., Straube, W.L., Baty, J.D., and Roti Roti, J.L., 2002, The effect of 835.62 MHz FDMA or 847.74 MHz CDMA modulated radiofrequency radiation on the induction of micronuclei in C3H 10T(1/2) cells. *Radiat Res* 157: 506–515.

Buttiglione, M., Roca, L., Montemurno, E., Vitiello, F., Capozzi, V., and Cibelli, G., 2007, Radiofrequency radiation (900 MHz) induces Egr-1 gene expression and affects cell-cycle control in human neuroblastoma cells. J Cell Physiol 213: 759–767.

Cain, C.D., Thomas, D.L., and Adey, W.R., 1997, Focus formation of C3H/10T1/2 cells and exposure to a 836.55 MHz modulated radiofrequency field. *Bioelectromagnetics* 18: 237–243.

Chauhan, V., Mariampillai, A., Bellier, P.V., Qutob, S.S., Gajda, G.B., Lemay, E., Thansandote, A., and McNamee, J.P., 2006, Gene expression analysis of a human lymphoblastoma cell line exposed in vitro to an intermittent 1.9 GHz pulse-modulated radiofrequency field. Radiat Res 165: 424–429.

Czyz, J., Guan, K., Zeng, Q., Nikolova, T., Meister, A., Schönborn, F., Schuderer, J., Kuster, N., and Wobus, A.M., 2004, High frequency electromagnetic fields (GSM signals) affect gene expression levels in tumor suppressor p53-deficient embryonic stem cells. Bioelectromagnetics 25: 296–307.

D'Ambrosio, G., Lioi, M.B., Scarfi, M.R., and Zeni, O., 1995, Genotoxic effects of amplitude-modulated microwaves on human lymphocytes exposed in vitro under controlled conditions. Electro Magnetobiol 14: 157–164.

Dabrowski, M.P., Stankiewicz, W., Kubacki, R., Sobiczewska, E., and Szmigielski S., 2003, Immunotropic effects in cultured human blood mononuclear cells pre-exposed to low-level 1300 MHz pulse-modulated microwave field. Electromagn Biol Med 22: 1-13.

Eberle, P., Erdtmann-Vourliotis, M., Diener, S., Finke, H.G., Löffelholz, B., Schnor, A., and Schräder, M., 1997, Cell proliferation, sister-chromatid exchange, chromosomal aberrations, micronuclei and mutation rate of the HGPRT locus following the exposure of human peripheral lymphocytes to electromagnetic high-frequency fields (440 MHz, 900 MHz, and 1.8 GHz). In: Brinkmann K., and Freidrich, G., Eds., Electromagnetic Compatibility of Biological Systems. VDE-Verlag GMBH, Berlin-Offenbach, pp. 135–155.

Garaj-Vrhovac, V., Horvat, D., and Koren, Z., 1991, The relationship between colony-forming ability, chromosome aberrations and incidence of micronuclei in V79 Chinese hamster cells exposed to microwave radiation. Mutat Res 263: 143–149.

Garaj-Vrhovac, V., Fucić, A., and Horvat, D., 1992, The correlation between the frequency of micronuclei and specific chromosome aberrations in human lymphocytes exposed to microwave radiation in vitro. Mutat Res 281: 181–186.

Garaj-Vrhovac, V., Vojvodic, S., Fucic, A., and Kubelka, D., 1996, Effects of 415 MHz frequency on human lymphocyte genome. In: Proceedings of IRPA9 Congress, Vienna, Austria, vol. 3, pp. 604–606.

Goswami, P.C., Albee, L.D., Parsian, A.J., Baty, J.D., Moros, E.G., Pickard, W.F., Roti Roti, J.L., and Hunt, C.R., 1999, Proto-oncogene mRNA levels and activities of multiple transcription factors in C3H 10T 1/2 murine embryonic fibroblasts exposed to 835.62 and 847.74 MHz cellular phone communication frequency radiation. Radiat Res 151: 300–309.

Gurisik, E., Warton, K., Martin, D.K., and Valenzuela, S.M., 2006, An in vitro study of the effects of exposure to a GSM signal in two human cell lines: monocytic U937 and neuroblastoma SK-N-SH. Cell Biol Int 30: 793–799.

Hirose, H., Nakahara, T., Zhang, Q.M., Yonei, S., and Miyakoshi, J., 2003, Static magnetic field with a strong magnetic field gradient (41.7 T/m) induces c-Jun expression in HL-60 cells. In Vitro Cell Dev Biol Anim 39: 348–352.

Hirose, H., Sakuma, N., Kaji, N., Suhara, T., Sekijima, M., Nojima, T., and Miyakoshi, J., 2006, Phosphorylation and gene expression of p53 are not affected in human cells exposed to 2.1425 GHz band CW or W-CDMA modulated radiation allocated to mobile radio base stations. Bioelectromagnetics 27: 494–504.

Hirose, H., Sakuma, N., Kaji, N., Nakayama, K., Inoue, K., Sekijima, M., Nojima, T., and Miyakoshi, J., 2007, Mobile phone base station-emitted radiation does not induce phosphorylation of Hsp27. Bioelectromagnetics 28: 99–108.

Ivaschuk, O.I., Jones, R.A., Ishida-Jones, T., Haggren, W., Adey, W.R., and Phillips, J.L., 1997, Exposure of nerve growth factor-treated PC12 rat pheochromocytoma cells to a modulated radiofrequency field at 836.55 MHz: effects on c-jun and c-fos expression. Bioelectromagnetics 18: 223–229.

Joubert, V., Leveque, P., Cueille, M. Bourthoumieu, S., and Yardin, C., 2007, No apoptosis is induced in rat cortical neurons exposed to GSM phone fields. Bioelectromagnetics 28: 115–121.

Kerbacher, J.J., Meltz, M.L., and Erwin, D.N., 1990, Influence of radiofrequency radiation on chromosome aberrations in CHO cells and its interaction with DNA-damaging agents. Radiat Res 123: 311–319.

Komatsubara, Y., Hirose, H., Sakurai, T., Koyama, S., Suzuki, Y., Taki, M., and Miyakoshi, J., 2005, Effect of high-frequency electromagnetic fields with a wide range of SARs on chromosomal aberrations in murine m5S cells. Mutat Res 587: 114–119.

Koyama, S., Nakahara, T., Wake, K., Taki, M., Isozumi, Y., and Miyakoshi, J., 2003, Effects of high frequency electromagnetic fields on micronucleus formation in CHO-K1 cells. Mutat Res 541: 81–89.

Koyama, S., Isozumi, Y., Suzuki, Y., Taki, M., and Miyakoshi, J., 2004, Effects of 2.45-GHz electromagnetic fields with a wide range of SARs on micronucleus formation in CHO-K1 cells. ScientificWorldJournal 20: 29–40.

Koyama, S., Takashima, Y., Sakurai, T., Suzuki, Y., Taki, M., and Miyakoshi, J., 2007, Effects of 2.45 GHz electromagnetic fields with a wide range of SARs on bacterial and HPRT gene mutations. J Radiat Res (Tokyo) 48: 69–75.

Lagroye, I., Hook, G.J., Wettring, B.A., Baty, J.D., Moros, E.G., Straube, W.L., and Roti Roti, J.L., 2004a, Measurements of alkali-labile DNA damage and protein-DNA crosslinks after 2450 MHz microwave and low-dose gamma irradiation in vitro. Radiat Res 161: 201–214.

Lagroye, I., Anane, R., Wettring, B.A., Moros, E.G., Straube, W.L., Laregina, M., Niehoff, M., Pickard, W.F., Baty, J., and Roti Roti, J.L., 2004b Measurement of DNA damage after acute exposure to pulsed-wave 2450 MHz microwaves in rat brain cells by two alkaline comet assay methods. Int J Radiat Biol 80: 11–20.

Lantow, M., Viergutz, T., Weiss, D.G., and Simkó, M., 2006, Comparative study of cell cycle kinetics and induction of apoptosis or necrosis after exposure of human mono mac 6 cells to radiofrequency radiation. Radiat Res 166: 539–543.

Leszczynski, D., Joenväärä, S., Reivinen, J., and Kuokka, R., 2002, Non-thermal activation of the hsp27/p38MAPK stress pathway by mobile phone radiation in human endothelial cells: molecular mechanism for cancer- and blood-brain barrier-related effects. Differentiation 70: 120–129.

Maes, A., Verschaeve, L., Arroyo, A., De Wagter, C., and Vercruyssen, L., 1993, In vitro cytogenetic effects of 2450 MHz waves on human peripheral blood lymphocytes. Bioelectromagnetics 14: 495–501.

Maes, A., Collier, M., Slaets, D., and Verschaeve, L., 1996, 954 MHz microwaves enhance the mutagenic properties of mitomycin C. Environ Mol Mutagen 28: 26–30.

Maes, A., Collier, M., Van Gorp, U., Vandoninck, S., and Verschaeve, L., 1997, Cytogenetic effects of 935.2-MHz (GSM) microwaves alone and in combination with mitomycin C. Mutat Res 393: 151–156.

Maes, A., Collier, M., and Verschaeve, L., 2001, Cytogenetic effects of 900 MHz (GSM) microwaves on human lymphocytes. Bioelectromagnetics 22: 91–96.

Malyapa, R.S., Ahern, E.W., Straube, W.L., Moros, E.G., Pickard, W.F., and Roti Roti, J.L., 1997a, Measurement of DNA damage after exposure to electromagnetic radiation in the cellular phone communication frequency band (835.62 and 847.74 MHz). Radiat Res 148: 618–627.

Malyapa, R.S., Ahern, E.W., Straube, W.L., Moros, E.G., Pickard, W.F., and Roti Roti, J.L., 1997b, Measurement of DNA damage after exposure to 2450 MHz electromagnetic radiation. Radiat Res 148: 608–617.

Mashevich, M., Folkman, D., Kesar, A. Barbul, A., Korenstein, R., Jerby, E., and Avivi, L., 2003, Exposure of human peripheral blood lymphocytes to electromagnetic fields associated with cellular phones leads to chromosomal instability. Bioelectromagnetics 24: 82–90.

McNamee, J.P., Bellier, P.V., Gajda, G.B., Lavallée, B.F., Lemay, E.P., Marro, L., and Thansandote, A., 2002a, DNA damage in human leukocytes after acute in vitro exposure to a 1.9 GHz pulse-modulated radiofrequency field. Radiat Res 158: 534–537.

McNamee, J.P., Bellier, P.V., Gajda, G.B., Miller, S.M., Lemay, E.P., Lavallée, B.F., Marro, L., and Thansandote, A., 2002b, DNA damage and micronucleus induction in human leukocytes after acute in vitro exposure to a 1.9 GHz continuous-wave radiofrequency field. Radiat Res 158: 523–533.

McNamee, J.P., Bellier, P.V., Gajda, G.B., Lavallée, B.F., Marro, L., Lemay, E., and Thansandote, A., 2003, No evidence for genotoxic effects from 24 h exposure of human leukocytes to 1.9 GHz radiofrequency fields. Radiat Res 159: 693–697.

Meltz, M.L., 2003, Radiofrequency exposure and mammalian cell toxicity, genotoxicity, and transformation. Bioelectromagnetics Suppl 6: S196–S213.

Meltz, M.L., Eagan, P., and Erwin, D.N., 1990, Proflavin and microwave radiation: absence of a mutagenic interaction. Bioelectromagnetics 11: 149–157.

Merola, P., Marino, C., Lovisolo, G.A., Pinto,. R, Laconi, C., and Negroni, A., 2006, Proliferation and apoptosis in a neuroblastoma cell line exposed to 900 MHz modulated radiofrequency field. Bioelectromagnetics 27: 164–171.

Miyakoshi, J., Ohtsu, S., Tatsumi-Miyajima, J., and Takebe, H., 1994, A newly designed experimental system for exposure of mammalian cells to extremely low frequency magnetic fields. J Radiat Res (Tokyo) 35: 26–34.

Miyakoshi, J., Kitagawa, K., and Takebe, H., 1997, Mutation induction by high-density, 50-Hz magnetic fields in human MeWo cells exposed in the DNA synthesis phase. Int J Radiat Biol 71: 75–79.

Miyakoshi, J., Koji, Y., Wakasa, T., and Takebe, H., 1999, Long-term exposure to a magnetic field (5 mT at 60 Hz) increases X-ray-induced mutations. J Radiat Res (Tokyo) 40: 13–21.

Miyakoshi, J., Yoshida, M., Shibuya, K., and Hiraoka, M., 2000a, Exposure to strong magnetic fields at power frequency potentiates X-ray-induced DNA strand breaks. J Radiat Res (Tokyo) 41: 293–302.

Miyakoshi, J., Yoshida, M., Yaguchi, H., and Ding, G.R., 2000b, Exposure to extremely low frequency magnetic fields suppresses x-ray-induced transformation in mouse C3H10T1/2 cells. Biochem Biophys Res Commun 271: 323–327.

Miyakoshi, J., Mori, Y., Yaguchi, H., Ding, G., and Fujimori, A., 2000c, Suppression of heat-induced HSP-70 by simultaneous exposure to 50 mT magnetic field. Life Sci 66: 1187–1196.

Miyakoshi, J., Yoshida, M., Tarusawa, Y., Nojima, T., Wake, K., and Taki, M., 2002, Effects of high-frequency electromagnetic fields on dna strand breaks using comet assay method. Electr Eng Jpn 141: 9–15.

Miyakoshi, J., Takemasa, K., Takashima, Y., Ding, G.R., Hirose, H., and Koyama, S., 2005, Effects of exposure to a 1950 MHz radio frequency field on expression of Hsp70 and Hsp27 in human glioma cells. Bioelectromagnetics 26: 251–257.

Nakahara, T., Yaguchi, H., Yoshida, M., and Miyakoshi, J., 2002, Effects of exposure of CHO-K1 cells to a 10 T static magnetic field. Radiology 224: 817–822.

Natarajan, M., Nayak, B.K., Galindo, C., Mathu,r S.P., Roldan, F.N., and Meltz, M.L., 2006, Nuclear translocation and DNA-binding activity of NFKB (NF-kappaB) after exposure of human monocytes to pulsed ultra-wideband electromagnetic fields (1 kV/cm) fails to transactivate kappaB-dependent gene expression. Radiat Res 165: 645–654.

Nylund, R., and Leszczynski, D., 2006, Mobile phone radiation causes changes in gene and protein expression in human endothelial cell lines and the response seems to be genome- and proteome-dependent. Proteomics 6: 4769–4780.

Phillips, J.L., Ivaschuk, O., Ishida-Jones, T., Jones, R.A., Campbell-Beachler, M., and Haggren, W., 1998, DNA damage in Molt-4 T-lymphoblastoid cells exposed to cellular telephone radiofrequency fields in vitro. Bioelectrochem Bioener 45: 103–110.

Qutob, S.S., Chauhan, V., Bellier, P.V., Yauk, C.L., Douglas, G.R., Berndt, L., Williams, A., Gajda, G.B., Lemay, E., Thansandote, A., and McNamee, J.P., 2006, Microarray gene expression profiling of a human glioblastoma cell line exposed in vitro to a 1.9 GHz pulse-modulated radiofrequency field. Radiat Res 165: 636–644.

REFLEX, Risk Evaluation of Potential Environmental hazards From Low Frequency Electromagnetic Field Exposure Using Sensitive in vitro Methods, Final Report, 2004 (http://www.verumfoundation.de)

Remondini, D., Nylund, R., Reivinen, J., Poulletier de Gannes, F., Veyret, B. Lagroye, I., Haro, E., Trillo, M.A., Capri, M., Franceschi, C., Schlatterer, K., Gminski, R., Fitzner, R., Tauber, R., Schuderer, J., Kuster, N., Leszczynski, D., Bersani, F., and Maercker, C., 2006, Gene expression changes in human cells after exposure to mobile phone microwaves. Proteomics 6: 4745–4754.

Roti Roti, J.L., Malyapa, R.S., Bisht, K.S., Ahern, E.W., Moros, E.G., Pickard, W.F., and Straube, W.L., 2001, Neoplastic transformation in C3H 10T(1/2) cells after exposure to 835.62 MHz FDMA and 847.74 MHz CDMA radiations. Radiat Res 155: 239–247.

Sagripanti, J.L., and Swicord, M.L., 1986, DNA structural changes caused by microwave radiation. Int J Radiat Biol Relat Stud Phys Chem Med 50: 47–50.

Sakuma, N., Komatsubara., Y, Takeda, H., Hirose, H., Sekijima, M., Nojima, T., and Miyakoshi, J., 2006, DNA strand breaks are not induced in human cells exposed to 2.1425 GHz band CW and W-CDMA modulated radiofrequency fields allocated to mobile radio base stations. Bioelectromagnetics 27: 51–57.

Sanchez, S., Haro, E., Ruffié, G., Veyret, B., and Lagroye, I., 2007, In vitro study of the stress response of human skin cells to GSM-1800 mobile phone signals compared to UVB radiation and heat shock. Radiat Res 167: 572–580.

Speit, G., Schütz, P., and Hoffmann, H., 2006, Genotoxic effects of exposure to radiofrequency electromagnetic fields (RF-EMF) in cultured mammalian cells are not independently reproducible. Mutat Res 626: 42–47.

Stankiewicz, W., Dabrowski, M.P., Kubacki, R., Sobiczewska, E., and Szmigielski, S., 2006, Immunotropic influence of 900 MHz microwave GSM signal on human blood immune cells activated in vitro. Electromagn Biol Med 25: 45–51.

Stodolnik-Barańska, W., 1967, Lymphoblastoid transformation of lymphocytes in vitro after microwave irradiation. Nature 214: 102–103.

Stronati, L., Testa, A., Moquet, J., Edwards, A., Cordelli, E., Villani, P., Marino, C., Fresegna, A.M., Appolloni, M., and Lloyd, D., 2006, 935 MHz cellular phone radiation. An in vitro study of genotoxicity in human lymphocytes. Int J Radiat Biol 282: 339–346.

Takashima, Y., Hirose, H., Koyama, S., Suzuki, Y., Taki, M., and Miyakoshi, J., 2006, Effects of continuous and intermittent exposure to RF fields with a wide range of SARs on cell growth, survival, and cell cycle distribution. Bioelectromagnetics 27: 392–400.

Thorlin, T., Rouquette, J.M., Hamneriu, Y., Hansson, E., Persson, M., Björklund, U., Rosengren, L., Rönnbäck, L., and Persson, M., 2006, Exposure of cultured astroglial and microglial brain cells to 900 MHz microwave radiation. Radiat Res 166: 409–421.

Tian, F., Nakahara, T., Wake, K., Taki, M., and Miyakoshi, J., 2002, Exposure to 2.45 GHz electromagnetic fields induces hsp70 at a high SAR of more than 20 W/kg but not at 5W/kg in human glioma MO54 cells. Int J Radiat Biol 78: 433–440.

Tice, R.R., Hook, G.G., Donner, M., McRee, D.I., and Guy, A.W., 2002, Genotoxicity of radiofrequency signals. I. Investigation of DNA damage and micronuclei induction in cultured human blood cells. Bioelectromagnetics 23: 113–126.

Tuschl, H., Novak, W., and Molla-Djafari, H., 2006, In vitro effects of GSM modulated radiofrequency fields on human immune cells. Bioelectromagnetics 27: 188–196.

Verschaeva, L., 2005, Genetic effects of radiofrequency radiation (RFR). Toxicol Appl Pharmacol 207: S336–S341.

Vijayalaxmi, 2006, Cytogenetic studies in human blood lymphocytes exposed in vitro to 2.45 GHz or 8.2 GHz radiofrequency radiation. Radiat Res 166: 532–538.

Vijayalaxmi, and Obe, G., 2004, Controversial cytogenetic observations in mammalian somatic cells exposed to radiofrequency radiation. Radiat Res 162: 481–496.

Vijayalaxmi, Mohan, N., Meltz, M.L., and Wittler, M.A., 1997, Proliferation and cytogenetic studies in human blood lymphocytes exposed in vitro to 2450 MHz radiofrequency radiation. Int J Radiat Biol 72: 751–757.

Vijayalaxmi, Leal, B.Z., Meltz, M.L., Pickard, W.F., Bisht, K.S., Roti Rot, J.L., Straube, W.L., and Moros, E.G., 2001a, Cytogenetic studies in human blood lymphocytes exposed in vitro to radiofrequency radiation at a cellular telephone frequency (835.62 MHz, FDMA). Radiat Res 155: 113–121.

Vijayalaxmi, Bisht, K.S., Pickard, W.F., Meltz, M.L., Roti Roti, J.L., and Moros, E.G., 2001b, Chromosome damage and micronucleus formation in human blood lymphocytes exposed in vitro to radiofrequency radiation at a cellular telephone frequency (847.74 MHz, CDMA). Radiat Res 156: 430–432.

Wang, J., Sakurai, T., Koyama, S., Komatubara, Y., Suzuki, Y., Taki, M., and Miyakoshi, J., 2005, Effects of 2450 MHz electromagnetic fields with a wide range of SARs on methylcholanthrene-induced transformation in C3H10T1/2 cells. J Radiat Res (Tokyo) 46: 351–361.

Wang, J., Koyama, S., Komatsubara, Y., Suzuki, Y., Taki, M., and Miyakoshi, J., 2006, Effects of a 2450 MHz High-Frequency Electromagnetic Field With a Wide Range of SARs on the Induction of Heat-Shock Proteins in A172 Cells. Bioelectromagnetics 27: 479–488.

Whitehead, T.D., Moros, E.G., Brownstein, B.H., and Roti Roti, J.L., 2006, Gene expression does not change significantly in C3H 10T(1/2) cells after exposure to 847.74 CDMA or 835.62 FDMA radiofrequency radiation. Radiat Res 165: 626–635.

Yaguchi, H., Yoshida, M., Ejima, Y., and Miyakoshi, J., 1999, Effect of high-density extremely low frequency magnetic field on sister chromatid exchanges in mouse m5S cells. Mutat Res 440: 189–194.

Yaguchi, H., Yoshida, M., Ding, G.-R., Shingu, K., and Miyakoshi, J., 2000, Increased chromatid-type chromosomal aberrations in mouse m5S cells exposed to power-line frequency magnetic fields. Int J Radiat Biol 76: 1677–1684.

Zeng, Q., Chen, G., Weng, Y., Wang, L., Chiang, H., Lu, D., and Xu, Z., 2006, Effects of global system for mobile communications 1800 MHz radiofrequency electromagnetic fields on gene and protein expression in MCF-7 cells. Proteomics 6: 4732–4738.

Zeni, O., Chiavoni, A.S., Sannino, A., Antolini, A., Forigo, D., Bersani, F., and Scarfì, M.R., 2003, Lack of genotoxic effects (micronucleus induction) in human lymphocytes exposed in vitro to 900 MHz electromagnetic fields. Radiat Res 160: 152–158.

Zeni, O., Romanò, M., Perrotta, A., Lioi, M.B., Barbieri, R., d'Ambrosio, G., Massa, R., and Scarfì, M.R., 2005, Evaluation of genotoxic effects in human peripheral blood leukocytes following an acute in vitro exposure to 900 MHz radiofrequency fields. Bioelectromagnetics 26: 258–265.

Zeni, O., Di Pietro, R., d'Ambrosio, G., Massa, R., Capri, M., Naarala, J., Juutilainen, J., and Scarfì, M.R., 2007, Formation of reactive oxygen species in L929 cells after exposure to 900 MHz RF radiation with and without co-exposure to 3-chloro-4-(dichloromethyl)-5-hydroxy-2(5H)-furanone. Radiat Res 167: 306–311.

Zhang, M.B., He, J.L., Jin, L.F., and Lu, D.Q., 2002, Study of low-intensity 2450-MHz microwave exposure enhancing the genotoxic effects of mitomycin C using micronucleus test and comet assay in vitro. Biomed Environ Sci 15: 283–290.

Zhao, R., Zhang, S., Xu, Z., Ju, L., Lu, D., and Yao, G., 2007, Studying gene expression profile of rat neuron exposed to 1800MHz radiofrequency electromagnetic fields with cDNA microassay. Toxicology 235: 167–175.

Zotti-Martelli, L., Peccatori, M., Scarpato, R., and Migliore, L., 2000, Induction of micronuclei in human lymphocytes exposed in vitro to microwave radiation. Mutat Res 472: 51–58.

Chapter 2

Carcinogenic Effect of Wireless Communication Radiation in Rodents

James C. Lin

ABSTRACT

The potential health effects of radio frequency (RF) radiation associated with cellular mobile telephones and related wireless devices remain a focus of concern. Although our knowledge regarding the health effects of RF radiation has increased considerably, the scientific evidence on biological effects of RF radiation associated with these wireless devices is still tentative. The uncertainties persist, in part, because of the limited number and scope of studies that have been conducted. Aside from the lack of a scientific consensus on experimental studies that provide clear evidence either refuting or supporting the cancer induction or promotion potential of RF radiation from cell phones, there is a concern that an established effect from wireless radiation, however small, could have a considerable impact in terms of public health. This chapter provides an updated review on recent research results on cancer induction and promotion in normal and transgenic mice and rats subjected to prolonged or life-long exposure to modulation schemes such as GSM, TDMA, CDMA, UMTS, and others. A majority of the laboratory mouse and rat studies did not exhibit a significant difference in carcinogenic incidences between exposed and sham-exposed animals. Although this observation may be comforting from the

J. C. Lin University of Illinois-Chicago, Suite 1020 SEO (MC 154), 851 South Morgan Street, Chicago, IL, 60607-7053, USA, e-mail: lin@uic.edu

J.C. Lin (ed.), *Advances in Electromagnetic Fields in Living Systems, Health Effects of Cell Phone Radiation, Volume 5*,
DOI 10.1007/978-0-387-92736-7_2, © Springer Science+Business Media, LLC 2009

perspective of safety evaluation, most of the studies are one-of-a-kind investigations – only three mouse and perhaps four rat studies were designed as replication or confirmation studies. It is noteworthy that the findings of these studies have not been consistent, making it difficult to arrive at a definitive conclusion. It could be a major flaw that in a majority of the investigations, cage-control animals were not part of the investigation or were not included in the data analyses. Moreover, restraining the experimental animals during exposure could have introduced a stress factor, which further complicates interpretation of the results since stress has often been associated with cancer induction in these animals.

1. INTRODUCTION

The number of cellular mobile telephone subscribers worldwide is in the billions and continues to increase. It is very likely that the market penetration is such that more people have access to cellular mobile radio telephone service than electricity for power and light in some territories. At the same time, the use of cordless telephones, which emit radio frequency (RF) or microwaves, are gaining popularity in the home and office to the extent that they are replacing cord telephones. The ubiquity of wireless systems has raised concerns about the safety of human exposure to radio waves emitted by these telecommunication devices.

While the biological effect of RF radiation has been an important research topic for more than half a century, there are two aspects of this technology prodding the resurgence of research interest related to human health. First, the proliferation of base-station antennas across many urban, suburban, and rural landscapes, and the rise of ambient RF radiation levels in residential and office environments. Second, for the first time in human history, a RF source is located in proximity to the brain or central nervous system (CNS) of a large number of users. The antenna of some devices, e.g., cellular telephones and Bluetooth devices, is typically located next to the user's head, thus creating a potential for RF interaction with brain tissues.

It is well known that at sufficiently high intensity, RF radiation can interact thermally with the human body and produce deleterious effects. However, biological responses from gross tissue heating would be a minor consideration for exposure to RF fields emitted by these wireless communication devices, where the maximum permitted specific absorption rate (SAR) of RF energy is between 1.6 and 2.0 W/kg in biological tissue. Accordingly, recent attention has converged on possible effects that may occur following prolonged or lifelong RF exposure at low levels. There is a need to provide a better understanding of the health effects to safeguard the general population against possible harm from RF radiation.

This chapter provides an update of recent research results on the carcinogenic effects of RF radiation from cellular mobile and personal communication devices. Specifically, the topics included are experimental studies involving cancer induction and promotion, and long-term survival of exposed laboratory animals. Of particular interest is tumorigenesis in the brain, tumors that start in the brain.

The most aggressive malignant brain tumors are astrocytoma and glioblastoma multiforme. They lack distinct borders, reproduce rapidly, and invade and infiltrate widely. These tumors also induce the formation of new blood vessels, so they can maintain their aggressive growth. They have a necrotic core, areas of dead cells in their center that are hypoxic, deficient in oxygen. At present, the prognosis or prediction about the future course of most aggressive brain tumors is not very encouraging. The survival rate is about 1 month for watchful waiting, about 1 year with surgery and radiation therapy, and is improved when combined with some form of chemotherapy. Many slow-growing primary brain tumors are benign or the least malignant, and could take decades for symptoms to emerge in humans. They are usually associated with long-term survival.

The incidence rate for brain tumors in US is currently 16.5 per 100,000 person-years (CBTRUS, 2008). The rate is slightly higher in females than males. An estimated 51,000 new cases of primary nonmalignant and malignant brain and CNS tumors are diagnosed each year. Note that the prevalence rate for all pediatric (ages 0–19) primary brain and CNS tumors was estimated at 9.5 per 100,000 with more than 26,000 children estimated to be living with this diagnosis in US in 2000.

It is estimated that the worldwide incidence rate of primary malignant brain and CNS tumors, age-adjusted using the world standard population, is 3.7 per 100,000 person-years in males and 2.6 per 100,000 person-years in females (Ferlay et al., 2004). This represents an estimated 108,277 males and 81,305 females who were diagnosed with a primary malignant brain tumor in 2002, an overall total of 189,582 individuals. The incidence rates are higher in more developed countries (males: 5.8 per 100,000 person-years; females: 4.1 per 100,000 person-years) than in less developed countries (males: 3.0 per 100,000 person-years; females: 2.1 per 100,000 person-years).

1.1. Some Early Cancer Studies in Laboratory Animals

The potential for cancer induction has been a major cause of concern. However, until recently, there were only a few studies involving frequencies in the spectral bands used for wireless communication. These reports showed an accelerated development of spontaneous mammary tumors in mice or promotion of tumor growth in animals, if the tumor was first initiated by other means, following exposure to 800–2,500 MHz radiations (Szmigielski et al., 1982; Szudinski et al., 1982; Wu et al., 1994). Some of these studies used relatively high average SARs (6–12 W/kg) that can induce appreciable temperature increases in the animal body. Since chemical action is facilitated by thermal energy, RF-induced heating could have influenced the action of such chemical agents as benzopyrene or 12-*O*-tetradecanoylphorbol-13-acetate (TPA). However, the potential for thermal enhancement apparently did not have any influence on the action of dimethylhydrazine (DMH).

An investigation by Kunz et al. (1985) was designed to study the effects of pulsed microwave exposure on a large number of animals throughout their life-span, with special emphasis on general health status and longevity. (The Kunz et al. report contains full details of the study on which the Chou et al. (1992) paper was based.) Beginning at 8 weeks of age, Sprague–Dawley rats were irradiated by pulsed microwaves (10-μs rectangular pulses modulated at 8 Hz and pulsed at 800 pps, 0.15–0.4 W/kg SAR) for 25 months. A statistically significant increase was observed in primary malignancies at death in irradiated rats (18) vs. sham-irradiated controls (5). However, lifelong exposure did not reveal any significant effects on the general health of exposed rats. Furthermore, the survival curves were virtually identical for microwave and sham-exposed rats, and there was no difference during any phase of the rats' lifetime.

1.2. Studies in the Spectral Bands Used for Wireless Communication

One of the first studies using frequencies and modulations specific to mobile phones involved the use of implanted rat brain tumors (Salford et al., 1993). Like most scientific inquiries, this study began as a rational discovery of any potential causality between exposure to mobile phone radiation and brain tumor promotion. This study was followed by the use of an experimental animal model, Eμ-Pim1 transgenic mice, in a first-of-a-kind experiment to systematically investigate a dose–response relationship for any risk of cancer associated with cell phone RF exposure (Repacholi et al., 1997). The Eμ-Pim1 transgenic mice carry a lymphoma oncogene and are predisposed to developing lymphomas spontaneously. Although there are physiological differences, test results in rodent studies have often shown that the same organs are affected in humans and in rodents by known carcinogens (NTP, 1999). Since then, to help evaluate the possible health risk of cellular mobile telecommunication devices and systems, a substantial number of investigations have been conducted using mice and rats under controlled or good-laboratory-practice (GLP) conditions. These experiments generally adhere to prescribed protocols, akin to product or drug testing. A summary and analysis of recent results is presented in what follows.

2. CANCER IN MICE EXPOSED TO RF RADIATION FROM CELL PHONES

2.1. Lymphomas in Genetically Prone Mice: GSM Exposure

Lymphomas are a type of cancer that affects the lymphatic system, which is part of the body's immune system. Specifically, the lymphatic system is the body's blood-filtering tissue that helps fight infection and disease. As other cancers, lymphomas occur when cells divide too much and too fast. Symptoms of lymphomas include swelling in one or more groups of lymph nodes, fever, weakness, weight loss, and an enlarged liver and spleen (Cotran et al., 1994). There are two major types of lymphomas: Hodgkin's disease and non-Hodgkin's lymphoma. Moreover, a lymphoblastic lymphoma – medium-sized lymphoid cells with a high nucleo-cytoplasmic ratio – is the most common type of non-Hodgkin's lymphoma, especially in children. Lymphoblastic lymphomas are the less predictable type, and they are more likely to spread to areas beyond the lymph nodes. Because lymphomas impair the immune system, there is the risk of death from infection. An estimated 60,000 people a year in the United States are diagnosed with lymphomas: 53,000 with non-Hodgkin's lymphoma and 7,000 with Hodgkin's disease, according to the Lymphoma Research Foundation of America. In most cases, the cause is not known.

The clinical course for non-lymphoblastic lymphomas is less rapid than for lymphoblastic lymphomas. In mice, lymphoblastic lymphomas are usually seen in animals less than 10 months of age as a mediastinal mass with attendant respiratory distress and rapid clinical decline when the enlarging mass compresses the thorax. Non-lymphoblastic lymphomas occur predominantly in mice older than 10 months, generally cause progressively increasing abdominal distension, and are readily palpable. A variety of factors has been associated with an increased risk of developing lymphomas; specifically: congenital status, infectious agents such as viral and bacterial infections, and chemical and physical agents such as pesticides, solvents, arsenate, paint thinners, lead, hair dyes, and high-dose ionizing radiation exposure. These have all been shown to increase the incidence of lymphomas in humans.

2.1.1. Plane Wave Exposure of GenPharm Eμ-Pim1 Mice

A study was conducted in Australia in which the incidence of lymphomas in female Eμ-Pim1 transgenic mice was shown to be significantly higher (OR = 2.4; $P = 0.006$, 95% CI = 1.3–4.5) in the exposed mice (43%) than in the sham controls (22%), following two 30-min periods per day exposure to 900 MHz plane-wave radiation repeated at 217 Hz (signals that mimic global system for mobile communication (GSM) digital mobile phones) (Repacholi et al., 1997). Follicular lymphomas were the major contributor to the increased tumor incidence. At the end of the experiment, 53% of the exposed mice had lymphomas, compared with 22% of the unexposed controls. The exposed transgenic mice also recorded a faster onset of lymphomas. In this study, 100 mice were sham-exposed and 101 were exposed for up to 18 months. The pulse width was 0.6 ms. The average incident power density and SAR were 2.6–13 W/m^2 and 0.13–1.4 W/kg, respectively.

It should be noted that the Eμ-Pim1 transgenic mice were genetically engineered for a predisposition to lymphoma. Thus, the extrapolation of results found in a very sensitive animal model to possible carcinogenesis in humans is not well established. Moreover, this study suffered from two general types of identifiable deficiencies. One type was dosimetric in nature; specifically, the plane-wave-equivalent exposure system used in this study allowed mice to roam and huddle freely during exposure to incident power densities of 2.6–13 W/m^2. Consequently, there was a wide variation of SARs (0.008–4.2 W/kg, averaging 0.13–1.4 W/kg). Only an average response could be inferred from an average SAR, not an individual SAR. Moreover, it is conceivable that the higher incidence of lymphomas was associated with the higher SAR instead of the reported average SAR. Further, mice selected for necropsy during the experiment were not replaced with either other mice or tissue-equivalent phantoms, thus altering dosimetry in the remaining animals. There are also some critical shortcomings concerning the biological assay, methods, and procedures. The study lacked any standardized assessment criteria for deciding which mice would be selected for necropsy and surviving mice were disposed of without performing necropsy to ascertain whether there was infection and/or other relevant diseases, such as kidney failure, in those animals. Apparently, cage-control animals were not included as part of the experiment.

2.1.2. Ferris Wheel Exposure of Eμ-Pim1 Mice

Subsequently, another study (Utteridge et al., 2002) was set up to test the same central hypothesis as that of the earlier (Repacholi et al., 1997) study, but with refinements to overcome some of the perceived shortcomings. For example, the variation in SAR was reduced by restraining the mice and by using tissue-equivalent phantoms to replace autopsied mice. The new exposure system, supplied by Motorola, consisted of 15 lossy, radial, parallel-plate electromagnetic cavities (Ferris Wheel), configured for far-field operation. Each cavity had 40 mice restrained individually in clear Perplex tubes, cylindrically arranged around a dipole antenna. The tubes were constructed to prevent each mouse from changing its orientation relative to the field to facilitate SAR determination. The exposed groups were divided into four SAR levels: 0.25, 1.0, 2.0, and 4.0 W/kg. A standardized set of criteria (10% reduction in body mass over a week) was used for selecting mice for necropsy, and all surviving animals were necropsied. A total of 120 lymphoma-prone Eμ-Pim1 mice and 120 wild-type mice were exposed for 1 h/day, 5 days/week, at each of the four SAR levels, for up to 24 months. In addition, 120 Eμ-Pim1 and 120 wild-type mice were sham-exposed; there was also an unrestrained negative (cage) control group.

The paper concluded that the results of the double-blind study did not show an increase in lymphomas following a 2-year exposure to GSM cell phone radiation (Utteridge et al., 2002). Furthermore, there was no significant difference in the incidence of lymphomas between exposed and sham-exposed groups at any of the exposure levels (with one exception). A dose–response effect was not detected. The findings showed that long-term exposures of lymphoma-prone mice to 898.4 MHz (referred to as 900 MHz) GSM RF radiation at SARs of 0.25, 1.0, 2.0, and 4.0 W/kg had no significant effects when compared with that of sham-irradiated animals. This was in contrast to the previous study, which reported that long-term (18 months) exposure of lymphoma-prone mice significantly increased the incidence of nonlymphoblastic lymphomas when compared with sham-irradiated animals.

Because this study was designed to test the same central hypothesis as that of the earlier study (Repacholi et al., 1997) but with refinements to overcome some of the perceived shortcomings, the study deserves close examination.

To be sure, the latter was not a replication of the earlier study. A replication, as a standard practice of the scientific approach, requires that the same methods and materials are followed as in the earlier study. Given that there are major differences in materials and methods (beyond refinements), the design of the latter is more appropriately characterized as an attempt to confirm or refute, rather than replicate.

More significantly, close examination of the source of mice, exposure regime, animal restraint, and the omission of data from analysis in the later study could lead to a different conclusion than that stated in the publication. It was stated in the paper that the mice were supplied from the same source used in the earlier study, and listed Taconic Farms, New York, as the source. However, mice for the earlier study came from GenPharm International of Mountainview, California. Thus, the Eμ-Pim1 mice appear not to be the same after all. Even the same strain of mice, from different suppliers, may have different characteristics and may respond differently, a factor to be considered further.

Mice in the later study were exposed to daily 1-h sessions, while those in the earlier study were exposed for two 30-min periods per day. The biological effect of fragmenting exposure duration is not well known. However, diurnal variations and the temporal dependence of physiologic, cellular and molecular processes are well established. The use of free-roaming vs. restrained animals by themselves is not a problem so long as the effects on these mice are characterized, with appropriate cage controls. Unfortunately, data for the cage-control mice were missing from the publication (Utteridge et al., 2002). Restraining the animal in a tight tube during the exposure session constitutes a continuing stress to the animal, which may lead to significant stress responses that potentially could obscure any effect from the exposure to cell phone radiation.

There are also some rather glaring inconsistencies in the published data (see Lin, 2002). For example, some or all of the mice were dead after 18 or 20 months, according to one figure (Fig. 1), but they still had weight gains up to 26 months, according to another figure (Fig. 2) [Figs. 1 and 2 in Utteridge et al., 2002]. The study design included equal numbers of freely moving mice for negative controls (cage controls). However, data for the cage-control group were not given in the paper and appear to have been excluded from the statistical analyses. By not having the free-moving mice form a part of the statistical analysis group, the report was deprived of the pathophysiology of cage-control mice for comparison. The cage controls can and should serve as valuable background materials, which potentially might be masked by stress response induced by the restraining tube used for sham

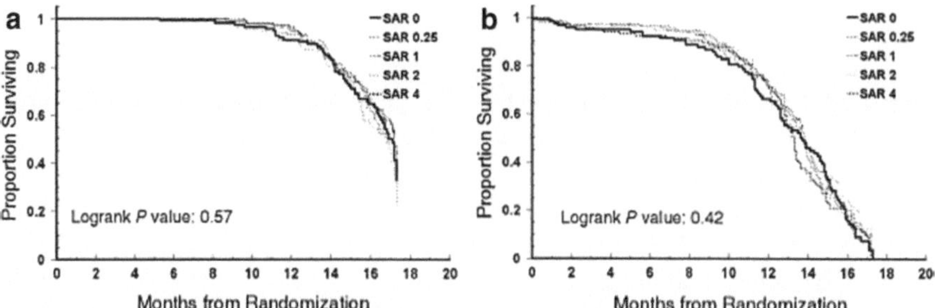

Figure 1. Survival curves for death from any cause for (**a**) wild-type mice and (**b**) heterozygous (transgenic) mice (Utteridge et al., 2002).

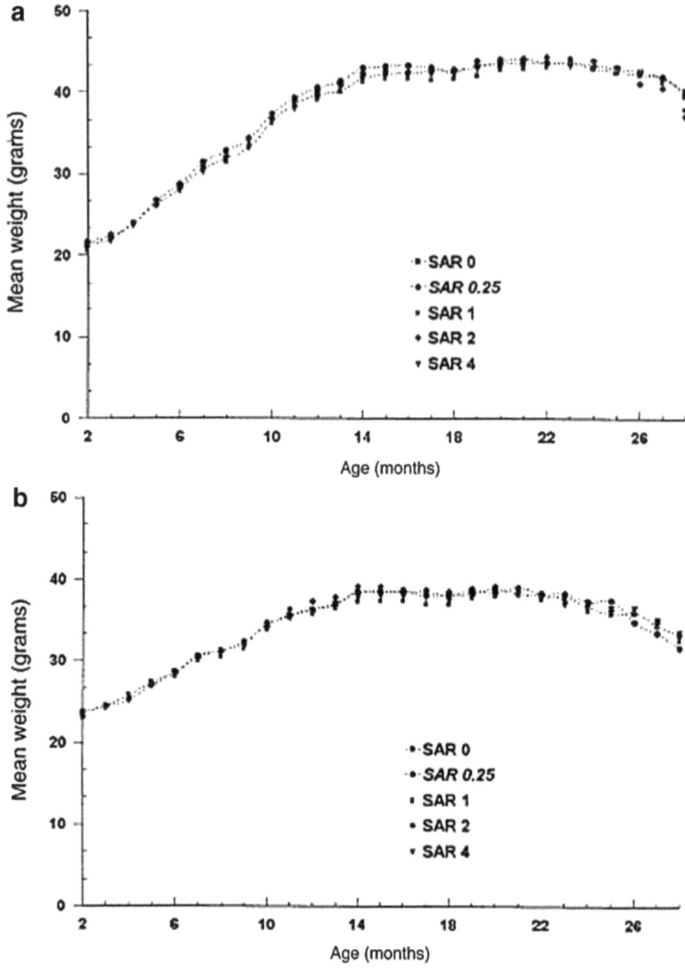

Figure 2. Distribution curves for weight gain by (**a**) wild-type mice and (**b**) heterozygous (transgenic) mice (Utteridge et al., 2002).

control. It is noteworthy that the incidence of lymphomas among the sham controls (SAR=0; mice are restrained but not exposed) was very high in this study. Specifically, among the transgenic mice, the incidence of lymphomas was 75% for the sham-control group (89 out of 120 mice developed lymphomas: 15 with lymphoblastic lymphomas, 74 with nonlymphoblastic lymphomas). In contrast, the incidence of lymphomas in the earlier study (Repacholi et al., 1997) was 22% for the sham-control mice (22 out of 100 mice developed the disease: 3 with lymphoblastic lymphomas, 19 with nonlymphoblastic lymphomas). The high degree of incidence in the sham controls (75 vs. 22%) makes the experimental protocol impractical.

It could have masked an effect from cell phones, or any other agent for that matter. It is unfeasible to come to any firm conclusions about lymphomas in transgenic mice exposed to cell phone radiation. These flaws – possibly in the sourcing or handling of mice, the statistical analysis of the data, or in the fundamental design of the experiment – limit the conclusions that can be drawn for the outcome of the Utteridge et al. study, despite the paper's claim.

Utteridge et al. (2003) have published a response to several comments (Kundi, 2003a; Lerchl, 2003; Goldstein et al., 2002, 2003) on their original article (Utteridge et al., 2002). Unfortunately, acceptability of results of the Utteridge et al. study has not been enhanced and clear, unambiguous data and information remain elusive for an unequivocal interpretation of the Utteridge et al. study (Kundi, 2003b). The need for other investigators to replicate or confirm these two studies (Repacholi et al., 1997; Utteridge et al., 2002) and to help appraise the acceptability and reliability of the reported results persisted for some time (Lin, 2008).

Later, a dosimetric evaluation of the Ferris-wheel exposure system used by Utteridge et al. (2002) for exposure of the Eμ-Pim1 transgenic mice to pulsed radiofrequency energy at 898.4 MHz was reported by Faraone et al. (2006). Twin-well calorimetry was used to measure the whole-body SAR of exposed mice. One major conclusion was that since the average lifetime weight was slightly higher than originally projected (30 g), the lifetime exposure received by the mice was somewhat less than anticipated. In particular, the mean lifetime exposure levels were lower by about 18% than the original targets for the wild-type mice and about 10% for the transgenic mice. Specifically, the lifetime average whole-body SARs were 0.21, 0.86, 1.7, and 3.4 W/kg for the four exposure groups. Infrared thermography showed SAR peaks in the abdomen, neck and head in thermograms taken over the sagittal plane of mouse cadavers. The peak local SAR (1-g) at these locations, determined by thermometric measurements, showed peak-to-average SAR ratios with typical values around 3:1, but some are close to 6:1. Thus, the average and peak SARs were slightly lower than originally reported in Utteridge et al. (2002).

2.1.3. Ferris Wheel Exposure of Taconic Pim1 Mice

The potential effect of chronic exposure to GSM-modulated 900 MHz fields and tumor development in mouse strains genetically predisposed to lymphoma development was the subject of a recent publication (Oberto et al., 2007). It was intended as a follow-up to the study by Repacholi et al. (1997) with improvements in dosimetry and methodology. The exposure system consisted of four Ferris Wheels and each wheel was composed of two parallel, circular, stainless-steel metal plates with a conical antenna in its center. Dosimetry was improved by restraining the mice in plastic tubes to obtain more uniform exposure. The incident field was adjusted as a function of body mass to obtain an age-independent exposure dose. Tissue-equivalent phantoms were used to replace necropsied mice to maintain a more consistent and symmetrical absorption profile. The study used identical RF signals as the previous study, i.e., animals were exposed to 217 Hz pulsed 900 MHz fields, but at average whole-body SARs of 0.5, 1.4, or 4.0 W/kg. In addition to whole body, dosimetric information about organ and spatial-average-peak SARs as well as their lifetime variations were reported. It is interesting to note that the ratio of organ or tissue average SAR to whole-body average SAR varied between 0.18 and 1.90. Moreover, the spatial peak SAR relative to the whole-body average SAR was as high as 62 and 85 for tissue mass of 5 and 0.5 mg, respectively.

At variance with Repacholi et al. (1997) and Utteridge et al. (2002), who used only female Eμ-Pim1 transgenic mice in their studies, this blinded study presented data on 500 female and male Eμ-Pim1 mice (250 females and 250 males purchased from Taconic Farms, New York). The animals were housed in a limited-access barrier rodent facility during the 20-day acclimatization period. The mice were trained to the exposure system before exposure started. Fifty female and 50 male mice were randomly selected for exposure at each SAR level (0.5, 1.4, or 4.0 W/kg), for sham exposure or as cage controls. The exposure was performed 1 h/day, 7 days/week for 18 consecutive months. Necropsy was performed on-site both for animals that died and for those that survived up to termination of the study.

The results of this study showed a large gender difference in the overall incidence of lymphomas in these Eμ-Pim1 transgenic mice. The incidence in females is two to three times higher than in males. The overall incidence of lymphomas did not show any relationship to GSM-900 exposure according to authors. In females, incidence was 52% in cage controls, 44% in sham-exposed controls, 36% at 0.5 W/kg, 60% at 1.4 W/kg, and 40% at 4.0 W/kg (Table 1). The corresponding incidences for males were 16%, 18%, 20%, 20%, and 6%, respectively. The results for malignant lymphoma (lymphoblastic and non-lymphoblastic) did not show any relationship to GSM-900 exposure in either sex. Specifically, in females, the individual group and combined incidence of malignant lymphoma, 46.4% (116/250) was substantially higher than the corresponding incidence for males, 16% (40/250). With the exception of males exposed at 0.5 W/kg, for which the incidence of lymphoblastic lymphoma was 50% of the total cases (5 out of 10), in all the other groups of both sexes, non-lymphoblastic lymphoma (mainly pleomorphic and follicular) was the prevailing type of lymphoma, similar to that of the Repacholi et al. (1997) and Utteridge et al. (2002) studies.

It was reported that for all tumors, there was no significant difference in the number of animals with tumors (incidence of tumors), regardless of malignancy or gender. However, the number of mice with tumors was about 20% higher in the cage controls than in the sham or any of the exposed groups (Table 2). The incidence of benign tumors in females did not show any significant differences among the various groups, while in males it was higher in the cage controls and in the 4.0 W/kg group than in the other groups. The incidence of malignant tumors did not vary significantly between the cage control and the exposed groups. However, the incidence was reduced by 34 and 57% at 4.0 W/kg for females and males, respectively.

At the end of the experiment, the incidence of lymphomas in decedents was 42% (cage controls), 41% (sham controls), 16.6% (0.5 W/kg), 37.5% (1.4 W/kg), and 37.5% (4.0 W/kg) in females and 9% (cage controls), 20% (sham controls), 25% (0.5 W/kg), 17.6% (1.4 W/kg), and 5.8% (4.0 W/kg) in males, respectively. Thus, these data did not show any increase in lymphomas in the exposed animals.

Table 1. Incidence of lymphomas (number of animals with tumor) in Eμ-Pim1Mice exposed to GSM-900 radiation (from Oberto et al., 2007)

		Cage control	Sham control	0.5 W/kg	1.4 W/kg	4.0 W/kg	Combined
Females	Total number of lymphomas (% of total)	26/50 (52%)	22/50 (44%)	18/50 (36%)	30/50 (60%)	20/50 (40%)	116/250 (46.4%)
	Pleomorphic/follicular	23	18	17	27	16	
	Small lymphocyte	0	2	0	0	2	
	Lymphoblastic	3	2	0	3	2	
	Plasma cells	0	0	1	0	0	
Males	Total number of lymphomas (% of total)	8/50 (16%)	9/50 (18%)	10/50 (20%)	10/50 (20%)	3/50 (6%)	40/250 (16%)
	Pleomorphic/follicular	5	6	3	5	3	
	Marginal zone	3	3	2	3	0	
	Lymphoblastic	0	0	5	1	0	
	Not specified	0	0	0	1	0	

Table 2. Overall incidence of tumors (number of animals with tumor) at any site in Eμ-Pim1 transgenic mice exposed to GSM-900 radiation (from Oberto et al., 2007)

		Cage control	Sham control	0.5 W/kg	1.4 W/kg	4.0 W/kg
All tumors	Females	41	34	33	38	33
	Males	23	16	16	14	18
	Total	64	50	49	52	51
Benign	Females	21	12	16	15	15
	Males	13	3	4	6	12
	Total	34	15	20	21	27
Malignant	Females	35	27	29	34	23
	Males	14	13	13	11	6
	Total	49	40	42	45	29

Compared with sham-exposed animals, mortality was higher in all the male groups exposed to GSM-900 radiation than in control groups. There was a significant ($P<0.05$) variation in the probability of death before the end of the study; however, it was not dose-dependent. In females, the only significant finding on survival was a reduction in time to death at 0.5 W/kg ($P<0.05$). Oberto et al. indicated that their study did not confirm the finding of a 2.0- to 2.4-fold increase in lymphomas (43% of exposed compared with 22% of sham control) by Repacholi et al. Indeed, they consider the finding by Repacholi et al. (1997) as incidental. Oberto et al. (2007) claimed that the culprit was the low tumor rates of the female Eμ-Pim1 transgenic mice used for sham controls. In the study by Repacholi et al., only 22% of the sham-control mice had lymphomas, whereas 44% of the sham-control female mice in their study had lymphomas.

2.1.4. Radial Waveguide Exposure of AKR/J Mice

The AKR/J mice genome carries the AK-virus, which leads within one year to spontaneous development of thymic lymphoblastic lymphoma. To investigate the effects of chronic exposure to GSM-modulated 900 MHz fields, a study using this strain of mice genetically predisposed to lymphoma development was chosen by Sommer et al. (2004). The unrestrained female mice were exposed for 24 h, 7 days a week at an average whole-body (10 g) SAR of 0.4 W/kg in radial waveguide, plane-wave-equivalent exposure systems, except for about 1 h per week for weighing and palpation, during which time the cages were cleaned. Animals without signs of disease were sacrificed and necropsied at about 46 weeks of age, but earlier for animals with signs of disease. The experimental design allowed the 160 exposed and 160 sham-exposed animals to be housed in the same room. The sham-exposed in this case had much lower field exposure values and therefore SARs than exposed mice, i.e., –65 dB; they are not true shams.

Since mice can move freely, the whole-body SAR varies with the animals' postures and positions inside their cages. The SAR was analyzed by numerical computation of field distributions inside the radial waveguide for five different configurations of the animals, which were assumed to be uniformly distributed in time. The five configurations are for groups of mice in the front and rear portions of the cage as well as for mice with heads, and left/right sides oriented toward the incident wave in an upright posture. The whole-body SAR was computed using simple homogeneous muscle phantoms (ellipsoids, 6 cm in length, 3 cm in diameter, and about 32 g in mass). The standard deviation of the whole-body SAR was found to be ±40%. Groups of anatomically shaped mice were used to evaluate the maximum localized SAR, which showed a maximum of value of 5.9 W/kg for 35 W of input power to the radial waveguide system.

The results of this 46-week study showed that compared with "sham-exposure," lymphoma-prone, female AKR/J mice exposed to 0.4 W/kg average whole-body SAR, 900 MHz GSM type radiation did not affect the incidence of lymphoma development. The median time for lymphoma development was 183 days for exposed mice or 193 days for "sham-exposed" mice, which was not statistically different. Cage controls were not included in this blinded study. Also, the high incidence in lymphoma development (~90%) for both the exposed and "sham-exposed" makes it a challenge to come to any firm conclusions about lymphomas in AKR/J mice exposed to mobile phone radiation. Further, the present experiment does not allow any conclusions about the onset of lymphomas or the kinetics of lymphoma development, since the animals were not sacrificed or examined at predetermined intervals, irrespective of clinical symptoms.

It is interesting to note that the exposure to GSM-900 RF fields had no influence on the absolute body mass of the female AKR/J mice. The rapid development of lymphoma in these mice was associated with a loss of individual body mass of about 9.2% in the exposed and 8.5% in the "sham-exposed" mice, but the group difference was not statistically significant. However, the relative gain in body mass of the female AKR/J mice was more pronounced in exposed than in "sham-exposed" animals and was statistically significant ($P<0.001$). If confirmed, this observation raises the intriguing question of potential trade-off between RF energy absorption and metabolism in the exposed or "sham-exposed" mice.

The plane-wave-equivalent exposure system, used in this study, has prompted some questions about whether the SAR might be higher than reported.

2.1.5. A Summary of Lymphomas in Transgentic Mice and GSM-900 Exposure

As discussed earlier, since the publication of Repacholi et al. in 1997, reporting a 2.0- to 2.4-fold increase of lymphomas in lymphoma-prone Eμ-Pim1 transgenic mice exposed to GSM-900 RF radiation compared with control animals, there have been two studies using the same strain of transgenic mice and one study using a different lymphoma-prone (AKR/J) strain of transgenic mice. However, in addition to the obvious difference in mouse strain, the latter study varies from the other three in exposure regimes and involved a single SAR value. Some of the key features of these studies are given in Table 3. It is obvious that there are major differences

Table 3. Tumor (lymphoma) incidence in transgenic female mice following exposure to GSM-900 fields

Reference	Exposure regime	Experimental animal (number) source	Control animal (number)	Exposure system	Whole-body SAR (W/kg)	Tumor (lymphoma) incidence[a]	Study design
Repacholi et al., 1997	(2) 30 min/day for 5-day/week; 18 months	Eμ-Pim1 Mice (101); GenPharm	Sham (100); cage (0)	Plane wave chamber (unrestrained)	0.008–4.2 (0.13-to-1.4 ave)	Ex – 43%; Sc – 22%; Cc – no data	Blind
Utteridge et al., 2002	1 h/day for 5 day/ week; 24 months	Eμ-Pim1 Mice (120); Taconic	Sham (120); cage (120)	Radial, parallel-plate cavities (Ferris Wheel) (restrained)	0.25, 1.0, 2.0, or 4.0 (Peak-to-average ratio 3:1–6:1)	Ex – 76% (0.25–73%; 1.0–72%; 2.0–77%; 4.0–82%); Sc – 75%; Cc – no data	Double-blind
Oberto et al., 2007	1 h/day, 7 day/week for 18 months	Eμ-Pim1Mice (50 for each SAR level); Taconic	Sham (50); cage (50)	Radial, parallel-plate cavities (Ferris Wheel) (restrained)	0.5, 1.4 or 4.0 (peak-to-average 62–85 times)	Ex– 45%; (0.5 – 36%; 1.4 – 60%; 4.0 – 40%); Sc – 44%; Cc – 52%	Blind
Sommer et al., 2004	23 h/day, 7 day/week for 46 weeks	AKR/J Mice (160); Jackson Lab	Sham (160); cage (0)	Plane-wave-equivalent radial waveguide; (unrestrained)	0.4 ± 40% SD; Local maximum 5.9	Ex – 85–90%; Sc – 85–90%; Cc – no data	Blind

Ex Exposed, Sc Sham control, Cc cage control

[a]Numbers in this (tumor) refer to SARs

among these studies. This summary will highlight some of the salient features of the three studies using Eμ-Pim1 transgenic female mice.

While all three studies used Eμ-Pim1 transgenic female mice and GSM-900 RF fields, they may be characterized at best as attempts to confirm or refute, rather than replicate, the earlier study. First, the exposure systems and protocols were different. Mice were free-roaming, not restrained, in a plane-wave exposure field for the initial study, but the Utteridge et al. and Oberto et al. studies used restrained animals in plastic tubes placed in radial waveguides for exposure. The resulting whole-body average SAR not only differed among the three studies but also varied between the two studies using restrained animals. Although the medium range of whole-body average SARs attained in the two subsequent studies was in the range of the average SARs reported in the earlier study, the tissue-specific exposure levels and peak-to-average SARs differed to a much larger degree (close to 100 times). Moreover, they varied even between the two studies using restrained animals in Ferris Wheel exposure systems.

The tumor incidence varied among all three studies. Cage-control data are available only from the Oberto et al. study, which exhibited a tumor incidence of 52%. The reported incidences of lymphomas in the sham controls are 22, 74, and 44% for the Repacholi et al., Utteridge et al., and Oberto et al. studies, respectively. (Since sham-control mice in Repacholi et al. were free-roaming, not restrained, it might be reasonably compared with the 52% in cage controls of Oberto et al.) Clearly, the incidence of lymphomas among the sham controls varied widely. Moreover, the restraining and sham exposure of mice are supposedly the same for the Utteridge et al. and Oberto et al. studies, but they presented totally different rates of tumor incidence, thus rendering a realistic comparison between and among these studies difficult, if not impossible. These flaws – possibly in the sourcing or handling of mice or in the fundamental design of the experiments – limit the conclusions that can be drawn.

It is noteworthy that the 46-week blinded study involving a different strain of female mice (AKR/J), which are also genetically predisposed to developing lymphomas cannot be regarded as a potential confirmation study. Specifically, the AKR/J mice were exposed for 24 h per day, 7 days per week at a single SAR of 0.4 W/kg. Cage controls were not included; the study was deprived of the pathophysiology of cage-control mice for comparison (Sommer et al., 2004). Further, the high incidence in lymphoma development makes it a challenge to come to any firm conclusion about lymphomas in AKR/J mice exposed to GSM-900 mobile-phone radiation.

2.2. Cancer Studies in Other Genetically Prone Mice

There are two reported investigations where transgenic mice were the experimental subjects. In one case, ODC transgenic K2 mice were used to study skin cancer induction and the other to study lymphomas from a 3G system.

2.2.1. Skin Cancers in ODC Transgenic Mice: GSM and DAMPS Exposures

The ODC transgenic K2 mice carries the human ornithine decarboxylase (ODC) gene in their genome. In one study, the effect of RF radiation from GSM-900 (operating at 902.5 MHz, 0.577 ms pulses, and 217 Hz modulation) and DAMPS on ultraviolet (UV)-induced skin cancer in female ODC transgenic mice was investigated (Heikkinen et al., 2003).

The DAMPS (digital advanced mobile phone system) is a second generation cell phone system developed for use in the US market; it has now been superseded by other technologies. It operates in the 800 and 900 MHz frequency bands with 30 kHz channels. Similar to GSM, DAMPS is a digital wireless communication system. However, it employs a different, noncompatible version of the Time Division Multiple Access (TDMA) technology. The frequency was 849 MHz for this DAMPS-849 study; the pulse duration was 6.67 ms and the pulse repetition frequency was 50 Hz.

In this study, groups of 50 transgenic female 12- to 15-week-old ODC-K2 mice were exposed for 1.5 h/day, 5 days a week, during the 52-week study (Heikkinen et al., 2003). Identical rectangular waveguide chambers were used for the RF and sham exposures. The mice were kept in small cylindrical acrylic restrainers that allowed the animals to turn around except for some larger ones toward the end of the experiment. Further, the placement of the restrainers in the transverse orientation of waveguide chambers prevented the mice from aligning their longitudinal axis parallel to the electric field. Each chamber accommodated the exposure of 25 mice at a time (additional animals from the same litters as the study animals were used to makeup for the capacity of chambers). The order of RF and sham exposures was changed weekly.

The whole-body average SAR was reported to be 0.5 W/kg in both the GSM and DAMPS groups; the whole-body average SARs were 4.0 and 1.5 W/kg, for a given pulse in the two respective groups. The maximum deviation of the SAR was estimated to be 30% both for the GSM-900 and DAMPS-849 groups.

The UV radiation was administered three times a week at a dose of 240 J/m^2 (1.2 times the human minimum erythemal dose) using lamps simulating the solar spectrum, except for the cage-control group. The protocol required UV exposures to precede RF exposures on Mondays and Fridays, and on Wednesdays the animals were exposed to RF first. Benign and malignant primary skin cancers developed in 6 (32%) of the transgenic animals, which underwent UV exposure and served as sham-exposed. Only one transgenic animal in the cage-control group developed a macroscopic skin tumor.

Among the number of mice available for histopathology, 12 were cage controls, and 21, 20, and 19 animals were in the GSM-, DAMPS-, and sham-exposed groups, respectively. The results showed that 5 (24%) and 8 (40%) of the GSM- and DAMPS-exposed mice developed macroscopic skin tumors, but neither the GSM nor DAMPS exposures had a statistically significant effect on the development of skin tumors in ODC transgenic mice. Moreover, GSM-900 and DAMPS-849 exposures did not appear to act as a cocarcinogenic to UV-induced skin tumors.

In spite of the small number of animals in this study, the results could be interpreted as comforting from the perspective of safety evaluation. Other limitations include the waveguide chamber exposure system, which likely produced highly selective absorption among the animals and, in principle, would have allowed the mouse closest to the source of RF energy to absorb most of the incident power. Although randomization of group assignment and daily placement of mice into the exposure chamber helped to ensure comparable long-term average exposure, they do not mitigate against the selective absorption that occurred during each exposure session. The selective absorption could have a confounding influence especially given the growth and maturation these mice experienced during the course of the study. Further, the dosimetric determinations are estimations of time and spatial average absorptions and they bear little relation to daily exposure or individual SAR or their distribution inside the animal body. It should be noted that the histology slides were evaluated in the blind except for the cage controls. This is also the case for other studies by this group of investigators (Heikkinen et al., 2001, 2003).

2.2.2. Lymphomas in Genetically Prone Mice: UMTS Exposure

The Universal Mobile Telecommunication System (UMTS) is a technical standard for third generation (3G) wireless communication. It uses a pair of 5 MHz channels in the frequency bands of 1,885–2,025 MHz and 2,110–2,200 MHz, for uplink (from user to base station) and downlink (from base station to user), respectively. It supports up to 2 Mbit/s data transfer rates, although the performance is around 64 kbit/s in the most heavily loaded system, but it is still higher compared with the typical 14.4 kbit/s of a GSM data channel and offers the prospect of practical, inexpensive access to the Internet on a mobile device. In the most commonly applied frequency division access mode, users are separated by different codes, a high data rate modulation (3.84 Mbit/s chiprate) on top of the basic 5 MHz information rate. This fact influences the total radiated power from base station antennas.

For the UMTS system, signals from all users must arrive at the base station with approximately the same power level. Thus, strict and fast power control is enforced at a rate of 1,500 Hz with steps as small as 1 dB. This means that the power radiated from a handset (and thus the SAR) will have a 1,500 Hz component. The maximum power radiated from a handset is governed by different classes. The most common is class IV with a maximum radiated mean power of 125 mW. (This is a factor of 2 less than the maximum mean power for GSM). In practice, the power radiated may be much less if the distance to the base station is short. For a small urban cell, the mean value could be as low as −6 dBm (6 dB below 1.0 mW, i.e., 0.25 mW). For a larger rural cell, a much higher fraction of the powers would be near the maximum value. However, SAR may vary with chip rates and the rates of power fluctuations associated with the automatic power level control feature (APC).

The effect of chronic exposure to UMTS fields on the development of lymphoma was investigated in a blind study using lymphoma-prone transgenic AKR/J female mice by the same group that reported on GSM exposures of AKR/J mice

(Sommer et al., 2007). The animals were obtained from the Jackson Laboratory (Bar Harbor, USA) at an age of about 7 weeks and were acclimated for 1 week before random assignments into the experimental groups. Unrestrained mice were exposed (160) or sham-exposed (160) in the same room in two identical radial waveguide exposure systems. The cage controls (30) were also kept in the same room. The female AKR/J mice were exposed or sham-exposed for 43 weeks to a modulated 1.966-GHz UMTS test signal for 24 h per day, 7 days per week at an average whole-body SAR of 0.4 W/kg. The UMTS fields received were different by more than –65 dB between the exposed and sham-exposed mice. Animals visibly diseased or older than 43 weeks were killed, and tissue slices were examined for metastatic infiltrations and lymphoma type.

Authors have reported that the 43-week-long exposure to UMTS-modulated fields did not have a negative influence on growth or lymphoma development in female AKR/J mice compared with sham-exposed animals. Indeed, as shown in Table 4, there is a nonsignificant trend toward a lower percentage in the incidence of lymphomas for the exposed mice when compared with the sham-exposed and cage-control animals. However, cage control AKR/J mice had a significantly lower mean body mass than those exposed in the radial waveguides. The median survival times were comparable among all experimental groups. However, the percentages of mice that survived to the end of the experiment were 17.5, 8.8, and 3.3, respectively, for exposed, sham-exposed and cage controls. Thus, a significantly higher percentage of the survivors were exposed mice.

It is difficult to arrive at a firm conclusion concerning lymphomas in AKR/J mice exposed to mobile phone radiation since the incidence of lymphoma development for the AKR/J strain of lymphoma-prone transgenic female mice is extremely high (88–96%) and not be obscured by it. Although a given set of data may show no negative effects from the mobile-phone radiation exposure, it is not obvious to what extent of increase or decrease in the incidence of lymphomas would constitute a significant change in the tumor incidence.

Apparently, the exposure was fairly uniform since the overall spatial variation of the field in the cage regions was 17.7%. While not restraining the animals minimizes the potential stress response induced by restraining, it also complicates dosimetry. It is well known that the distribution of absorbed energy varies with body

Table 4. Mean body mass, final survival, lymphoma incidence and median survival time of female AKR/J mice chronically exposed to UMTS fields (from Sommer et al., 2007)

	RF-exposed mice	Sham-exposed mice	Cage-control mice
Mean body mass			
Beginning of study (g)	24.6±2.4 SD	24.4±2.6 SD	24.7±2.5 SD
End of study (g)	40.4±4.8	38.9±4.6	27.2±0.0
Final survival	28/160 (17.5%)	14/160 (8.8%)	1/30 (3.3%)
Lymphoma incidence	141/160 (88.1%)	149/160 (93.1%)	29/30 (96.7%)
Median survival time (days)	172	165	166

posture and from location to location inside the animal's body, even though the exposure field is uniform. Thus, a standard deviation of mean whole-body SAR of 50%, while comforting could mean as much as sixfold variations in peak SAR in local tissues and organs. The actual SAR could be much higher or lower than reported. This observation would also apply to the other study using AKR/J mice and radial waveguide exposure systems by the same group of investigators (Sommer et al., 2004).

2.3. Cancer Promotion and Induction in Normal or Nontransgenic Mice

There are four reported cancer studies in normal or nontransgenic mice: skin cancer promotion in CD-1 female mice, X-ray-induced tumors in mice, cocarcinogenesis of skin cancer in nontransgenic ODC mice, and carcinogenic potential in female and male B6C3F1 mice.

2.3.1. Skin Cancer in DMBA Treated CD-1 Mice: TDMA Exposure

The CD-1 mouse model for cancer initiation/promotion has been used to examine the potential for cell phone fields to promote skin cancer after a single dose of the carcinogen 7,12-dimethylbenz[α]anthracene (DMBA) in a medium-term bioassay (Imaida et al., 2001). Since the combination of DMBA initiation and 12-*O*-tetradecanoylphorbol-13-acetate (TPA) promotion is routinely used to study carcinogenesis, TPA was used for positive control. In this study, 10-week-old female CD-1 mice were treated with a single application of DMBA on shaved dorsal skin. A week later, mice were divided into four groups: 48 for sham exposure (DMBA-sham), 48 for RF exposure (DMBA-RF), 30 for positive controls (DMBA-TPA), and 30 as cage controls (DMBA-control).

Mice were exposed dorsally to 1,439 MHz RF radiation in individual chambers lined with absorbing materials in the near field of a monopole antenna using TDMA-1500 signals of the personal digital cellular (PDC) phone. The 19-week exposure was carried on for 1.5 h/day, 5 day/week, at a dorsal skin local peak SAR of 2.0 W/kg, with a whole-body average SAR of 0.084 W/kg. The fact that the ratio of peak to average SAR was 24 is irrelevant and misleading because of localized exposure in the near field of the antenna. It was not a whole-body exposure scenario.

The results showed that the incidences of skin cancers in DMBA-RF, DMBA-Sham, DMBA-TPA, and DMBA-Control groups were 0/48 (0%), 0/48 (0%), 29/30 (96.6%), and 1/30 (3.3%), respectively. As expected, the incidence in the DMBA-TPA group was nearly 100%; tumor response sensitivity of CD-1 mouse skin to this pair is well known. These results indicate that near-field exposure to TDMA-1500 fields did not indicate a promotional effect on skin tumorigenesis initiated by DMBA.

2.3.2. X-Ray-Induced Tumors in Mice: GSM and NMT Exposures

The capacity of cell phone RF radiation to act as a cancer promoter was the subject of an investigation examining the cell phone's effect on the development of cancers initiated in mice by ionizing radiation (Heikkinen et al., 2001). Ionizing radiation was selected as an initiator because it is known to induce leukemia and lymphomas as well as several other types of cancers in mice. Young female CBA/S mice (3- to 5-week old) were randomized into four groups of 50 mice: cage control, sham, and two groups of RF-exposed animals. Except for the cage-control group, all mice were irradiated by X-rays at the beginning of the study and then to cell phone RF radiation for 1.5 h per day, 5 days a week for 78 weeks.

The total-body X-ray dose was 4 Gy delivered as three equal fractions of 1.33 Gy at 1-week intervals with linear accelerators. Appropriate steps were taken to ensure uniform irradiations of the whole body. The cell phone exposure started on the day of exposure to the ionizing radiation.

The two types of RF exposures were signals from the analog NMT (Nordic Mobile Telephony) system at 902.5 MHz used mainly in North European countries, and the digital GSM system operating at 902.4 MHz. The exposures involved three identical rectangular waveguide chambers; the same as those used by this group in another study mentioned above (see Heikkinen et al., 2003). The average whole-body SAR was 0.35 and 1.5 W/kg for the GSM-900 and NMT-900 groups, respectively.

The survival rate of mice in the cage-control group was significantly higher at 96% compared with 68% in the sham-exposed group; cage controls were not exposed to ionizing radiation. The survival rates of 68%, 66%, and 68% in the GSM, NMT and sham-exposed groups, respectively, were similar in the exposed and sham-exposed groups. Specifically, the results showed that the proportion of X-ray irradiated mice with any neoplasm were 94%, 98%, and 98% in the GSM, NMT, and sham-exposed groups, respectively. Exposure to cell phone radiation did not significantly increase the incidence of any primary neoplasm in the tissues examined. The overall incidence of primary malignant neoplasm was 50%, 56%, and 40% in the GSM, NMT, and sham-exposed groups, respectively. The corresponding incidences of benign neoplasm were 82%, 76%, and 84%.

Although the results of this study do not suggest cancer promotion by RF radiation from GSM-900 or NMT-900 cell phones, the proportion of X-ray irradiated mice with any neoplasm was as high as 100% in all exposed groups, irrespective of exposure conditions. It should also be mentioned that a particular limitation or uncertainty surrounding this study is use of the waveguide chamber exposure system, which likely produced highly selective absorption among the animals. Further, the dosimetric determinations are estimations of time and spatial average absorptions and they bear little relation to daily exposure or individual SAR or their distribution inside the animal body. Some of the animals may have encountered either considerably lower or higher SARs during a given exposure session, which would be washed out in the averaged responses.

2.3.3. Skin Cancer in Nontransgenic Mice: GSM and DAMPS Exposures

A parallel study of the potential cocarcinogenic effect of GSM-900 and DAMPS-849 fields in ODC nontransgenic mice was conducted by Heikkinen et al. (2003). The study design and protocol were the same as those for the UV-induced skin cancer work in transgenic female ODC mice described above, except for the use of ODC nontransgenic mice. Female mice were exposed for 1.5 h/day, 5 days a week, during the 52-week study at a whole-body average SAR of 0.5 W/kg in rectangular waveguide chambers. Among the mice available for histopathology, 8 were cage controls, and 27, 26, and 26 were in the GSM-, DAMPS-, and sham-exposed groups, respectively. Microscopic skin tumors developed in 3 (11.5%) mice that were subjected to UV exposure and served as sham-exposed. None in the cage-control group developed a macroscopic skin tumor. The exposure results showed that 4 (15%) and 5 (19%) of the GSM-900 and DAMPS-849 exposed mice developed macroscopic skin tumors, but neither the GSM nor DAMPS exposures had a statistically significant effect on the development of skin tumors in ODC nontransgenic mice. Further, GSM and DAMPS fields did not appear to act as a cocarcinogenic to UV-induced skin tumors.

2.3.4. Cancer Induction in B6C3F1 Mice: GSM and DCS Exposures

The carcinogenic potential from exposure to GSM and digital cellular system (DCS) fields operating at 902 and 1,747 MHz, respectively, was studied by Tillmann et al. (2007). The study involved a large number (1,170) of female and male B6C3F1 mice. This strain of mice is a first-generation hybrid strain produced by crossing C57BL/6 females with C3H males. The animals were 8–9 weeks of age at the start of RF exposures. The DCS system is commonly known as DCS 1800 and is a mobile communication system that operates in the 1,710–1,880 MHz region of the radio frequency spectrum. It uses the spectrum between 1,710 and 1,785 MHz for uplink and 1,805 and 1,880 MHz for downlink operations, respectively. Standard signaling schemes were used in this study. The study design included groups of 50 B6C3F1 mice of each sex for cage control, sham, GSM-900 and DCS-1800 exposures at whole-body averaged SARs of 0.4, 1.3, and 4.0 W/kg for 2 h per day, 5 days per week for 2 years. The sham- and RF-exposed groups were housed in the same room. It should be noted that while 100 mice were designated as cage controls, they were not included in any comparison among various study groups. Instead, the publication included the statement, "comparison to published tumor rates in untreated mice revealed that the observed tumor rates were within the range of historical control data."

The RF exposure was conducted using "Ferris Wheel" chambers developed for the two frequencies of interest. Each chamber supported the simultaneous exposure of up to 65 mice restrained in plastic tubes. The GSM-900 and DCS-1800 exposures were conducted during the same time of the year, under essentially the same technical, laboratory, and environmental conditions. Corresponding to the whole-body average SARs of 0.4, 1.3, and 4.0 W/kg, the maximum average SAR during an active burst was 3.7, 11.1, 33.2 W/kg, respectively. The average absorption in the brain of a mouse was 2.5 W/kg for GSM and 5 W/kg for DCS. It should be noted that while the incident field was adjusted to maintain the same exposure level, independent of the animal's mass or age, the average uncertainty for SAR was ±400 and 200% for GSM and DCS, respectively. Moreover, the spatial peak SAR at

4 W/kg may be as high as 250 W/kg for GSM and 30 W/kg for DCS. Obviously, the SARs varied widely under both GSM and DCS exposures.

For GSM-900 exposures, the results showed that while the number of tumor-bearing B6C3F1 female mice (77%) at all levels was about 18% higher than in males (65%), they were not significantly different from the sham exposure group in either females or males (Tillmann et al., 2007). Also, the results did not show any significant increase in the incidence of any particular organ-specific tumor type in the GSM-exposed compared to the sham-exposed. Likewise, the incidence of hepatocellular carcinomas was similar in GSM- and sham-exposed groups. However, there appeared to be a dose-dependent decrease of the incidence of hepatocellular adenomas in males. Further, the decrease of hepatocellular adenomas in males exposed to 4.0 W/kg was significantly different ($P=0.048$) from that in the sham-exposed males.

In DCS-exposed mice, the incidence of tumor-bearing females was highest (37/50, 74%) in the sham-exposed group, but it was not significantly different from the 31/50 (62%), 35/50 (70%), and 33/50 (66%) for 0.4, 1.3, and 4.0 W/kg groups, respectively. However, there was a distinct dose-dependent decrease in the incidence of tumor-bearing males compared with the sham-exposed group. Specifically, the incidence was 37/50 (74%) in the sham-exposed group and 30/50 (60%; $P=0.202$), 25/50 (50%; $P=0.023$), and 24/50 (48%; $P=0.013$) in the three respective SAR levels. Again, while the incidence of hepatocellular carcinomas was similar in DCS and sham-exposed groups; in male B6C3Fl mice, there was a dose-dependent decrease. Moreover, the decrease in males exposed to 4.0 W/kg was significantly different ($P=0.015$) from that in the sham-exposed.

2.3.5. A Summary of Cancer Studies in Other Genetically Prone and Nontransgenic Mice

The two reports in which other strains of transgenic mice were the experimental subjects differed in nearly every aspect of the experiments: the strain of mice, RF field, exposure regime study design, and tumor type, but they used comparable SARs (Table 5). The overall results from these studies showed no difference in cancer incidence from prolonged GSM or UMTS fields except for a nonsignificant trend toward a lower incidence of lymphomas for the UMTS-exposed AKR/J mice when compared with the cage controls.

Among the studies of cancers in nontransgenic or normal mice, only one was a 2-year or life-long study, others varied from 19 to 78 weeks. There were two investigations on the promotional or cocarcinogenic potential for DMBA- and UV-induced skin cancers in CD-1 mice for 19-week exposures to TDMA fields, and ODC mice for 52-week exposures to GSM and DAMPS modulations, respectively. In both cases, the animals were partially restrained. These experiments did not indicate a promotional or cocarcinogenic effect on skin tumorigenesis (Table 6).

The 2-year study on the carcinogenic potential in female and male B6C3F1 mice is especially worthy of note in several regards. While exposure of male and female B6C3F1 mice to wireless GSM-900 and DCS-1800 fields did not show any overall

Table 5. Cancer studies in genetically prone ODC-K2 and AKR/J mice

Reference	Signal, modulation, exposure regime	Mice (number)	Cocarcinogen	Exposure system	SAR(W/kg) (whole-body, local peak)	Tumor type or location (results)	Study design
Heikkinen et al., 2003	GSM-900 DAMPS-849; 1.5 h/day, 5 day/week, 52 week	ODC-K2 (200 female; 50 exposed, 50 sham)	UV radiation, 3 times/week at 240 J/m^2	Waveguide chamber (partially restrained)	0.5 (1.5–4.0)	Skin (no difference)	Cocarcinogen; RF, sham, cage control
Sommer et al., 2007	UMTS-1.97; 24 h/day, 7 day/week, 43 weeks	AKR/J (350 female; 160 exposed 160 sham)	None	Radial waveguide (unrestrained)	0.4	Lymphoma (no difference; trend toward lower incidence for exposed)	RF, sham, cage control

Table 6. Cancer promotion and induction in normal or nontransgenic mice

Reference	Signal, modulation, exposure regime	Mice (number)	Cocarcinogen	Exposure system	SAR (W/kg) (whole-body, local peak)	Tumor type or location (results)	Study design
Imaida et al., 2001	TDMA-1500 1.5 h/day, 5 day/week, 19 weeks	CD-1 (156 female; 48 and 30 per group)	DMBA on dorsal skin	Monopole (near field, partially restrained)	0.8 (2.0)	Skin (no difference)	Cocarcinogen RF, sham, cage control
Heikkinen et al., 2001	GSM-900 NMT-900; 1.5 h/day, 5 day/week, 78 weeks	CBA/S (200 female; 50 each in 4 groups)	X rays, 4 Gy total body in 3 weeks	Waveguide chamber (partially restrained)	0.35 GSM-900; 1.5 NMT-900	Neoplasm (no difference)	Cocarcinogen; RF, sham, cage control (no X-ray)
Heikkinen et al., 2003	GSM-900 DAMPS-849; 1.5 h/day, 5 day/week, 52 weeks	ODC-Nontransgenic (200 female; 50 exposed, sham, cage)	UV radiation, 3 times/week at 240 J/m^2	Waveguide chamber (partially restrained)	0.5 (1.5–4.0)	Skin (no difference)	Cocarcinogen; RF, sham, cage control
Tillmann et al., 2007	GSM-900; DCS-1800; 2 h/day, 5 day/week, 2 years	B6C3F1 (1170 female/male; 50/group)	None	Ferris Wheel chambers (restrained)	0.4, 1.3, and 4.0 (brain ave 2.5 for GSM and 5 for DCS)	Lymphoma (no general difference; SAR-dependent decrease in males)	RF, sham, cage control

increase in the incidence of tumors, there was a dose-dependent decrease in the number of tumor-bearing males and more so for incidence of hepatocellular carcinomas. The SARs in restrained mice varied widely (by as much as 85-fold) for both GSM and DCS exposure in "Ferris Wheel" chambers, although the incident field was adjusted to maintain the same exposure level, independent of the animal's mass or age.

3. TUMOR INDUCTION AND PROMOTION IN RATS

The carcinogenic and cocarcinogenic potentials of RF electromagnetic fields employed for cellular mobile telephone systems have been the subject of several investigations using three different strains of laboratory rats. To date, the published reports include 8, 3, and 5 studies using Fischer 344, Wistar, and Sprague–Dawley rats, respectively. In some cases, the animals were restrained during exposure and others were not, under either plane-wave equivalent or near-zone exposure conditions. These tests were typically two years in duration. However, there was a 6-week liver bioassay study by Imaida et al. (1998), and an implanted brain tumor study in rats irradiated for 2–3 weeks following glioma cell implantation; these animals typically die of glioma 2–3 weeks after glioma cell implantation (Salford et al., 1993). The following section will begin with a summary of the short-term studies using Fischer 344 rats.

3.1. Implanted Brain Tumors in Fischer 344 Rats

The first study using frequencies and modulations specific to cellular mobile phones and implanted brain tumors did not show any significant difference in tumor growth between microwave- and sham-exposed rats (Salford et al., 1993). In particular, the study used pulse-modulated 915 MHz RF fields and two rat glioma models of central nervous system tumors (RG2 and N32). It should be noted that gliomas, including astrocytomas and glioblastomas, are the most common malignancy of the central nervous system in adult humans, and the prognosis is extremely poor. The growth rate of N32, a glioma cell line, is approximately one-half that of RG2 tumor type. (The RG2 tumor model is an ethylnitrosourea-induced cell line, which grows in cell culture in vitro, and provides a reproducible glioma model when inoculated into the brain.) In both cases, tumor cells were injected stereotaxically into the right caudate nucleus of male and female rats (37 experimental and 37 matched-sham-control Fisher 344 rats, 150–250 g). Starting on the fifth day after inoculation, intact (unanesthetized) animals were either RF- or sham-irradiated in individual TEM exposure chambers for 7 h/day, 5 days/week for 2–3 weeks. The modulation characteristics were 0.57 ms wide, 1 W pulses repeated at 0, 4, 8.33, 16, 50, or 217 Hz. The reported SARs were 0.008–0.4 W/kg. At 50 Hz modulation, the pulse width was 6.67 ms and peak power was 2 W, which produced SARs of 1.0 W/kg. Results from histopathological examinations indicate that repeated exposure to mobile phone RF fields did not promote growth of either the faster or the slower growing implanted gliomas beyond their normal course. Note that these animals typically die of glioma 2–3 weeks after glioma cell implantation.

3.2. Promotion of Chemically Induced Rat Liver Cancer

The potential for cancer promotion by local exposure to pulse modulated fields was investigated in a medium-term bioassay employing chemically-induced rat liver carcinogenesis (Imaida et al., 1998). Male Fischer 344 rats (48 exposed and 48 sham-exposed, 6-week old initially, at week 0) were given a single dose of diethylnitrosamine (DEN, 200 mg/kg body mass, I.P.). Exposure began 2 weeks later and lasted for 6 weeks. The exposure to the near-field 929.2 MHz TDMA signal for PDC (PDC, Japanese cellular telephone standard) was directed to the lateral mid-section of the rat body through a quarter-wavelength monopole antenna. The maximum local SARs (temporal average) were 6.6–7.2 W/kg for the whole body and 1.7–2.0 W/kg within the liver, the target organ. Temporal peak SARs were three times higher due to the duty ratio of the PDC signal. (Although less relevant, the whole-body average SARs were 0.58–0.80 W/kg.) The animals were exposed for 90 min a day, 5 days a week, for 6 weeks. At week 3, all rats were subjected to a 2/3 partial hepatectomy. At the end of the 6-week exposure period when these young animals were 14 weeks of age, the experiment was terminated and all animals were killed. Carcinogenic potential was scored by comparing the numbers and areas of the induced glutathione *S*-transferase placental form (GST-P) positive foci in the livers of the exposed and sham-exposed rats. Another group of 24 rats, given only DEN and partial hepatectomy, served as the controls. The numbers (no./cm^2) of GST-P positive foci were 4.61 ± 1.77, 5.21 ± 1.92 ($P<0.05$, vs. control), and 4.09 ± 1.47 and the areas (mm^2/cm^2) were 0.30 ± 0.16, 0.36 ± 0.21, and 0.28 ± 0.15, for the exposed, sham-exposed and control groups, respectively. There are no significant differences between the exposed and sham-exposed groups. These findings showed that local body exposure to a 929.2 MHz field with a PDC modulation does not have a significant effect on rat liver carcinogenesis under the experimental conditions employed. It should be noted that these are young animals and there did not appear to be any positive controls.

3.3. Tumor Induction or Promotion in Chronically Exposed Fischer 344 Rats

In a study that included fetal exposure, offsprings of pregnant Fischer 344 rats were tested for spontaneous tumorigenicity and the incidence of induced CNS tumors after a single dose of the carcinogen, *N*-ethylnitrosourea (ENU) in utero, followed by exposure to 836 MHz TDMA signals pulse-modulated at 50 Hz. The protocol involved both plane-wave-like far-field and near-field exposures (Adey et al., 1999). Far-field exposure of pregnant dams began on gestational day 19, and later with offspring in their cage up to weaning at 21 days of age. RF exposure was suspended until all pups were weaned. Near-field exposures began after weaning and continued for the next 22 months, with each rat in individual restraints for four consecutive days weekly, 2 h/day.

For far-field exposures, rat cages were positioned in a vertically oriented 3×3 matrix at the square aperture of a large tapered horn radiator (2.0 m on a side). Sham exposures were made in a square chamber of identical dimensions and materials. The power density at the center of the horn aperture was 26 ± 5.0 W/m^2, and it was within 1.6 dB across the cage exposure area. However, no SAR was given. Circular polarization was used to reduce possible orientation-dependent coupling to the animals, because dams and pups were free to move about their cages. Apparently, cage-control animals were not included in this study.

Near-field exposure was provided by a carousel-type exposure system with 10 rats oriented radially around a central antenna. To accommodate 120 rats simultaneously (60 exposed, 60 sham), 12 exposure carousels were used. A plastic tubular restraint confined each rat for the duration of the exposure to facilitate dosimetry. The animals faced the antenna at a fixed distance from the tip of the nose (30 mm from weaning to 120 days, 45 mm thereafter). Exposures were conducted in three shifts to accommodate the 360 exposed/sham-exposed rats in this study. Dosimetry was obtained using two different techniques, each of which was verified by an independent method: numerical modeling verified by electric probe measurements, and infrared thermography verified by thermometric probes. Numerical modeling was based on magnetic resonance imaging data sets of a rat cadaver with a resolution of 0.125 mm^3 in the brain and 1.0 mm^3 in the rest of the body. The results were validated at 30 specific points within a cadaver brain, using an electric field probe. In thermography, bisected rat cadavers were exposed to a 235 W field at 836 MHz for ≤90 s and a series of infrared images of the cut surfaces was acquired for 2 min. Thermographic readings were compared with measurements made using a Vitek thermistor probe. The average brain SAR was 1.0–1.6 W/kg (time-averaged SAR of 0.33–0.53 W/kg) for rats ranging in size from 250 to 450 g.

This study demonstrated that exposure of Fischer 344 rats to TDMA-modulated 836 MHz RF fields from late gestation through 24 months of age did not change the incidence of either spontaneous primary or ENU-induced CNS tumors. All animals did not survive to the end of the experiment; the 182 (77%) that survived were sacrificed for detailed histopathological examination. There was no evidence of tumorigenic effects in the CNS from the field exposure; however, some evidence of tumor-inhibiting ("protective") effects of TDMA field was observed. Overall, the TDMA field-exposed animals exhibited trends toward a reduced incidence of spontaneous CNS tumors ($P < 0.16$) and ENU-induced CNS tumors ($P < 0.16$). In the 54 rats (23%) that died during the study ("preterm rats") where primary CNS tumors were determined to be the cause of death, the TDMA-field exposure significantly reduced the incidence of ENU-induced tumors ($P < 0.03$).

The observed tumor-inhibiting ("protective") effects of TDMA exposure were apparent but unexpected. Moreover, both the numbers of rats and tumors were small. The observation was confounded by such issues as stress introduced by the restraint device and the absence of cage controls. Furthermore, the incidence of spontaneous CNS tumors was several times higher than historical data reported for this strain of rats. A plausible explanation is that the historical data were based on gross examination rather than the detailed histopathology used in this study. To help assess the uncertainty of the observed protective effect, it would be desirable to conduct additional dose–response relationship experiments with a large number of animals.

In a related study using the same exposure systems and protocols, Fisher 344 rats were exposed to frequency-modulated (FM), 836 MHz RF radiation from simulated cellular telephone operations during talk. Exposure-related changes were neither detected in number, incidence, or histological type of either spontaneous or ENU-induced CNS tumors, nor were gender differences observed in tumor numbers (Adey et al., 2000). Thus, these two studies seem to suggest a relationship between the observed tumor reduction and the modulation scheme used for the cell phone RF field.

The protocol involving exposure of pups from Fischer 344 dams subjected to a single dose of ENU in utero was used to study TDMA-modulated 1.44 GHz field with Japanese PDC cellular phone operating standards (Shirai et al., 2005). The exposure apparatus was a carousel-type system in an environmentally controlled chamber. Rats with their nose direction toward the antenna in the center of the carousel were restrained individually in plastic holders; four different size holders were used to accommodate the animals' growth throughout the 2-year experiment. Brain average SARs of 0.67 and 2.0 W/kg were selected for a low and a high level exposure; the whole-body average SAR was less than 0.4 W/kg. A total of 500 pups were divided into five groups, each composed of 50 males and 50 females: untreated cage controls; ENU alone; 3 groups of ENU + RF (sham exposure and 2 at 0.67 and 2.0 W/kg exposure levels, respectively). Furthermore, an additional 63 rats for each sex were used as dummy subjects to cover any vacancy in the RF exposure boxes due to interim death to maintain the same exposure conditions.

The results showed that the growth rates of treated rats were not significantly different from those of untreated controls in both the females and males. Restraining the animals was associated with curtailed growth in the males (and apparently in females after the age of 1.5 years). Otherwise, there were no inter-group differences in body mass, food consumption, or survival rates. Increase in the incidence or number of brain or spinal cord tumors was not observed in the RF-exposed groups (Fig. 3). In addition, no clear changes in tumor types were detected. Thus, TDMA-modulated 1.44 GHz RF exposure at 0.67 and 2.0 W/kg to the heads of rats for a 2-year period did not exhibit any promotional effect on ENU-initiated brain tumorigenesis. It should be noted that in contrast to the Adey et al. assay, the protocol of the present experiment used four different sized restraining holders during the experimental period to accommodate the animals' growth. This approach prevented the smaller animals from turning around in the holder and reduced the associated dosimetric uncertainties. However, this procedure may have contributed to increased stress on the restrained animals.

This exposure system and protocol were applied to investigate whether chronic (2-year) exposure to wide-band code division multiple access (W-CDMA) RF fields has any effect on promotion of ENU-induced tumorigenesis. The monopole antenna was adjusted for the 1.95 GHz cellular operation. W-CDMA signal is a feature of the International Mobile Telecommunication 2000 (IMT-2000) wireless communication system. Pregnant Fischer 344 rats were administered a single dose of ENU on gestational day 18 and a total of 500 pups was divided into five groups as in the other study, each composed of 50 females and 50 males. In general, no significant increase

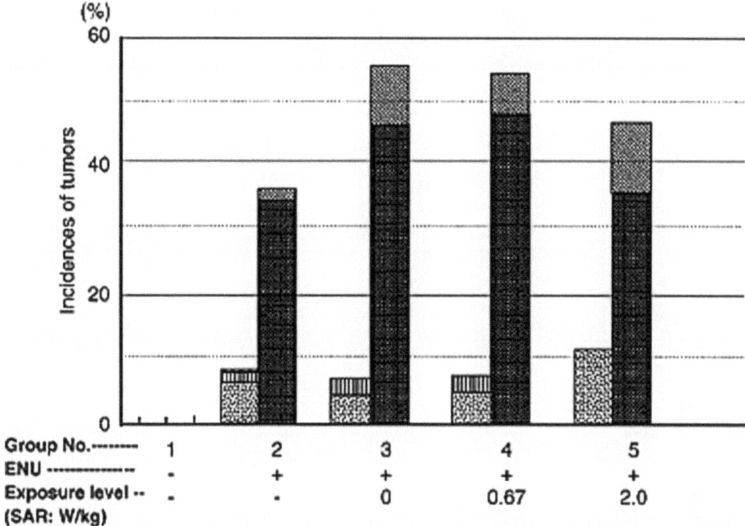

Figure 3. Incidences of CNS tumors among exposure groups in female and male F344 rats. Brain tumors: "white dots on black –" moribund and killed; "diagonal lines –" end of the experiment. Spinal cord tumors: "black dots on white –" moribund and killed; "vertical lines –" end of the experiment (Shirai et al., 2005).

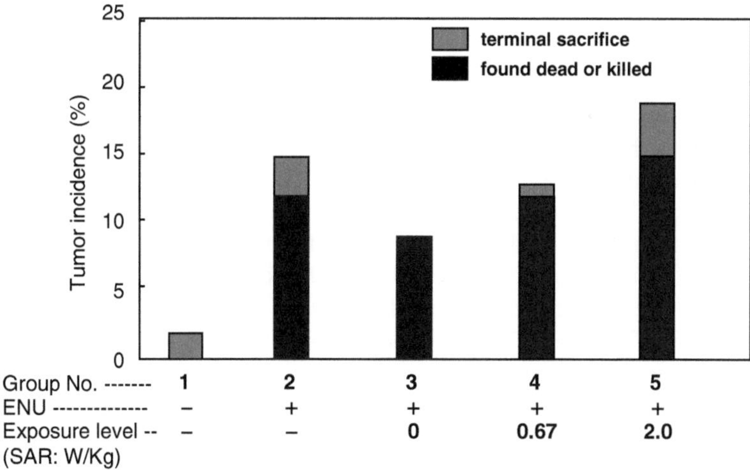

Figure 4. Incidences (%) of brain tumors among the five exposure groups; females and males are combined (Shirai et al., 2007).

in the incidence or number of tumors was observed in the experimental animals (Fig. 4). Moreover, the results showed no clear changes in tumor types in the brain. However, there was a tendency of slight increase in brain tumor development in the females exposed to 1.95 GHz W-CDMA modulated field (Table 7).

Table 7. Incidence and number of CNS tumors in Fischer 344 females (Shirai et al., 2007)

Organ and findings	Group	1	2	3	4	5
	ENU	–	+	+	+	+
	EMF exposure (SAP: W/kg)	–	–	0	0.67	2.0
Brain	No. of animals	50	50	50	50	50
Astrocytoma		1	6	3	5	9
Oligodendroglioma		0	0	1	0	1
Mixed glioma		0	1	0	0	1
Ependymoma		0	0	1	0	0
Meningioma		0	0	0	0	0
Granular cell tumor		0	0	0	0	0
No. of rates with tumor		1 (2)[a]	7 (14)	5 (10)	5 (10)	11 (22)
No. of total tumor		1	7	5	5	11
Spinal cord						
Astrocytoma		0	0	0	0	0
Mixed glioma		0	0	0	0	0
Reticulosis, malignant		0	0	0	0	0
No. of rates with tumor		0 (0)	0 (0)	0 (0)	0 (0)	0 (0)
No. of total tumor		0	0	0	0	0

[a]The numbers in the parenthesis represents percent incidences

The spontaneous tumorigenesis of Fischer 344 rats, without the use of ENU initiation, was the subject of another investigation (La Regina et al., 2003) using frequency-modulated continuous-wave (836 MHz) radiation in the form of frequency division multiple access (FDMA). In addition, an experiment was conducted using CDMA-modulated 848 MHz in carousel-type exposure systems. A total of 480 young female and male Fischer 344 rats, 80 female and 80 male, was placed randomly in each of three experimental groups: sham, FDMA and CDMA groups exposed to 847.74 MHz CDMA. Exposure began when the animals were 6-weeks old. Rats were placed in their respective chambers and exposed for a total of 4 h each day, 5 days a week during the subsequent 2-year study period. Although it appeared that cage-control animals did not form a part of this study, sentinel rats were kept in the room to monitor for infectious disease. Results showed exposure to 835.62 MHz FDMA or 847.74 MHz CDMA RF radiation had no effect on spontaneous tumor development in brain or other organs of either male or female Fischer 344 rats.

The Fischer 344 rats were used as subjects of a study on the effect of Iridium signal modulation, which uses differentially encoded quaternary phase shift keying (DEQPSK). The Iridium system is a satellite-based, digital, wireless, personal communication network. In this study, pregnant Fischer 344 rats from 19th day of gestation and their offspring were exposed to a far-field 1.6 GHz Iridium fields for

2 h/day, 7 days/week until weaning (Anderson, et al., 2004). Far-field whole-body exposures were conducted in a parallel-plate system with a field intensity of 4.3 W/m^2 and whole-body average SAR of 0.036–0.077 W/kg (0.10–0.22 W/kg in the brain). This was followed by chronic, head-only exposures of female and male offspring to a near-field produced in a carousel system for 2 h/day, 5 days/week for 2 years. Near-field exposures were conducted at a SAR of 0.16 or 1.6 W/kg in the brain. Concurrent sham-exposed and cage-control rats were also included in the study.

A total of 150 female rats were divided into 3 groups: 42 untreated cage controls, 36 sham control, and 72 RF exposure. They remained singly housed, or with their pups in the same cage during far-field exposure until the weaning of the offspring. For the near-field exposure phase of the study, three rats of the same gender from the same exposure group were housed per cage. The 700 pups were divided into 4 groups composed of 80 females and 80 males as untreated cage controls; 3 groups each of 90 females and 90 males for sham, 0.16 and 1.6 W/kg, respectively, in the rat brain. Neither statistically significant differences were observed among treatment groups for number of live pups/litter, survival index, and weaning mass, nor were there differences in clinical signs or neoplastic lesions among the treatment groups. It should be noted that the reporting of clinical histopathology was not consistent in this study. In particular, the incidence of brain tumors in untreated cage controls was not reported. Instead, incidences of brain tumors was compared with and found to be comparable to published historical control incidences for Fischer 344 rats. The percentages of animals surviving at the end of the near-field exposure were not different among the male groups. In females, a significant decrease in percentage of survival and survival time was observed for the cage-control group.

3.4. Carcinogenic and Cocarcinogenic Potentials in Wistar Rats

In addition, Wistar rats were the subject of two studies on the carcinogenic and cocarcinogenic potential of cell phone RF electromagnetic fields, the first of which investigated tumorigenesis induced by the mutagen 3-chloro-4-(dichloromethyl)-5-hydroxy-2(5H)-furanone (MX) given in drinking water. Female Wistar rats aged 7 weeks at the beginning of the experiments were randomly divided into four groups of 72 animals: a cage-control group and three MX-treated groups (a daily average dose of 1.7 mg MX/kg body mass for two years) (Heikkinen et al., 2006). MX is known to be a potent bacterial mutagen and a multisite carcinogen in Wistar rats. In this case, MX rats were exposed to RF radiation for 2 h per day, 5 days per week for 104 weeks to GSM-modulated 900 MHz fields at whole-body average SARs of 0.0 (sham), 0.3 and 0.9 W/kg. Unrestrained animals were exposed to GSM radiation in individual cages installed in a radial transmission line system. The rats were able to move freely in the cages. Food was available at all times, but water bottles were removed for the RF field exposure sessions. Histopathological examination performed on the rats showed that GSM exposure did not affect tumor types and incidences observed in the MX-exposed animals. There were no statistically significant changes in mortality or organ-specific incidence of any tumor type.

A more recent publication reported two sets of carcinogenic results from Wistar rats exposed to GSM at 902 Hz and DCS at 1,747 MHz, respectively (Smith et al., 2007). The RF exposure took place in a waveguide wheel – a circular array of waveguides excited by a common quarter-loop circularly-polarized antenna located in the center. In addition to cage and sham controls, for each frequency, 500 rats (7-week old in 5 groups of 50 females and 50 males per group) were exposed for 2 h/day, 5 days/week for up to 104 weeks at target SARs of 0.44, 1.33, and 4.0 W/kg. These two double-blinding studies did not produce any evidence that RF field exposure at GSM-900 or DCS-1747 had any effect on the incidence or severity of any primary tumors or the type, incidence, multiplicity, and latency of any neoplastic lesion (Table 8).

It is interesting to note that while the combined female and male incidence of palpable mass was similar across all groups, the incidence in females was higher than in males, with the highest incidence occurring in the sham control females for both GSM-902 and DCS-1474 (Table 9). The macroscopic findings showed several statistically significant gross lesions. Compared with sham control, the incidence of foci in the liver of males of the 1.33 W/kg GSM group and of skin nodules in males of the 0.44 W/kg DCS group were higher ($P<0.05$), while incidence of foci in the lachrymal glands of males of the 1.33 and 4.0 W/kg GSM group was lower. Also, the incidence of cysts in the liver of females of 9.44 W/kg GSM group was lower compared with an incidence of 9% in the corresponding sham control group. Similar to the observation of 4/50 prostate adenomas in the 4.0 W/kg DCS group compared with 0/50 in the sham-exposed controls, these observations were considered isolated, incidental findings unrelated to RF exposure by authors (Smith et al., 2007). It is noted that histopathology was not performed for the cage-control rats in this study.

In addition to whole-body-averaged SARs, this study provided detailed dosimetry including the spatial peak and organ-averaged SAR values for the GSM and DCS systems. Because of the differences in frequency, the distribution of the induced fields at the same whole-body averaged exposure is significantly different between the GSM and DCS experiments. For example, the brain-averaged exposure differed by a factor of 5 (i.e., 1.5 W/kg at GSM compared with 7.6 W/kg at DCS), whereas the SARs of other organs such as liver, kidneys, etc. were similar. It should be mentioned that this study employed an exposure protocol with a targeted whole-body SAR averaged over the entire exposure period. For example, the whole-body SAR 4 W/kg was achieved in the DCS study, but the SAR levels had to be decreased in the GSM study since the body mass increase of the rats was greater than predicted such that the available power was insufficient to maintain 4 W/kg. In fact, the whole-body SAR averaged over the entire exposure period was 3.7 W/kg for the GSM study.

3.5 Induction or Promotion of Cancer in Sprague–Dawley Rats

The Sprague–Dawley strain of rats has been used to investigate the carcinogenic potential of cell phone radiation especially with regard to neural and mammary tumors. In one study (Zook and Simmens, 2001), Sprague–Dawley rats were exposed in a carousel-type system to a FM (CWRF) or a pulsed RF (PRF) field generated by

Table 8. The type, incidence, and multiplicity of primary neoplastic lesion in Wistar rats (Smith et al., 2007)

	GSM							
	Males				Females			
	Sham control	Low dose	Mid dose	High dose	Sham control	Low dose	Mid dose	High dose
No. of animals	50	50	50	50	50	50	50	50
No. of animals with neoplasms	32	34	31	35	44	42	45	49
No. of animals with more than one primary neoplasm	14	13	8	11	23	24	25	28
No. of animals with benign neoplasms	30	29	31	32	43	39	40	47
No. of animals with malignanat neoplasms	9	10	1	8	3	11	10	11
No. of animals with metastases	1	0	0	0	1	1	0	2
No. of primary neoplasms	51	53	39	50	74	81	81	91

	DCS							
	Males				Females			
	Sham control	Low dose	Mid dose	High dose	Sham control	Low dose	Mid dose	High dose
No. of animals	50	50	50	50	50	50	50	50
No. of animals with neoplasms	36	32	33	33	43	43	43	49
No. of animals with more than one primary neoplasm	45	14	10	11	28	24	24	24
No. of animals with benign neoplasms	35	29	28	29	42	42	41	46
No. of animals with malignanat neoplasms	3	7	9	5	12	3	9	13
No. of animals with metastases	0	1	1	2	1	1	0	1
No. of primary neoplasms	55	53	49	51	88	76	85	93

The low, mid, and high doses correspond to SARs of 0.44, 1.33, and 4.0 W/kg for both GSM-902 and DCS-1474

Table 9. Incidence of palpable mass in RF-exposed Wistar rats (Smith et al., 2007)

	Cage control	Sham control	0.44 W/kg	1.33 W/kg	4.0 W/kg
GSM					
Males	10/50 (20)[a]	3/50 (6)	8/50 (16)	2/50 (4)	9/50 (18)
Females	14/50 (28)	20/50 (40)	17/50 (34)	18/50 (36)	15/50 (30)
DCS					
Males		2/50 (4)	8/50 (16)	5/50 (10)	10/50 (20)
Females		21/50 (42)	12/50 (24)	15/50 (30)	16/50 (32)

[a]Number of rats with palpable mass/number of rats per group (%)

Table 10. Number of malignant brain and nerve tumors in Sprague–Dawley rats exposed to MiRS sources (Zook and Simmens, 2001)

Group (60/rats/group)[a]			Brain tumors		Nerve tumors		
No.	ENU (mg/kg)	RF-field exposure	No. of brain tumors	No. of rats with brain tumors	Spinal (number)	Cranial (number)	Spinal cord tumors
1	0	PRF	5	5	0	0	0
2	0	Sham	3	3	0	0	0
9	0	CWRF	3	3	0	0	1
10	0	Sham	5	5	0	0	0
13	0	Cage	6	6	0	0	0
5	2.5	PRF	10	7	2	5	2
6	2.5	Sham	10	9	1	2	0
7	2.5	PRF	9	9	2	1	0
8	2.5	Sham	11	10	3	2	0
11	2.5	CWRF	3	3	6	2	2
12	2.5	Sham	7	6	3	2	0
14	2.5	Cage	5	5	4	1	1
3	10.0	PRF	58	36	15	5	2
4	10.0	Sham	48	35	12	7	2
15	10.0	Cage	52	41	12	6	0

[a]There were only rats necropsided in group 6

a Motorola Integrated Radio Services (MiRS) source. The 860 MHz RF exposure at a SAR of 1.0 W/kg averaged over the brain took place for 6 h/day, 5 days/week from 2 up to 24 months of age. The rats were assigned to 15 groups. Each group consisted of 60 rats (30 males and 30 females). Every group exposed to an RF field had a matching sham-exposed group held in identical exposure units for the same periods. These offspring were injected i.v. with 0, 2.5, or 10 mg/kg of ENU to induce brain tumors. Three groups of cage controls were killed at the same time as the rats given corresponding ENU doses. All rats but 2, totaling 898, were necropsied, and major tissues were histopathologically examined. Table 10 gives the number of malignant brain and nerve tumors and the number of animals with tumors. Overall, there was no statistically significant indication that the pulsed (PRF) or FM (CWRF) exposure induced cancer in the Sprague–Dawley rats. Additionally, there was no significant indication of promotion of CNS or spinal cord tumors. The PRF or CWRF had no statistically significant effect on the number, volume, location, multiplicity, histological type, malignancy, or fatality of brain tumors. However, authors suggest there was a trend for the group that received a high dose of ENU and was exposed to the PRF to develop fatal brain tumors at a higher rate than its sham group. Indeed, the result showed a 50% reduction in numbers for sham or CWRF compared with cage controls in the low or zero ENU-dose groups. In contrast, for PRF, the numbers either doubled or were the same compared with cage controls in the low or zero ENU-dose groups.

In several studies, the RF field employed in cellular mobile communication was tested using 7,12-dimethylbenz[α]anthracene (DMBA)-induced mammary tumors in female Sprague–Dawley rats as a model for human breast cancer.

Bartsch et al. (2002) conducted three experiments using female Sprague–Dawley rats under standardized conditions that were replicated twice by starting the two subsequent experiments on the same day of the two following years. In each experiment, 120 rats (60 for sham) were injected with a single 50 mg/kg dose of DMBA and continuously exposed to 900 MHz GSM fields in two separate plane wave chambers, except for brief servicing and house-keeping periods, until practically all animals had developed mammary tumors. The animals had freedom to move within their cages. Circularly-polarized RF fields in the exposure chambers had an average power density of 100 μW/cm^2 at the bottom of the animal cages. For an adult female Sprague–Dawley rat weighing 300 g, the whole-body SAR was 0.017–0.070 W/kg. Note that the whole-body SAR declined continuously during the course of the experiment due to body-resonant energy absorption. At the beginning of the experiment (51-day old, 150 g), animals had whole-body SARs between 0.033 and 0.13 W/kg. The overall results of the three studies are that low-level GSM-900 RF field exposure did not have any significant effect on tumor latency and that the cumulative DMBA-initiated mammary tumor incidence at the end of the experiment was unaffected by the exposure. However, in the first experiment, the median latency for the development of malignant tumors was statistically significantly extended for RF field-exposed rats compared to sham controls (278 days compared with 145 days). This difference was not detected in the two subsequent experiments. Cage controls were not included in this study. The results show that low-level GSM-900 RF radiation did not appear to have a cancer-promoting effect on DMBA-induced mammary tumors.

The promotion of DMBA-initiated mammary tumors in Sprague–Dawley rats subchronically exposed to GSM-900 radiation over a wide range of whole-body SARs was investigated in one study involving two separate experiments (Anane et al., 2003). Mammary tumors were induced by ingestion of a single 10 mg dose of DMBA in 55-day-old female Sprague–Dawley rats. RF exposure started 10 days later for 2 h/day, 5 days/week for 9 weeks. Rats (128) were exposed to plane waves with the electric field parallel to the long axis of the body at whole-body SARs of 0.0 (sham), 0.1, 0.7, 1.4, 2.2, and 3.5 W/kg in 8 groups of 16 animals. Among these were two groups at 0.4 W/kg, separated by one month in time. Another 8 rats served as an untreated cage-control group, but were not included in the data analysis. Rats were killed 3 weeks after the end of exposure. The results obtained indicated that there were no differences in latency, multiplicity, or tumor volume among the groups. With regard to tumor incidence (Table 11), while these data showed both increases and decreases compared with sham exposure, overall the results are rather inconsistent. Nevertheless, there seems to be a trend toward reduced rate of incidence of DMBA-initiated mammary tumor for rats exposed to GSM-900 RF fields at 1.4 W/kg or lower. Note that the number of animals per group (16) is relatively small in this study. A smaller number of cage controls (8) were mentioned but data were not presented in this study.

Another study designed to test the carcinogenic or promotional potential of GSM-modulated 900 MHz fields in female Sprague–Dawley rats involved the use of a different exposure system (Yu et al., 2006). The "exposure wheel" consisted of a circular array of 17 sectored waveguides, excited by a single loop antenna located in the center. To enhance homogeneity of field exposure, each week the exposure position of each rat was rotated one position to the right on the wheel so that the position and exposure of individual rats varied throughout the 26-week exposure duration. Individual rats were administered a single 35 mg/kg dose of DMBA and a total of 500 rats were divided into five groups: cage control and four exposure groups, including sham and three RF exposure groups for SARs of 0.0, 0.44, 1.33, and 4.0 W/kg, respectively. The 26-week exposure started one day after DMBA administration for 4 h/day, 5 days/week. Rats were palpated weekly for the presence of mammary tumors and were killed at the end of the 26-week exposure period. The results showed no significant differences in body mass between sham- and GSM 900-exposed groups. No significant differences in overall mammary tumor incidence, latency to tumor onset, tumor multiplicity, or tumor size were observed between GSM 900- and sham-exposed groups. There were significant differences in body mass and benign mammary tumors between the cage control and experimental groups (sham and exposure). Specifically, body mass and mammary tumor incidence, especially benign tumors in the cage-control group are significantly

Table 11. Number of malignant mammary tumors detected by palpation at week 11 and confirmed at necropsy at week 12 in DMBA treated Sprague–Dawley rats (Anane et al., 2003)

		Number of tumors per group		
First experiment	Sham	1.4 W/kg	2.2 W/kg	3.5 W/kg
Week 11	14	18	22	19
Week 12	21	24	24	29
Second experiment	Sham	0.1 W/kg	0.7 W/kg	1.4 W/kg
Week 11	15	6	10	4
Week 12	17	8	13	4

higher than in the sham- and GSM 900-exposed groups. The latency to mammary tumor onset was also significantly shorter in the cage-control group than in the other groups.

For rats in exposure groups, including the sham control group, food and water were not available during exposure. The duration of food and water deprivation was 4.5–5.0 h per experiment day. In contrast, for rats in the cage-control group, food and water were available ad libitum for the 6-month experimental period. Given that many reports indicate chronic food restriction inhibits the development of mammary tumors in mice and rats, the observed difference in DMBA-induced mammary tumors in sham and exposed female Sprague–Dawley rats is most likely associated with dietary restriction.

A parallel study of DMBA-induced mammary tumors in female Sprague–Dawley rats has been published recently (Hruby et al., 2008). This study used the same protocol and "waveguides in a wheel" exposure system as the Yu et al. study. Rats in the cage-control group had in most aspects the highest incidence and malignancy of tumors or neoplasms among all groups. In particular, when compared with the sham-exposed group the cage-control group had significantly more palpable tissue masses, more benign and malignant tumors, perhaps for the same reasons as mentioned previously in connection with the Yu et al. study. In addition, the results showed several significant differences among the various exposure groups: all GSM-exposed groups had, at different times, significantly more palpable tissue masses. There were fewer rats with benign tumors, but more with malignant tumors or neoplasms in the 4.0 W/kg group (Table 12, where SARs of 0.4, 1.33 or 4.0 W/kg are designated as low, mid, or high dose). In addition, there were more adenocarcinomas in the 0.4 W/kg group, more malignant tumors in the 0.4 and 4.0 W/kg groups, more Sprague–Dawley rats with adenocarcinomas in the 4.0 W/kg group, and fewer rats with fibroadenomas in the 0.4 and 4.0 W/kg groups. None of the above findings in GSM-exposed rats produced a clear dose–response relationship. The significant

Table 12. DMBA-induced mammary gland tumors in Sprague–Dawley rats exposed to GSM fields (Hruby et al., 2008)

	Cage control	Sham exposure	Low does	Mid dose	High dose
Total number of animals	100	100	100	100	100
Animals with malignant or benign neoplasms	73	60	57	50	65
Animals with malignant neoplasms	45	30	40	35	47
Animals with benign neoplasms	28	30	17	15	18
Animals with hyperplasia	12	11	19	22	9
Animals with hyperplasia or neoplasia	85	71	76	72	74
Mean number of tumors per tumor-bearing animal	1.73	1.42	1.74	1.72	1.57

differences between the sham-exposed animals and one or more GSM-exposed groups may be interpreted as evidence of an effect of GSM exposure. However, authors of the paper had opined that the differences between the groups are incidental because of the high variability in results.

3.6. A Summary of Studies on Cancer and Cell Phone RF-Exposed Rats

Among the 2-year cancer promotion studies using Fischer 344 rats (Table 13), four involved ENU induction. They each used a different carrier frequency or modulation scheme specific to wireless communication, but none gave any indication of an increase in the promotion of ENU-induced brain or CNS cancer. Likewise, the two spontaneous tumor induction studies did not show any significant difference in CNS tumor growth or incidence between RF- and sham-exposed rats. As part of their ENU study, Adey et al. (2000) had included a non-ENU group, which yielded a reduction in tumor incidence for TDMA-modulated 836 MHz exposures. The interpretation of this finding becomes obscure since cage-control animals did not form a part of this investigation. Moreover, restraining the experimental animals during exposure in the carousel-type exposure system could have introduced a stress factor, which further complicates interpretation of the results.

The Wistar rats exposed to GSM-900 studies provided the same null results with regard to any tumor type. However, there were major differences in most aspects of the studies conducted in two different laboratories. One was a promotional study (Heikkinen et al., 2006) where unrestrained rats were exposed in a plane wave environment and the other studied the induction of cancer in restrained rats exposed in the near field of a waveguide-wheel exposure system (Smith et al., 2007). This study also reported on a DCS study at 1,747 MHz. As for GSM, the combined female and male incidence of palpable mass was found to be similar across all exposed groups. Histopathology was not performed for the cage-control rats in the Smith et al. study. The report showed that the incidence in females was higher than in males, with the highest incidence occurring in the sham control females for both GSM-900 and DCS-1474. The macroscopic findings showed several statistically significant gross lesions comparing sham control with GSM exposed groups.

As a model for human breast cancer, DMBA-induced mammary tumors in female Sprague–Dawley rats formed the objective in four studies employing RF radiation from GSM-900 cellular mobile communication systems (Table 13). The Bartsch et al. investigation was a self-replicated study using unrestrained rats and it found no difference between sham and plane-wave RF-exposed animals. Restraining the rats as in the Anane et al. study and somewhat higher SARs did not produce any statistical difference either. Note that neither the Bartsch et al. or the Anane et al. studies included cage controls. However, a parallel investigation involving frequencies and modulations specific to GSM-900 mobile telephones, and identical "waveguide-wheel" exposure systems producing the same SARs gave very different pictures in mammary tumor incidence. Although Yu et al. found no difference between RF and sham-exposed rats, benign tumors in the cage-control group are significantly higher than in the sham and GSM 900-exposed groups. The latency to mammary tumor

Table 13. Cancer induction or promotion in rats by continuous wave, frequency, and pulsed modulated cell phone RF exposure

Reference	Signal, modulation, exposure regime	Rats (number)	Cocarcinogen	Exposure system	SAR (W/kg) (whole-body, local peak)	Tumor type or location (results)	Study design
Salford et al., 1993	915 MHz GSM pulse modulation, 7 h/day, 5 day/week, 2–3 weeks	Fischer-344 (male and female); 37 exposed, 37 sham)	Implanted RG2 and N32	TEM chamber (partially restrained)	0.008–1.0	Brain (no difference)	Promotion; RF and sham control only
Imaida et al., 1998	929.2 MHz TDMA 90 min/day, 5 day/week, 6 weeks	Fischer-344 (male; 48 exposed, 48 sham; 24 DEN control)	Diethyl-nitrosamine (DEN)	Monopole Near-field (partially restrained)	0.58–0.8 wb ave; 6.6–7.2 peak wb: 1.7–2.0 peak in liver	Liver (no difference)	Promotion; RF and sham only
Adey et al., 1999	836 MHz TDMA 2 h/day, 4 day/week, for 2 years	Fischer 344 (236 offspring; in 4 groups of 60:30 female/30 male, ENU/RF (26 female/30 male)	*N*-ethyl-nitrosourea (ENU, 4 mg/kg) in utero	Plane Wave Horn in utero and offsprings; Carousel after weaning (restrained)	0.27–0.72 whole-body; (0.74–1.6 in brain)	Brain or CNS (no overall difference; reduction w/o ENU)	Spontaneous tumorigenicity and promotion; RF, sham, no cage controls

(continued)

Table 13. (continued)

Reference	Signal, modulation, exposure regime	Rats (number)	Cocarcinogen	Exposure system	SAR (W/kg) (whole-body, local peak)	Tumor type or location (results)	Study design
Adey et al., 2000	836 MHz FM 2 h/day, 4 day/week for 2 years	Fischer 344 (102 pregnant dams; 540 offspring in 6 groups: 45 female and 45 males/group)	ENU 4 mg/kg in utero	Plane wave horn for pregnant dams and offspring; carousel after weaning (restrained)	0.27–0.72 whole-body; (0.74–1.6 in brain)	Brain or CNS (no overall difference)	Spontaneous tumorigenicity and Promotion; RF, sham, and cage control
Shirai et al., 2005	1.44 GHz TDMA, 90 min/day, 5 day/week for 2 years	Fischer 344 (500 5-week old pups; 50 female and 50 male per group)	ENU 4 mg/kg in utero	Carousel (restrained with 4 sizes of holders)	<0.4 whole-body (0.67 and 2.0 brain ave)	Brain or CNS (no overall difference)	Tumor promotion; RF, sham, cage control
Shirai et al., 2007	1.95 GHz W-CDMA, 90 min/day, 5 day/week for 2 years	Fischer 344 (500 5-week old pups; 50 female and 50 male per group)	ENU 4 mg/kg in utero	Carousel (restrained with 4 sizes of holders)	<0.4 whole-body (0.67 and 2.0 brain ave)	Brain or CNS (no overall difference)	Tumor promotion; RF, sham, cage control
La Regina et al., 2003	836 MHz FM or 848 MHz CDMA, 4 h/day, 5 day/week for 2 years	Fischer 344 (480 6-week old; 80 female and 80 male per group)	None	Carousel (restrained)	1.3 brain ave (0.5 and 2.5 local brain)	Brain or organs (no difference in any tumor in any organ)	Spontaneous tumorigenicity RF, sham, no cage controls

(continued)

Table 13. (continued)

Reference	Signal, modulation, exposure regime	Rats (number)	Cocarcinogen	Exposure system	SAR (W/kg) (whole-body, local peak)	Tumor type or location (results)	Study design
Anderson et al., 2004	1.62 GHz Iridium 2 h/day, 4 day/week for 2 years	Fischer 344 (102 pregnant dams; 540 offspring in 6 groups: 45 female and 45 males/group)	None	Parallel plate, plane wave for pregnant dams and offspring; Carousel after weaning (unrestrained)	0.036–0.077 whole-body (0.16 and 1.6 brain ave)	Brain or organs (no difference in treatment groups), not reported for cage control	Spontaneous tumorigenicity RF, sham, cage controls
Heikkinen et al., 2006	900 MHz GSM 2 h/day, 5 day/week for 2 years	Wistar (288 female 7-week old in 4 groups of 72)	Cocarcinogen 3-chloro-4-(dichloromethyl)-5-hydroxy-2(5H)-furanone (MX), 1.7 mg/kg Oral daily	Radial transmission line TEM plane wave (unrestrained)	0.3 (0.07–1.2) and 0.9 (0.21–3.6) whole-body ave	No effect	Cocarcinogenesis RF, sham, cage control
Smith et al., 2007	902 MHz GSM 2 h/day, 5 day/week for 2 years	Wistar 500 (7-week old in 5 groups of 50 females and 50 males/group)	None	Exposure wheel waveguides (restrained)	0.44, 1.33, and 4.0 whole-body ave	No effect	Carcinogenesis RF, sham, cage control

(continued)

Table 13. (continued)

Reference	Signal, modulation, exposure regime	Rats (number)	Cocarcinogen	Exposure system	SAR (W/kg) (whole-body, local peak)	Tumor type or location (results)	Study design
Smith et al. 2007	1,747 MHz DCS 2 h/day, 5 day/week for 2 years	Wistar 500 (7-week old in 5 groups of 50 females and 50 males/group)	None	Exposure wheel waveguides (restrained)	0.44, 1.33, and 4.0 whole-body ave	No effect	Carcinogenesis RF, sham, cage control
Zook BC, Simmens, 2001	860 MHz FM or MiRS pulsed, 6 h/day, 5 day/week for 2 years	Sprague–Dawley (900 pups in 15 groups of 30 females and 30 males/group)	ENU (0, 2.5, 10 mg/kg) in utero	Carousel with different size wedge or tube holders (restrained)	0.27–0.42 whole-body ave (1.0 brain ave)	Brain or organs (no difference), but trend in high ENU	Spontaneous tumorigenicity and Promotion; RF, sham, Cage controls
Bartsch et al., 2002	900 MHz GSM CW exposure until tumors developed in 150–280 days	Sprague–Dawley (360 female 51-day old in 6 groups of 60)	DMBA (50 mg/kg)	Plane wave exposure chamber (unrestrained)	0.017–0.13 whole-body ave (low-level)	Mammary (no difference)	Promotion; RF, sham, No cage control

(continued)

Table 13. (continued)

Reference	Signal, modulation, exposure regime	Rats (number)	Cocarcinogen	Exposure system	SAR (W/kg) (whole-body, local peak)	Tumor type or location (results)	Study design
Anane et al., 2003	900 MHz GSM 2h/day, 5day/week for 9 weeks	Sprague–Dawley (128 female 55-day old in 8 groups of 16)	DMBA (10 mg total)	Plane wave exposure chamber (restrained)	0.1, 0.7, 1.4, 2.2, and 3.5 whole-body ave	Mammary (no difference) but trend in low SAR	Promotion; RF, sham, no cage control
Yu et al., 2006	900 MHz GSM 4 h/day, 5 day/week for 26 weeks	Sprague–Dawley (500 female 48-day old in 5 groups of 100)	DMBA (35 mg/kg)	Exposure Wheel waveguides (restrained)	0.44, 1.33, and 4.0 whole-body ave	Mammary (no sham/exposed difference; but in cage control)	Promotion; RF, sham, Cage control
Hruby et al., 2008	902 MHz GSM 4 h/day, 5 day/week for 26 weeks	Sprague–Dawley (500 female 47-day old in 5 groups of 100)	DMBA (17 mg/kg, oral)	Exposure wheel waveguides (restrained)	0.44, 1.33, and 4.0 whole-body ave	Mammary (many differences but no dose response relation)	Promotion; RF, sham, cage control

onset was also significantly shorter in the cage-control group than in the exposed groups. This difference in DMBA-induced mammary tumors was thought to be associated with dietary restrictions imposed on the sham and exposed female Sprague–Dawley rats. Similar to the Yu et al. report, cage-control rats in the Hruby et al. study had in most cases the highest incidence and malignancy of neoplasms. However, the results showed several significant differences among the various exposure groups: All GSM-exposed groups had, at different times, significantly more palpable tissue masses. Although it may serve as evidence of an effect of GSM field exposure, the fact that none of the findings in GSM-exposed rats produced a clear dose–response relationship makes it difficult to arrive at a definitive conclusion, especially since the DMBA dose and manner of administration were different. Moreover, the DMBA-mammary tumor model seems prone to produce variable results in some cases.

4. CONCLUDING REMARKS

The carcinogenic investigations reviewed have included 10 studies in laboratory mice and 16 studies in rats exposed to RF fields from a variety of wireless communication schemes. The investigations using mice have involved three strains of genetically prone mice: Eμ-Pim1, AKR/J, and ODC-K2. The three studies using Eμ-Pim1 lymphoma prone mice all employed GSM-900 RF field, but gave varying results. Moreover, differences and uncertainties in the animal protocols and exposure systems limit the conclusions that can be drawn. There are two studies using the AKR/J lymphomas prone mice. One study was done for GSM-900 field exposure but it differed substantially in SAR and exposure durations, thus it cannot be regarded as a potential confirmation of the Eμ-Pim1 results. The other is somewhat isolated; the exposure was conducted with UMTS-1.97. Lastly, a small and shorter duration study using ODC-K2 mice showed that skin cancers were not changed by a 52-week exposure to DAMPS-TDMA-849 fields.

Cancer induction and promotion by wireless communication fields of differing frequencies and modulations were the subject of studies using four different strains of normal mice: CD-1, CBA/S, ODC-nontransgenic, and B6C3F1. For exposures of one year or less, experiments with the first three strains of mice did not show a promotional or cocarcinogenic effect on tumorigenesis. The 2-year study with female and male B6C3F1 mice showed while exposure to GSM-900 and DCS-1800 fields did not produce an overall increase in the incidence of tumors, there was a dose-dependent decrease in the number of tumor-bearing males and more so for incidence of hepatocellular carcinomas.

The 16 published reports on carcinogenesis in rats include three different strains: Fischer 344 (8), Wistar (3), and Sprague–Dawley (5), respectively. In some cases the animals were restrained during exposure and others were not but under either plane-wave equivalent or near-zone exposure conditions. These investigations were typically 2 years in duration. However, there was an implanted brain tumor study with Fischer 344 rats irradiated using GSM-900 fields for 2–3 weeks following glioma cell implantation in Salford et al., and a 6-week liver bioassay study also with Fischer 344 rats by Imaida et al. for TDMA-900 fields. Neither study attained any overall significant difference in the experiment animals.

With few exceptions in the 2-year studies, the solitary studies of Fischer 344 rats exposed to a variety of carrier frequency or modulation scheme specific to wireless communication did not provide indications of an increase in the promotion of ENU-induced brain or CNS cancer or spontaneous tumor induction compared with sham-exposed rats. The two GSM-900 exposed Wistar rat studies provided the same null results with regard to any tumor type. However, there were major differences in most aspects of the studies conducted in two different laboratories. Nonetheless, the macroscopic findings from one study showed several statistically significant gross lesions comparing sham control with GSM-exposed groups.

The four DMBA-induced mammary tumors in female Sprague–Dawley studies are especially interesting because they all used GSM-900 RF radiation. One investigation (Bartsch et al., 2002) was a self-replicated study using unrestrained rats and it found no difference between sham and plane-wave RF-exposed animals. However, two parallel investigations (Yu et al., 2006; Hruby et al., 2008) involving restrained rats in identical "waveguide-wheel" exposure systems at the same SARs resulted in very different mammary tumor incidences. Although Yu et al. found no difference between RF and sham-exposed rats, benign tumors in the cage-control group are significantly higher than in the sham and GSM 900-exposed groups. The latency to mammary tumor onset was also significantly shorter in the cage-control group than in the exposed groups. In addition, all GSM-exposed groups had, at different times, significantly more palpable tissue masses and none of the findings in GSM-exposed female Sprague–Dawley rats produced a clear dose–response relationship.

In summary, a majority of the laboratory mouse and rat studies did not exhibit a significant difference in carcinogenic incidences between exposed and sham-exposed animals. Although this observation may be comforting from the perspective of safety evaluation, most of them are one-of-a-kind investigations – only three mouse and perhaps four rat studies were designed as replication or confirmation studies. It is noteworthy that the findings of these studies have not been consistent, making it difficult to arrive at a definitive conclusion. It could be a major flaw that in a majority of the investigations, cage-control animals were not part of the investigation or were not included in the data analyses. Moreover, restraining the experimental animals during exposure could have introduced a stress factor, which further complicates interpretation of the results since stress has often been associated with cancer induction in these animals.

REFERENCES

Adey, W.R., Byus, C.V., Cain, C.D., Higgins, R.J., Jones, R.A., Kean, C.J., Kuster, N., MacMurray, A., Stagg, R.B., Zimmerman, G., Phillips, J.L., and Haggren, W., 1999, Spontaneous and nitrosourea-induced primary tumors of the central nervous system in Fischer 344 rats chronically exposed to 836 MHz modulated microwaves. *Radiat Res* 152: 293–302.

Adey, W.R., Byus, C.V., Cain, C.D., Higgins, R.J., Jones, R.A., Kean, C.J., Kuster, N., MacMurray, A., Stagg, R.B., and Zimmerman, G., 2000, Spontaneous and nitrosourea-induced primary tumors of the central nervous system in Fischer 344 rats exposed to frequency-modulated microwave fields. *Cancer Res* 60: 1857–1863.

Anane, R., Dulou, P.E., Taxile, M., Geffard, M., Crespeau, F.L., and Veyret, B., 2003, Effects of GSM-900 microwaves on DMBA-induced mammary gland tumors in female Sprague–Dawley rats. *Radiat Res* 160(4): 492–497.

Anderson, L.E., Sheen, D.M., Wilson, B.W., Grumbein, S.L., Creim, J.A., and Sasser, L.B., 2004, Two-year chronic bioassay study of rats exposed to a 1.6 GHz radiofrequency signal. *Radiat Res* 162(2): 201–210.

Bartsch, H., Bartsch, C., Seebald, E., Deerberg, F., Dietz, K., Vollrath, L., and Mecke, D., 2002, Chronic exposure to a GSM-like signal (mobile phone) does not stimulate the development of DMBA-induced mammary tumors in rats: results of three consecutive studies. *Radiat Res* 157(2): 183–190.

CBTRUS, 2008, *Statistical Report: Primary Brain Tumors in the United States, 2000–2004.* Central Brain Tumor Registry of the United States, Hinsdale, IL, USA. Available at: http://www.cbtrus.org/reports//2007-2008/2007report.pdf

Chou, C.K., Guy, A.W., Kunz, L.L., Johnson, R.B., Crowley, J.J., and Krupp, J.H., 1992, Long term, low-level microwave irradiation of rats, *Bioelectromagnetics* 13: 469–496.

Cotran, R.S., Kumar, V., and Robbins, S.L., 1994, *Pathologic Basis of Disease*, 5th ed. Philadelphia, PA: Saunders.

Faraone, A., Luengas, W., Chebrolu, S., Ballen, M., Bit-Babik, G., Gessner, A.V., Kanda, M.Y., Babij, T., Swicord, M.L., Chou, C.K., 2006, Radiofrequency dosimetry for the Ferris-wheel mouse exposure system. *Radiat Res* 165(1):105–112.

Ferlay, J., Bray, F., Pisani, P., and Parkin, D.M., 2004, GLOBOCAN 2002: Cancer Incidence, Mortality and Prevalence Worldwide, Version 2.0. IARC Cancer Base No. 5. Lyon, IARC Press, 2004. Limited version available from: URL: http://www_depdb.iarc.fr/globocan2002.htm

Goldstein L.S., Kheifets L., van Deventer E., and Repacholi M., 2002, Comments on "Long-term exposure of Eμ-Pim1 transgenic mice to 898.4 MHz microwaves does not increase lymphoma incidence" by Utteridge et al., *Radiat Res* 158, 357–364.

Goldstein L.S., Kheifets L., van Deventer E., and Repacholi M., 2003, Comments on "Long-term exposure of Eμ-Pim1 transgenic mice to 898.4 MHz microwaves does not increase lymphoma incidence" by Utteridge et al., *Radiat Res* 159, 275–276

Heikkinen, P., Kosma, V.-M., Hongisto, T., Huuskonen, H., Hyysalo, P., Komulainen, H., Kumlin, T., Lahtinen, T., Lang, S., Puranen, L., and Juutilainen, J., 2001, Effects of mobile phone radiation on X-ray-induced tumorigenesis in mice. *Radiat Res* 156: 775–785.

Heikkinen, P., Kosma, V.M., Alhonen, L., Huuskonen, H., Komulainen, H., Kumlin, T., Laitinen, J.T., Lang, S., Puranen, L., and Juutilainen, J., 2003, Effects of mobile phone radiation on UV-induced skin tumourigenesis in ornithine decarboxylase transgenic and non-transgenic mice. *Int J Radiat Biol* 79(4): 221–233.

Heikkinen, P., Ernst, H., Huuskonen, H., Komulainen, H., Kumlin, T., Mäki-Paakkanen, J., Puranen, L., and Juutilainen, J., 2006, No effects of radiofrequency radiation on 3-chloro-4-(dichloromethyl)-5-hydroxy-2(5H)-furanone-induced tumorigenesis in female Wistar rats. *Radiat Res* 166(2): 397–408. Erratum in: *Radiat Res* 2007 167(1):124–125.

Hruby, R., Neubauer, G., Kuster, N., and Frauscher, M., 2008, Study on potential effects of "902-MHz GSM-type Wireless Communication Signals" on DMBA-induced mammary tumours in Sprague–Dawley rats. *Mutat Res* 649(1–2): 34–44.

Imaida, K., Taki, M., Yamaguchi, T., Ito, T., Watanabe, S., Wake, K., Aimoto, A., Kamimura, Y., Ito, N., and Shirai, T., 1998, Lack of promoting effects of the electromagnetic near-field used for cellular phones (929.2 MHz) on rat liver carcinogenesis in a medium-term liver bioassay. *Carcinogenesis* 19(2): 311–314.

Imaida, K., Kuzutani, K., Wang, J., Fujiwara, O., Ogiso, T., Kato, K., Shirai, T. 2001, Lack of promotion of 7,12-dimethylbenz[a]anthracene- initiated mouse skin carcinogenesis by 1.5 GHz electromagnetic near fields. *Carcinogenesis* 22(11): 1837–1841.

Kundi, M., 2003a, Comments on "Long-term exposure of Eμ-Pim1 transgenic mice to 898.4 MHz microwaves does not increase lymphoma incidence" by Utteridge et al., Radiat. Res. 158, 357–364 (2002). *Radiat. Res.* 159: 274.

Kundi, M., 2003b, Comments on the responses of Utteridge et al. (Radiat. Res. 159, 277–278, 2003) to letters about their paper (Radiat. Res. 158, 357–364, 2002). *Radiat Res* 160(5): 613–614.

Kunz L.L., Johnson R.B., Thompson D., Crowley J., Chou C-K., Guy A.W., 1985, Effects of long-term low-level radiofrequency radiation exposure on rats. Vol 8: Evaluation of longevity, cause of death, and histopathological findings.Texas, Brooks Air Force Base, *USAF School of Aerospace Medicine*, ASAFSAM-TR-85–11

La Regina, M., Moros, E.G., Pickard, W.F., Straube, W.L., Baty, J., and Roti Roti, J.L., 2003, The effect of chronic exposure to 835.62 MHz FDMA or 847.74 MHz CDMA radiofrequency radiation on the incidence of spontaneous tumors in rats. *Radiat Res* 160(2): 143–151.

Lerchl, A., 2003, Comments on the recent publication on microwave effects on Eμ-Pim1 transgenic mice (Utteridge et al., Radiat. Res. 158, 357–364, 2002). *Radiat Res* 159: 276.

Lin, J.C., 2002, Cellular-phone radiation effects on cancer in genetically modified mice, *IEEE Antenn Propag Mag* 44(6): 165–168.

Lin, J.C., 2008, Studies on tumor incidence in mice exposed to GSM cell-phone radiation, *IEEE Microw Mag* 9(3): 48–54.

NTP, 1999, Toxicology and carcinogenesis studies of 60-Hz magnetic fields in F344/N rats and B6C3F1 mice (whole-body exposure studies). *Natl Toxicol Program Tech Rep Ser* 488: 1–168.

Oberto, G., Rolfo, K., Yu, P., Carbonatto, M., Peano, S., Kuster, N., Ebert, S., and Tofani, S., 2007, Carcinogenicity study of 217 Hz pulsed 900 MHz electromagnetic fields in Pim1 transgenic mice. *Radiat Res* 168: 316–326.

Repacholi, M.H., Basten, A., Gebski, V., Noonan, D., Finnie, J., and Harris, A.W., 1997, Lymphomas in Eu-Pim1 transgenic mice exposed to pulsed 900 MHz electromagnetic fields, *Radiat Res*, 147: 631–640.

Salford, L.G., Brun, A., Persson, B.R.R., and Eberhardt, J.L., 1993, Experimental studies of brain tumour development during exposure to continuous and pulsed 915 MHz radiofrequency radiation, *Bioeletrochem Bioenerg* 30: 313–318.

Shirai, T., Kawabe, M., Ichihara, T., Fujiwara, O., Taki, M., Watanabe, S., Wake, K., Yamanaka, Y., Imaida, K., Asamoto, M., and Tamano, S., 2005, Chronic exposure to a 1.439 GHz electromagnetic field used for cellular phones does not promote *N*-ethylnitrosourea induced central nervous system tumors in F344 rats. *Bioelectromagnetics* 26: 59–68.

Shirai, T., Ichihara, T., Wake, K., Watanabe, S., Yamanaka, Y., Kawabe, M., Taki, M., Fujiwara, O., Wang, J., Takahashi, S., and Tamano, S., 2007, Lack of promoting effects of chronic exposure to 1.95-GHz W-CDMA signals for IMT-2000 cellular system on development of *N*-ethylnitrosourea-induced central nervous system tumors in F344 rats. *Bioelectromagnetics* 28(7): 562–572.

Smith, P., Kuster, N., Ebert, S., and Chevalier, H.J., 2007, GSM and DCS wireless communication signals: Combined chronic toxicity/carcinogenicity study in the Wistar rat. *Radiat Res* 168(4): 480–492.

Sommer, A.M., Streckert, J., Bitz, A.K., Hansen, V.W., and Lerchl, A., 2004, No effects of GSM-modulated 900 MHz electromagnetic fields on survival rate and spontaneous development of lymphoma in female AKR/J mice. *BMC Cancer* 4: 77–90.

Sommer, A.M., Bitz, A.K., Streckert, J., Hansen, V.W., and Lerchl, A., 2007, Lymphoma development in mice chronically exposed to UMTS-modulated radiofrequency electromagnetic fields. *Radiat Res* 168: 72–80.

Szmigielski, S., Szudzinski, A., Pietraszek, A., Bielec, M., Janiak, M., and Wrembel, J.K., 1982, Accelerated development of spontaneous and benzopyrene-induced skin cancer in mice exposed to 2450 MHz microwave radiation, *Bioelectromagnetics* 3: 179–191.

Szudinski, A., Pietraszek, A., Janiak, M., Wrembel, J.K., Kalczak, M., and Szmigielski, S., 1982, Acceleration of the development of benzopyrene-induced skin cancer in mice by microwave radiation, *Arch Dermatol Res* 274: 303–312.

Tillmann, T., Ernst, H., Ebert, S., Kuster, N., Behnke, W., Rittinghausen, S., and Dasenbrock, C., 2007, Carcinogenicity study of GSM and DCS wireless communication signals in B6C3F1 mice, *Bioelectromagnetics* 28: 173–187

Utteridge, T.D., Gebski, V., Finnie, J.W., Vernon-Roberts, B., and Kuchel, T.R., 2002, Long-term exposure of E-mu-Pim1 transgenic mice to 898.4 MHz microwaves does not increase lymphoma incidence. *Radiat Res* 158: 357–364.

Utteridge, T.D., Gebski, V., Finnie, J.W., Vernon-Roberts, B., and Kuchel, T.R., 2003, Response to the letters to the editor sent by (1) Kundi, (2) Goldstein/Kheifets/van Deventer/Repacholi, and (3) Lerchl. *Radiat Res* 159: 277–278.

Wu, R.Y., Chiang, H., Shao, B.J., Li, N.G., and Fu, Y.D., 1994, Effects of 2.45 GHz microwave radiation and phorbol ester 12-*O* tetradecanoylphorbol-13-acetate on dimethylhydrazine-induced colon cancer in mice. *Bioelectromagnetics* 15: 531–538.

Yu, D., Shen, Y., Kuster, N., Fu, Y., and Chiang, H., 2006, Effects of 900 MHz GSM wireless communication signals on DMBA-induced mammary tumors in rats. *Radiat Res* 165(2): 174–180.

Zook, B.C., and Simmens, S.J., 2001, The effects of 860 MHz radiofrequency radiation on the induction or promotion of brain tumors and other neoplasms in rats. *Radiat Res* 155(4): 572–583.

Chapter 3

Epidemiological Studies of Cellular Telephone Use and Risk of Cancer

Minouk J. Schoemaker and Anthony J. Swerdlow

ABSTRACT

The increasing worldwide use of cellular telephones has generated public concern about exposure to radiofrequency fields as a potential risk factor for cancer. Over the last decade, there has been substantial effort to investigate the potential health effects of cellular phone use, and this chapter examines the evidence about the relation to cancer risks obtained from epidemiological studies. Most studies have focused on brain tumors and other intracranial tumors, with a few studies having investigated other neoplasms such as parotid gland and ocular tumors. Nearly all have been case–control studies, the main exception being a large Danish cohort study based on record linkage of network operator and cancer registry data. Methodological strengths and weaknesses of individual studies are discussed, and the overall evidence is evaluated in terms of consistency between studies, timing and magnitude of associations observed, and dose–response associations. It is concluded that the epidemiological studies reviewed show, on balance, no convincing or consistent evidence for a raised risk of cancer in relation to cellular phone use. The overall evidence suggests that it is unlikely that there are large increases of risk in relation to cellular phone use in the lag period for which there are substantial data.

M. J. Schoemaker and A. J. Swerdlow Section of Epidemiology, Institute of Cancer Research, Sir Richard Doll Building, Sutton, Surrey, SM2 5NG, UK, e-mail: Minouk.Schoemaker@icr.ac.uk

J.C. Lin (ed.), *Advances in Electromagnetic Fields in Living Systems, Health Effects of Cell Phone Radiation, Volume 5*,
DOI 10.1007/978-0-387-92736-7_3,

Past studies have had limitations, however, in particular that exposure assessment has been crude, data on risk after lag periods of 10 years or more, prolonged use, high intensities of use and childhood exposures, are still limited, and the possibility of risk in relation to these remains open.

1. INTRODUCTION

The increasing worldwide use of cellular (mobile) telephones and microwave communications has generated public concern about exposure to radiofrequency (RF) fields as a potential risk factor for cancer. RF fields are nonionizing, and there is no established biological mechanism by which they are thought to cause or promote tumor growth (AGNIR, 2003). However, if such an effect were to exist, it would potentially have a large impact on health because of the widespread use of cellular phones and other devices based on wireless technology. In the year 2006, there were two billion cellular phone users in the world, and this number is expected to rise to three billion by the year 2010 (MobileTracker, 2006). Over the last decade or so, there has been substantial effort to investigate the potential health effects of cellular phone use, and in this chapter, we examine the evidence about the relation to cancer risks obtained from epidemiological studies.

Epidemiology is the study of the distribution and determinants of disease in human populations. Two main types of epidemiological study have been used to address the possible relationship between cellular phone use and adverse health outcomes: the case–control study and the cohort study. The case–control study involves comparing a group of patients with the health outcome under investigation, for example, a brain tumor, with an unaffected comparison group with regard to the exposure of interest. A main feature is therefore, that exposure is assessed retrospectively when subjects are already affected by the condition. In cohort studies, people are followed up over time and subsequent risks of the health outcome under investigation are compared between people with and without the exposure of interest. Exposure is assessed prior to development of the health outcome, but for rare diseases large numbers of subjects need to be followed up over a long period for a sufficient number of cases of disease to occur.

Most studies investigating cancer risks related to cellular phone use have focused on brain tumors and other tumors in the head, because these develop close to where the energy from the cellular handset is deposited. Studies have investigated the association of cellular phone use with glioma, the most common form of brain tumor, meningioma, and acoustic neuroma (a tumor of the vestibulocochlear nerve). Other types of neoplasm investigated include parotid (salivary) gland tumors, ocular tumors, non-Hodgkin's lymphoma, and testicular cancer, the latter because of the habit of men carrying phones in their trouser pockets. Reported studies on cellular phone use have largely been case–control studies as results can be obtained relatively fast, typically after a few years, and as the number of study subjects needed is more manageable in logistic and financial terms. A substantial number of published studies on brain tumors, acoustic neuroma, and parotid gland tumors were carried out under the umbrella of the Interphone study, a 13-country collaborative case–control study into cellular phone use and risk of several types of intracranial tumor, coordinated by the International Agency for Research on Cancer. Two cohort studies of cellular phone subscribers have been published to date, one from the United States and one from Denmark. Such cohort studies are informative because they circumvent several problems that are inherent to case–control studies, but also these studies have had some limitations, in particular with regard to exposure classification.

Published studies have analyzed cancer risks in relation to regular phone use, with various definitions of regular use. Risk associations have also been investigated by time since first exposure as a measure of induction time, and number of years of use, cumulative hours of use and number of calls as measures of duration and dose. Most reports have also investigated associations of risk with use of analogue and digital cellular phones separately, because analogue phones were the first phones to be introduced and have a higher power output than digital phones (Independent Expert Group on Mobile Phones (IEGMP), 2000). As exposure is largely on the side of the head where the phone is held (Dimbylow and Mann, 1999), studies have also analyzed risks of intracranial or parotid tumors by laterality of the tumor and reported side of phone use.

Case–control studies have reported the odds ratio as a measure of association, which indicates how many times more (or less) the risk is among exposed people (e.g., cellular phone users) compared with unexposed or infrequently exposed people. In cohort studies, the measure of risk generally used has been the standardized incidence ratio, which compares the number of subjects who developed the outcome of interest (e.g. cancer) in the cohort to the number expected if the age, sex, and calendar year-specific rates of the background, general population applied.

In this chapter, we describe published epidemiological studies into cellular phone use and risk of cancer, and discuss the overall body of evidence regarding a potential association.

2. AN OVERVIEW OF PUBLISHED STUDIES OF CELLULAR PHONE USE AND RISK OF CANCER

The following sections summarize chronologically published case–control studies through to August 31, 2007 and comments on their design and results. Studies were identified through searches on Pubmed using key words such as "cellular phone," "neoplasms," and "brain neoplasms" and through searches of the citations of the identified articles. We first describe published studies of risk of intracranial tumors (glioma, meningioma, acoustic neuroma), and then studies of other types of neoplasm. Studies of intracranial tumors that were conducted as part of the Interphone study are discussed separately because they were conducted under a shared protocol.

2.1. Case–Control Studies of Cellular and Cordless Phone Use and Risk of Intracranial Tumors

2.1.1. Studies of Intracranial Tumors Conducted Outside the Umbrella of the Interphone Collaboration

Sweden, Brain Tumor and Acoustic Neuroma Cases, 1994–1996: (Hardell et al., 1999, 2000, 2001)

Hardell et al. (1999) published the first case–control study of intracranial tumors in relation to cellular phone use. Cases were defined as people diagnosed with a histologically verified brain tumor or acoustic neuroma at ages 20–80 years. The study included cases diagnosed in the Uppsala-Örebro region of Sweden in 1994–1996 and in the Stockholm region in 1995–1996, retrospectively identified through the cancer registry. For each interviewed case, two controls were selected from the population register, with matching on sex, birth year, and region.

Information on cellular phone use was obtained by self-administered postal questionnaire. The questionnaire included questions on cellular phone use as well as a complete occupational history and information on chemical exposures. Use of a cellular phone during leisure time and work was assessed, as well as the average number of minutes of exposure per day, years of use, types of phone used and use of hands-free devices, and participants were asked which ear they used most for phone calls. The questionnaire was followed up by a telephone call by a nurse for clarification if answers were unclear. Models of telephones were classified as analogue or digital, based on the first digits of the telephone number. Interviews and coding of answers were reported to have been conducted "blind" to case–control status.

The analysis was based on 209 cases and 425 population controls (Hardell et al., 1999). The analysis was reportedly based on a conditional logistic regression model of matched case–control sets, but it is not entirely clear how this was done, as the designed two controls per case (even if all participated) would give 418 not 425 controls. Subjects were considered "exposed" if they reported a minimum of 8 h of cumulative phone use. Phone use in the year prior to diagnosis was not considered in the analyses. Controls were assigned the same year as their matched case for this censoring of phone use. Usage of a hands-free device was considered as no exposure to cellular phones; it was not stated what level of hands-free use was required for this.

A total of 78 cases (37%) and 161 controls (38%) were considered exposed to cellular phones (Hardell et al., 1999). Risk of an intracranial tumor was not raised in relation to exposure to cellular phones overall (Odds ratio (OR) = 0.98, 95% confidence interval (CI): 0.69–1.41), or to digital or analogue phones separately. There was no relation of risk to cumulative duration of use or use several years earlier. Analyses of tumor laterality in relation to side of phone use showed statistically nonsignificantly raised odds ratios, based on five cases on the left and eight cases on the right, for a temporal, occipital, or temporo-parietal tumor, an apparently ad hoc combination of sites, on the same side of the head as reported phone use. The risk of an ipsilateral tumor was reported as 2.40 (95% CI: 0.52–10.9) for the left, and 2.45

(95% CI: 0.78–7.76) for the right. It is not clear from the paper what the baseline group for these odds ratios was: e.g., nonphone users, or those with contralateral tumors. As Rothman (Rothman, 2000) subsequently argued, since there had been no increase in the overall risk of a tumor, the association between side of tumor and side of telephone use would require the implausible inference that telephone use does not affect the risk of whether a brain tumor will occur but only its location. More plausibly, it might be due to biased reporting by cases of side of phone use. Furthermore, the results had wide confidence intervals including 1.0 so chance as an explanation remains open.

The cases included 136 subjects with malignant and 62 with benign tumors, the latter group including 46 meningiomas, 13 acoustic neuromas, and three other benign cases (Hardell et al., 1999). Reported relative risks for phone use overall were similar for all malignant tumors, astrocytic tumors, all benign tumors together, or meningioma separately.

The authors reported that 90% of cases and 91% of controls took part in the study. This rate for cases appears to represent the number of participants among all cases contacted for the study, rather than the proportion among all eligible cases. The latter is, however, the conventional definition of participation (response) rate (Slattery et al., 1995), as the aim is to investigate all incident cases within a particular time period, to obtain as representative a study sample as possible. This conventionally calculated participation rate cannot be obtained from the paper because the number of eligible patients deceased at the time of the study was not reported, but as there were 270 subjects known to be alive at the start of the study, it is likely to be considerably lower than 77% (209/270).

Ahlbom and Feychting (1999) further reported that comparisons against cancer registry data showed that the number of reported patients in the study was considerably lower than the number of incident patients during the study period. They noted that the cancer registry had recorded 862 cases that met the study criteria on age, region of residence, and date of diagnosis, with the implication that the participation rate might be as low as 209/862, i.e. 24%. Mortality could not account for the disparity, suggesting the possibility that additional exclusion criteria were applied, ascertainment methods were incomplete, or both. Cancer registries are known not to have completely up-to-date information on all incident cases, with a lag period to achieve completeness, and it is unclear to what extent the higher number cited by Ahlbom and Feychting could be due to incident cases ascertained later in time. It later emerged that some of the above inconsistencies were due to lack of transparency in reporting; in a subsequent paper (Hardell et al., 2000), it was revealed that benign tumors were included only for the Stockholm area in 1996 (i.e., for only part of the study region and study period), which was not disclosed in the first paper (Hardell et al., 1999) and could have contributed substantially to this discrepancy in numbers.

Hardell et al. (1999) did not provide a breakdown of how the 91% participation rate for controls was derived, nor whether exclusion criteria were applied.

The data from this study were reanalyzed and published twice more (Hardell et al., 2000, 2001). In these subsequent papers, the authors reported intracranial tumor risks in relation to cellular phone use obtained from multivariate analyses incorporating variables for medical X-ray exposures to the head and neck and laboratory work, both variables significantly related to brain tumor risk in the study. The risk of temporal, occipital, or temporoparietal tumors was 2.62 (95% CI: 1.02–6.71) in relation to ipsilateral phone use, 0.97 (95% CI: 0.36–2.59) for contralateral use, and 0.71 (95% CI: 0.14–3.68) for use on both sides of the head (Hardell et al., 2000, 2001).

United States, Malignant Brain Tumors and Acoustic Neuroma: (Muscat et al., 2000, 2002)

Muscat et al. (2000) carried out a hospital-based case–control study in five US academic medical centers between 1994 and 1998. Eligible cases were those diagnosed with a primary brain cancer during the past year and with sufficient command of English to be interviewed. A total of 469 cases with primary malignant brain cancer, and 422 matched controls were interviewed using a structured questionnaire. Regular cellular phone use was defined as having had a subscription to a cellular phone service. Information was obtained on the number of years of use, minutes/hours of use per month, year of first use, manufacturer, and reported average monthly bill. Eligible control patients were inpatients from the same hospitals as the cases and identified from daily admission rosters. Controls were admitted for benign conditions, except at two centers, which predominantly treated cancer. The benign conditions included musculoskeletal disorders (24%), systemic disease (21%), skin conditions (16%), and other specified conditions (23%). Six percent of controls were admitted for malignant neoplasms, but excluding leukemia or lymphoma due to a possible link with RF fields. Additional matching criteria included age at diagnosis, sex, race, and month of admission. A matched control could not be found for all cases, in particular for cases under age 30 years. Unconditional logistic regression, with adjustment for the matching factors, type of respondent, years of education and month and year of interview was used to obtain odds ratios in order that unmatched cases could be included in the analyses. Cell phone use was evaluated up to the interview date, but it was reported that 70% of cases were interviewed within 2 months of diagnosis.

Eighty-two percent of approached cases and 90% of approached controls took part in the study. Fifty-five cases who were too ill were not approached; therefore, it can be derived that interviewed cases represented approximately 75% of all eligible cases. The proportion of proxy interviews was higher for cases than for controls (9.2% vs. 1.4%). Sixty-six cases (14%) and 76 controls (18%) reported having had a subscription to a cellular phone service. Risk of brain cancer was not raised in relation to ever having had a subscription (OR=0.8, 95% CI: 0.6–1.2), and was not raised for four or more years of cellular phone use (OR=0.7, 95% CI: 0.4–1.4), or for the highest category of hours per month of use (>10.1 h) or cumulative hours of use (>480 h). Relative risks in ever subscribers overall were 1.1 or less for each anatomical site of brain tumor. Among case users, there was a nearly significant excess of tumors ipsilateral to the side of phone use for brain cancer overall ($p=0.06$), but no ipsilateral excess for temporal lobe tumors ($p=0.33$). No significant association of risk was observed when analyses were stratified by tumor lobe or by histological group.

Muscat et al. (2002) also published results regarding acoustic neuroma risk in relation to cellular phone use, using a similar study design to that described earlier. The study included 90 cases diagnosed between 1997 and 1999 and 86 controls with nonmalignant conditions. Participation rates were not reported. Ever-subscription for a cellular phone was reported by 20% of cases and 27% of controls. The abstract states a corresponding odds ratio of 0.9, but this appears from the text to be the relative risk of a tumor ipsilateral to reported side of use (with $p=0.07$). There was no trend in risk of acoustic neuroma in relation to number of years of phone use, hours per month of use, total hours of use, or cumulative hours of use.

United States, Brain Tumors and Acoustic Neuroma: (Inskip et al., 2001)

Inskip et al. (2001) obtained data from 782 cases of intracranial tumors treated at participating hospitals in Phoenix, Arizona, Boston, and Pittsburgh in the US during 1994–1998. Eligible cases had to have been diagnosed with a first intracranial glioma or neuroepitheliomatous tumor within the 8 weeks preceding hospitalization at a participating hospital. Cases also had to be resident within 50 miles of the hospital or, for the Phoenix centre, within Arizona, and be aged 18 year and older at diagnosis. Participating cases included 489 patients diagnosed with glioma, 197 with meningioma, and 96 with acoustic neuroma. A total of 799 controls were recruited from the same hospitals and were admitted with nonmalignant conditions, mainly accidents, cardiovascular disease, and musculoskeletal disease. Controls were frequency-matched to the total group of cases on hospital, age, sex, ethnicity, and proximity of residence to the hospital. Data on cellular phone use were obtained by interview by a research nurse. Cellular phone use appeared to have been included up to the interview date of cases and controls, but 80% of the cases were interviewed within three weeks of diagnosis. Analyses were conducted using conditional logistic regression adjusting for the matching variables and several other factors including date of interview (as a continuous variable).

A total of 92% of eligible cases and 86% of contacted controls agreed to take part. Proportions of proxy responses were higher for glioma and meningioma cases than controls: for 16% of glioma cases, 8% of meningioma cases, 3% of acoustic neuroma cases and 3% of controls the interview was carried out with a proxy, usually the spouse. Twenty-nine percent of controls had used a handheld cellular phone more than five times in their lives, with the lowest percentage (22%) among controls with circulatory disease, likely due to their older age. Regular use of a handheld cellular phone, defined as use of a phone for at least two calls per week, was associated with an odds ratio of 0.8 (95% CI: 0.6–1.1) for intracranial tumors overall. The relative risk for five or more years of use was 0.9 (95% CI: 0.5–1.6). No significant increase in risk of glioma, meningioma, or acoustic neuroma was observed in relation to regular phone use, or for longest duration of use. For acoustic neuroma, the relative risk after 5 or more years of use was 1.9 (95% CI: 0.6–5.9), based on five cases. Tumor risk was not related to duration of phone use, daily frequency of use, or total cumulative hours of use. Among cases who had used a cellular phone for at least 6 months before diagnosis, tumor laterality was not related to reported side of phone use.

Finland, Brain and Salivary Gland Tumors: (Auvinen et al., 2002)

Auvinen et al. (2002) reported on a case–control study based on record linkage between cancer registry records and subscription records of the two cellular phone network providers in Finland. The researchers identified all brain tumors and salivary gland tumors diagnosed at ages 20–69 years in 1996 in Finland from the National Cancer Registry, resulting in 398 brain tumors and 34 salivary gland tumors, and selected five age and sex-matched controls for each case from the national population register. Thirteen percent of brain tumor cases, 12% of salivary gland tumor cases, and 11% of controls had ever had a personal subscription to a cellular phone network. Ever having a cellular phone subscription was associated with a relative risk of 1.3 (95% CI: 0.9–1.8) for brain tumors and 1.3 (95% CI: 0.4–4.7) for salivary gland tumors. A borderline significant association was shown for glioma (OR = 1.5, 95% CI: 1.0–2.4), but the relative risk was close to unity for meningioma or other brain tumors. Glioma risk showed a significant relationship to ever-subscription for an analogue phone (OR = 2.1, 95% CI: 1.3–3.4), but not for a digital phone (OR = 1.0). There was a weak increasing trend of glioma risk with duration of analogue subscription, but not of digital subscription. The average duration of subscription was 2–3 years for analogue phones, and less than 1 year for digital phones. There were no substantial differences in distribution of histology, tumor lobe, or laterality between gliomas occurring in phone users and in those who did not have a subscription. The results remained similar after adjustment for place of residence, occupation, and socio-economic status.

Although this study avoids problems with quality of recall from self-reported cellular phone use, exposure misclassification is a potential problem. The authors had no information whether those who held a personal subscription were also the users of the phone, and had no information about the frequency or duration of calls made by the subject, or about the use of cellular phones provided by an employer. They note that in the study period, there were more corporate subscriptions than private subscriptions in Finland, with the implication that many cellular phone users could have been classified as nonusers in the study. The authors stated, however, that a 50% unbiased sensitivity in exposure assessment would have attenuated any real effects by only 10%.

Sweden, Brain Tumor and Acoustic Neuroma Cases, 1997–2000: (Hardell et al., 2002a, b, 2003a, b, 2004b, 2005b)

A second case–control study by Hardell et al. (2002a), much larger than the first, included cases with intracranial tumors diagnosed during 1997 to mid-2000, and followed a similar study design to their first study (Hardell et al., 1999). Cases diagnosed with brain tumors and acoustic neuroma at ages 20–80 years were identified from the cancer registry in the Uppsala-Örebro, Stockholm, Linköping, and Göteborg medical regions of Sweden. Unlike in the previous study, in which the cases appeared to have been identified retrospectively, the researchers were notified by the cancer registry of newly incident cases when they occurred. One control per case was selected from the population register, matched on sex, age, and geographical area of residence. Exposures up to 1 year prior to diagnosis were included in the analyses. For controls, the same year was used as the case it was matched to.

The authors reported that 1,429 cases and 1,470 controls completed the questionnaire, corresponding to 88% and 91% of those mailed. If we calculated the case participation rate including patients in the denominator who were deceased (over 20% of all eligible patients), for whom their treating physician refused permission, or those who were not able to take part for medical reasons, it would be 63% for all cases combined (Hardell et al., 2002a) and 53% for malignant cases only (Hardell et al., 2002b), i.e., the participation rates were much lower than they appear to be from the article. No breakdown was given of reasons for nonparticipation of controls, and we can not establish how the 91% control participation rate was derived. The analyses were predominantly based on 1,303 matched case–control pairs in a conditional logistic regression model, including 529 malignant cases (Hardell et al., 2002b) and an undisclosed number of less than 611 meningiomas and 159 acoustic neuromas. A higher proportion of cases than controls were assisted by relatives in completing the questionnaire, 32 vs. 9%, respectively (Hardell et al., 2002a). Exposure was defined as "ever" use of cellular or cordless phones more than 1 year prior to diagnosis (Hardell et al., 2002b).

The relative risk of any intracranial tumor in relation to phone use was 1.3 (95% CI: 1.02–1.6) for analogue phone use, 1.0 (95% CI: 0.8–1.2) for digital phone use, and 1.0 (95% CI: 0.8–1.2) for cordless phone use (Hardell et al., 2002a). For analogue phones, start of use more than 5 years earlier was associated with a relative risk of 1.4 (95% CI: 1.04–1.8) and use more than 10 years earlier of 1.8 (95% CI: 1.1–2.9). Relative risks for cordless phones were similar to those for analogue phones: 1.3 (95% CI: 0.99–1.8) and 2.0 (95% CI: 0.5–8.0), for start of use more than 5 and 10 years earlier, respectively. No subjects reported digital phone use 10 or more years ago. In a multivariate analysis of all three phone types, there was no trend of greater risk with increasing induction time for analogue phone use, but the trend remained for cordless phones (Hardell et al., 2002a). Additional analyses in a further paper (Hardell et al., 2003b) showed a significant trend in risk for intracranial tumors overall in relation to duration of use for analogue (increase in OR per year = 1.04, 95% CI: 1.01–1.08), but not for digital or cordless phones. The trend analysis treated duration of use as a continuous variable and appears to have included the nonexposed group. The increased risk of acoustic neuroma in phone users overall might be responsible for this trend by itself, rather than a trend within exposed subjects only, but no categorized data were shown to examine this possibility. The paper restricted to malignant tumors (Hardell et al., 2002b) suggested a greater risk with longer induction period for digital and cordless but not analogue phones.

Risk of an intracranial tumor, of any type, in the temporal area was increased in relation to analogue phone use (OR=2.0, 95% CI: 1.3–3.1), but not in relation to digital or cordless phone use, and risks were not appreciably raised for tumors at other lobes, including tumors in the temporo-parietal area (OR=0.8) (Hardell et al., 2002a). Analyses of risk of temporal, occipital, or temporoparietal tumors combined were not presented, despite significant findings reported for such tumors in the first study by this research group (Hardell et al., 1999). For each type of phone, borderline significant or significantly raised risks were found for tumors ipsilateral to reported phone use, whereas contralateral risks were generally lower than 1.0. When risks were examined by histological tumor type and type of phone, it appeared that the raised risk of tumors in the temporal area in relation to analogue phone use was confined to acoustic neuroma (OR=3.5, 95% CI: 1.8–6.8), constituting the large majority of benign tumors considered under "temporal," and to a lesser extent meningioma (OR=4.5, 95% CI: 0.97–20.8), and that risk of malignant tumors in the temporal area was not raised. The findings for analogue phone use overall and for temporal tumors specifically were, therefore, largely due to an effect for acoustic neuroma. No effect of induction time was shown for acoustic neuroma, however; the odds ratio was 3.7 for 5 or more years and 3.5 for 10 or more years since first use. Hardell et al. (2003a) have also reported evidence of an increasing national trend in incidence rates of acoustic neuroma in Sweden, based on cancer registry data, but this increase was already apparent before the major increase in cellular phone use in Sweden.

Further papers based on the same dataset were published, with stratifications by urban/rural residence and by age at diagnosis (Hardell et al., 2004b, 2005b). Relative risks of an intracranial tumor in relation to analogue phone use were identical for rural or urban residence (OR=1.3 for each), but were somewhat higher for rural residence in relation to digital use (OR=1.4 vs. OR=0.9) and cordless use (OR=1.3 vs. OR=1.0) compared with urban residence, although with overlapping confidence intervals (Hardell et al., 2005b). These findings were attributed by Hardell et al. to a higher average output power in rural than in urban areas due to adaptive power control (APC) (Lonn et al., 2004b). Stratification of risk by age at diagnosis showed that risks of intracranial tumors were highest in the 20–29 year age group (Hardell et al., 2004b).

United States, Acoustic Neuroma and Intratemporal Facial Nerve Tumors: (Warren et al., 2003)

Warren et al. (2003) investigated risks of acoustic neuroma and intratemporal facial nerve tumors in relation to cellular phone use. The stated rationale for selection of these two types of tumors was that the intratemporal facial nerve would have higher RF exposure levels than the acoustic nerve. The study was small, 51 cases of acoustic neuroma and 18 of facial nerve tumor, and as they were recruited from a tertiary care center, the study was not population-based. Two control groups were recruited from the same hospital; 1 of 72 patients with rhinosinusitis and 1 of 69 patients with dysphonia or gastroesophageal reflux. The report did not state participation rates. Risks of either type of tumor was not raised in relation to regular handheld cellular phone use, defined as more than one call per week, compared with nontumor controls. No information was reported on risks by duration of phone use or other features of exposure.

Sweden, Brain Tumor and Acoustic Neuroma Cases, 2000–2003: (Hardell et al., 2005a, c)

The third case–control study by Hardell et al. (2005a) was based on patients diagnosed between mid-2000 and the end of 2003. It included 317 malignant brain tumor cases, 305 meningiomas, 84 acoustic neuromas, and 692 controls. Cases were ascertained from a smaller area than the previous study: from the Uppsala-Örebro and Linköping regions. The study design was similar to that of the second study by this research group (Hardell et al., 2002a), with incident cases ascertained consecutively from the cancer registries and controls selected from the population register frequency-matched on age and sex to cases. The median time between diagnosis and sending of the questionnaire was 79 days. As in previous studies, exposure was assessed by self-administered questionnaire, except that all subjects were subsequently contacted by telephone to supplement the questionnaire with extra details. This was reportedly to have been done "blind" to case–control status.

As with previous studies by this research group, relatively high participation rates were reported; 89% for benign tumor patients, 88% for malignant brain tumor patients, and 84% for controls. It can be derived from the article, however, that the participating cases represented 67% of eligible cases overall, and less than 58% of eligible malignant cases, after taking into account deceased patients as well as patients excluded for medical reasons and because the treating physician refused. Exposure was defined as ever use of cordless or cellular phones. Analyses were adjusted for sex, age, year of diagnosis, and socio-economic status, which was derived from the last occupation reported in the questionnaire.

Sixty-six percent of controls reported ever use of a cellular or cordless phone (11% for analogue, 50% for digital, and 44% for cordless). For benign tumors, there was a statistically significant positive association of acoustic neuroma risk with phone use more than 1 year prior to diagnosis for analogue phones (OR=4.2, 95% CI: 1.8–10) and digital phones (OR=2.0, 95% CI: 1.05–3.8) and a nonsignificant association for cordless phones (OR=1.5, 95% CI: 0.8–2.9). There was no clear trend of risk with increasing time since first use; in fact, for analogue and cordless phones risk was highest among those who started use less than 5 years ago. Risk of meningioma was borderline significantly raised overall with use of analogue phones (OR=1.7, 95% CI: 0.97–3.0), and was nonsignificantly raised for digital and cordless phones. Meningioma risk 10 or more years after start of use was significantly raised for analogue phones (OR=2.1, 95% CI: 1.1–4.3), and borderline significantly raised for cordless phones (OR=1.9, 95% CI: 0.97–3.6). Restriction of analyses to tumors of the temporal lobe, after excluding acoustic neuroma, showed higher odds ratios than for benign tumors overall, and than for tumors in the frontal lobe or other parts of the brain. Fivefold significantly increased risks were observed for benign temporal tumors in relation to analogue, digital, and cordless phones 10 or more years since first use.

In analyses restricted to malignant tumors, significantly raised risks were reported in relation to analogue (OR=2.6, 95% CI: 1.5–4.3), digital (OR=1.9, 95% CI: 1.3–2.7), and cordless (OR=2.1, 95% CI: 1.4–3.0) phone use (Hardell et al., 2005c). Risks were higher in subjects who started phone use 10 or more years prior to diagnosis, and in those in the highest categories of cumulative hours of use. Risks for phone use overall were higher for high-grade than for low-grade astrocytoma, but with overlapping confidence intervals, and based on a small number of cases with low-grade tumors.

Sweden, Pooled Analyses of Second and Third Hardell Studies of Brain Tumors and Acoustic Neuroma, 1997–2003: (Hardell et al., 2006a, b)

The second and third studies on brain tumors by Hardell et al. (2006a, b) described earlier have also been reported as a pooled analysis of patients diagnosed between 1997 and 2003.

The analyses of benign tumors included 916 cases of meningioma, 243 of acoustic neuroma, 96 other benign brain tumors, and 2,162 controls (Hardell et al., 2006a). Risk of acoustic neuroma was raised for phone use up to 1 year prior to diagnosis in relation to analogue phones (OR=2.9, 95% CI: 2.0–4.3), digital phones (OR=1.5, 95% CI: 1.1–2.1), and cordless phones (OR=1.5, 95% CI: 1.04–2.0), but was raised only for analogue phone use when considering first use 10 or more years earlier (OR=3.1, 95% CI: 1.7–5.7). There was an apparent trend of risk with number of cumulative hours of use, with the highest risk after 1,000 or more hours of analogue use (OR=5.1, 95% CI: 1.9–14). Risks of acoustic neuroma ipsilateral to reported side of phone use were only marginally higher than the overall risks regardless of laterality. For meningioma, risk was borderline significantly raised in relation to analogue phone use overall (OR=1.3, 95% CI: 0.99–1.7), but not substantially raised for digital or cordless phone use. Risk of meningioma 10 or more years after first use was raised for analogue phone use (OR=1.6, 95% CI: 1.02–2.5), and nonsignificantly raised for cordless phones (OR=1.6) and digital phones (OR=1.3). There was no clear trend of risk of meningioma with number of cumulative hours of use.

Pooled analyses of 905 malignant tumors and 2,162 controls showed raised risks in relation to analogue (OR=1.5, 95% CI: 1.1–1.9), digital (OR=1.3, 95% CI: 1.1–1.6), and cordless (OR=1.3, 95% CI: 1.1–1.6) phone use up to 1 year prior to diagnosis (Hardell et al., 2006b). Risks were 1.8 to 2.8-fold increased 10 or more years after first use. After separating high-grade from low-grade astrocytoma, relative risks were higher for high-grade than for low-grade tumors for phone use incorporating a 10-year induction time, and to a lesser extent also for phone use with a 1-year induction time, but with overlapping confidence intervals. Subjects with more than 2,000 lifetime hours of phone use had a relative risk of 5.9 (95% CI: 2.5–14) in relation to analogue use, 3.7 (95% CI: 1.7–7.7) in relation to digital use and 2.3

(95% CI: 1.5–3.6) in relation to cordless use, with statistically significant trends of increasing risk in relation to cumulative hours of use.

Somewhat disconcertingly, a later review paper by this research group detailing results of all their previous case–control studies of cancer in relation to cellular phone use (Hardell et al., 2006c) showed results for the above pooled studies that did not correspond with the original papers. The discrepancy was particularly prominent in relation to digital phone use with 10 or more years of later. For example, for malignant tumors, the original paper (Hardell et al., 2006b) showed an odds ratio of 2.8, whereas the review (Hardell et al., 2006c) gave 3.4. Neither mention was given of the existence of these disparate results, nor was an explanation given.

2.1.2. Studies of Intracranial Tumors within the Interphone Collaboration

The Interphone study is a large international case–control study involving 13 countries, coordinated by the International Agency for Research on Cancer. Participating countries include eight West European countries as well as Australia, Canada, Israel, Japan, and New Zealand. The number of cases for all centers combined was 2,765 gliomas, 2,425 meningiomas, 1,121 acoustic neuromas, and 109 malignant parotid gland tumors, as well as 7,658 controls (Cardis et al., 2007). Cases had to be aged 30–59 years at diagnosis to be included in the Interphone analyses, although several study centers increased this age range for their own analyses to include younger as well as older patients. Recruitment of participants for the Interphone analyses was finished in 2004. Several centers have published their national results, and two pooled studies of data from the Nordic countries and United Kingdom have also been reported. Pooled analyses based on data from all 13 countries have not yet been published, but are currently in progress. The results will be of interest because they will be based on larger numbers than yet published. All centers followed a largely shared protocol and used the same questionnaire. The general design features have been described previously (Cardis et al., 2007), and a brief description of the data collection is given here.

The Interphone study collected information on cellular phone use, use of other wireless communication devices, occupational exposures to electromagnetic fields, and other potential confounders, by a computer-assisted personal interview (CAPI) at all but one study centre, which used paper questionnaires. The interview was conducted by an interviewer recording responses directly into a laptop computer. This was intended to ensure that uniform data were collected between centers, but had the disadvantage that there was no written record of the interview to check against if errors were made in data entry. Subjects who answered that they had ever used a cellular phone on average once a week for 6 months or more, defined as regular use, were asked detailed questions about their phone use patterns. They were asked, with the help of a compendium of show-cards of cellular phones, to identify each of the phone models they had used. For each individual phone, the subject was asked about the network operator, the start and end date of use, the average amount of time the phone was used and the average number of calls that had been made with it.

If any substantial changes in the pattern of use between the start and end date were reported that lasted for more than 6 months, the same information was also collected for these periods of changed usage. Other exposure parameters collected were the proportion of usage time that the antenna was extracted, if extendable, a hands-free set had been used, and the phone had been used while moving in a vehicle, and whether the phone had mainly been used in rural areas, urban areas, or both. Finally, the interviewer asked about the side of the head on which the phone was mainly used and whether the subject was left or right-handed. Information on the use of digital enhanced cordless telecommunications (DECT) phones was collected in most centers, as well as information on other communication devices such as CB radios, walkie-talkies, other types of radios and transmitters, and ham radios. A validation study was carried out to try to assess the accuracy of recall of cellular phone use, including a study of short-term recall in volunteer subjects using software modified phones or network operator records (Vrijheid et al., 2006a) as well as a simulation study of the potential effects of random error and selection bias on the results (Vrijheid et al., 2006b). We present in this chapter, results of the published national and pooled studies.

Denmark, Brain Tumors and Acoustic Neuroma: (Christensen et al., 2004, 2005a)

Christensen et al. (2004) reported on the Danish Interphone study of acoustic neuroma, including 106 cases aged 20–69 years at diagnosis and 212 age- and sex-matched controls. Eligible cases had to be incident between September 2000 and August 2002 and resident in Denmark. Eighty percent of eligible cases (including those deceased before contact) took part in the study. The reported participation rate among controls was 64%, but an erratum published later suggested that the participation rate was closer to 52%, i.e., considerably lower than those reported by the other previously published studies, in Sweden and the United States (Christensen et al., 2005b). In total, 42% of cases and 46% of controls had used a cellular phone regularly, defined as at least one call per week for 6 months or more up to the date of diagnosis for cases or equivalent date for controls. The relative risk of acoustic neuroma in relation to regular use of a cellular phone was 0.90 (95% CI: 0.51–1.57). Tumor risk did not increase with increasing time since first regular use, lifetime cumulative number of calls, or cumulative hours of use; in fact, odds ratios were lowest for the highest categories of these variables. Only two cases and 15 controls had started using a phone 10 or more years ago, corresponding to a reduced relative risk (OR=0.22, 95% CI: 0.04–1.11). Analyses of tumor laterality vs. reported side of phone use among cases showed a statistically significantly reduced relative risk of 0.68 ($p=0.02$) for a tumor ipsilateral to reported side of phone use. Among 45 cases who were regular cellular phone users, 25 out of 35 (71%) cases who expressed a preferred side of use stated they had used the phone predominantly on the contralateral side, and ten cases expressed no preference. This observation might well have been due to unilateral hearing loss as an early symptom of the tumor (Matthies and Samii, 1997), which would have made subjects switch their preferred side of phone use to the unaffected ear.

Christensen et al. (2005a) also published their investigation of the relation of glioma and meningioma risk to use of cellular phones in Denmark. This analysis included 252 glioma cases, 175 meningioma cases, and 822 age and sex-matched controls. Information on 19 glioma cases (7.5%) and 3 meningioma cases (1.7%) was based on proxy interviews. Reported participation rates were 71% for glioma and 74% for meningioma. Subjects who had had cancer prior to selection for the study were excluded from the study. Risk of meningioma in regular cellular phone users was somewhat reduced (OR=0.83, 95% CI: 0.54–1.28), and risk was borderline-significantly reduced for glioma overall (OR=0.71, 95% CI: 0.50–1.01). This decrease was due to a significantly reduced relative risk for high-grade gliomas (OR=0.58, 95% CI: 0.37–0.90), whereas that for low-grade gliomas was around unity (OR=1.08, 95% CI: 0.58–2.00). For meningioma and low-grade glioma, risks were not significantly related to time since first use, lifetime number of calls, lifetime hours of use or daily frequency of use, whereas for high-grade glioma significantly reduced odds ratios were observed for several categories of time since first use and lifetime hours of use. Memory tests in cases and controls using the Folstein Mini-Mental State Examination showed that cases scored significantly lower than controls, with similar scores in meningioma and low-grade glioma patients, and significantly poorer scores in high-grade glioma patients. Exclusion of subjects with a poor score led to an odds ratio in relation to regular phone use somewhat closer to 1.0 for high-grade glioma. Comparison of reported usage data with data from network operators for 27 cases and 47 controls showed similar levels of poor agreement (kappa~0.30) for number of calls in cases and controls, and very poor agreement for hours of use.

Sweden, Brain Tumors and Acoustic Neuroma: (Lonn et al., 2004a, 2005)

Lonn et al. (2004a) conducted a case–control study of acoustic neuroma in the Stockholm, Göteborg, and Lund regions of Sweden, including 148 cases and 604 controls. Cases were patients diagnosed with acoustic neuroma between September 1999 and August 2002, aged 20–69 years. Reported participation rates were high; 93% for cases and 72% for controls. Sixty percent of cases and 59% of controls were classified as regular phone users, with an overall relative risk of 1.0 (95% CI: 0.6–1.5). Exposures within 12 months prior to diagnosis were not considered in the evaluation of regular use. Ten years after start of phone use this risk was nonsignificantly increased to 1.9 (95% CI: 0.9–4.1). There was no association of risk with cumulative number of calls or number of hours of use. Risk was somewhat raised in relation to regular use of analogue phones (OR=1.6, 95% CI: 0.9–2.8), but not for use of digital phones (OR=0.9, 95% CI: 0.6–1.4). The authors analyzed tumor risk on the same and opposite side to reported side of phone use. Risk of a tumor on the same side as phone use was not appreciably raised for regular use overall, or for less than 10 years since start of use, but was significantly raised 10 or more years after first use (OR=3.9, 95% CI: 1.6–9.5). The corresponding relative risk of a contralateral tumor was 0.8 (95% CI: 0.2–2.9). Odds ratios for ipsilateral and contralateral tumors were obtained by randomly assigning controls to the left or right side on a 50:50 basis. On the basis of these findings, the authors concluded that their findings did not indicate any increased risk of acoustic neuroma related to short-term cellular phone use, but that they did suggest an increased risk for cellular phone use of at least 10 years duration. Regular use of DECT phones was not associated with tumor risk.

A second report of the Swedish Interphone study was published in 2005, detailing the results for glioma and meningioma (Lonn et al., 2005). Cases and controls were identified from the same study areas as for the study of acoustic neuroma, but with the addition of the Umeå region. The study included 371 glioma cases, 273 meningioma cases, and 674 controls, with good participation rates (74%, 85%, and 71%, respectively). Proxy interviews were carried out for 9% of glioma and 3% of meningioma cases. Tumor risk in relation to regular cellular phone use was nonsignificantly reduced for glioma (OR=0.8, 95% CI: 0.6–1.0) and significantly reduced for meningioma (OR=0.7, 95% CI: 0.5–0.9). There was no relationship of risk with duration of use, time since first use, or cumulative number of hours of use or number of calls, and there was no appreciable difference between risk for regular use of analogue and digital phones separately. There were no indications of material variation in relative risks by tumor grade or tumor lobe. Laterality analyses showed no significant associations overall or after restriction to tumors of the temporal or parietal lobes only. A nonsignificantly raised ipsilateral relative risk of 1.8 (95% CI: 0.8–3.9) was observed after 10 or more years of use for glioma, and of 1.4 (95% CI: 0.4–4.4) for meningioma, but restriction of such analyses to temporal and parietal tumors only showed relative risks close to unity. No association of risk was observed with regular use of DECT phones.

Nordic Countries and United Kingdom, Acoustic Neuroma: (Schoemaker et al., 2005)

A pooled analysis of acoustic neuroma risk in relation to cellular phone use was reported based on six Interphone studies in five North European countries (Schoemaker et al., 2005). This report included 678 cases and 3,553 controls, and was therefore several times larger than previously reported studies. It included data from the already published studies in Denmark and Sweden (Christensen et al., 2004; Lonn et al., 2004a), as well as from the Interphone study in Finland and Norway and two Interphone studies in the United Kingdom. Cell phone use was evaluated up to 1 year prior to the date of diagnosis for cases. For controls, a similar censoring date was derived based on interview year and how far back subjects were asked to recall exposures. Analyses were conducted with conditional logistic regression with strata of centre, region, age group, and sex and adjusting for highest educational level and combinations of interview year and interview lag time.

Pooled participation rates were 83% for cases (range between centers 69 to 91%) and 51% for controls (range 42–69%). Risk of acoustic neuroma in relation to regular cellular phone use was not raised overall (OR=0.9, 95% CI: 0.7–1.1). There was no association of risk with duration of use, cumulative number of calls or hours of use, or for analogue or digital phones separately. Analyses of tumor laterality in relation to reported side of phone use using the method of Lönn et al. (2004a) showed no raised risk of an ipsilateral tumor in regular phone users overall (OR = 0.9, 95% CI: 0.7–1.1), but a statistically significantly raised ipsilateral risk in subjects with at least 10 years of use (OR = 1.8, 95% CI: 1.1–3.1), whereas the corresponding contralateral risks were 1.1 and 0.9, respectively. As this analysis was based on randomly allocating controls to a left and right-sided group, the extent of variability in this result was investigated. On the basis of 300 simulations, the mean odds ratio for ipsilateral long-term use was 1.80, and was within the boundaries of 1.46 ($p = 0.19$) and 2.10 ($p = 0.007$) in 95% of the simulations (Schoemaker, 2007).

The study also analyzed the proportion of right-sided tumors in relation to cellular phone use, because, as the majority of controls expressed a preference for right-sided phone use, one would expect an excess of right-sided tumors in regular or long-term phone users if there were an etiological relationship. A total of 49.3% of cases classified as never or nonregular users had a right-sided tumor, compared with 53.3% in regular users and 58.8% in regular analogue phone users overall ($p = 0.34$ and 0.13, respectively). However, there was a nonsignificant deficit of right-sided tumors in long-term users overall (48.8%) and in long-term analogue users (48.7%).

Japan, Acoustic Neuroma: (Takebayashi et al., 2006)

Results on acoustic neuroma have been published as part of the Interphone study from the Tokyo municipality and surrounding areas in Japan. The study included 101 acoustic neuroma cases recruited from participating neurosurgical departments and 339 controls individually matched to cases on age, sex, and area of residence, recruited through random digit dialing of fixed home phones. Reported participation rates were 84% for cases and 52% for controls. Fifty-three percent of cases and 58% of controls were classified as regular phone users, with a relative risk for acoustic neuroma of 0.73 (95% CI: 0.43–1.23). There was no increase in risk with cumulative duration of use or cumulative call time. The percentage of long-term users was low, with 4.1% of cases and 3.6% of controls having used a cellular phone for 8 or more years, and more than 90% of phone users only having used digital phones. No significant associations were observed in analyses restricted to cellular phone use on the same side of the head as the tumor. The relative risk was 1.09 (95% CI: 0.58–2.06) when only phone use 5 or more years prior to diagnosis was considered. Fewer cases than controls started using a cellular phone between 1 and 5 years before diagnosis; this was suggested to be evidence for latent disease bias, i.e., that the preclinical stage of the tumor might have affected cellular phone use patterns.

United Kingdom, Glioma: (Hepworth et al., 2006)

Hepworth et al. (2006) analyzed glioma data from the two Interphone case–control studies in the United Kingdom: one in Southeast England, and one in Central

Scotland, the West Midlands, West Yorkshire, and the Trent area. Cases were recruited from multiple clinical sources in these regions, and in the Southeast England were also identified through the cancer registry. The pooled study comprised 966 cases and 1,716 controls, aged 18–69 years. Controls were recruited through National Health Service general practitioners lists in the study region. Such lists are virtually a complete population register, because 98% of the population is registered with a general practitioner (OPCS, 1995). Exposures were evaluated up to 1 year prior to the date of diagnosis, and equivalent date for controls, derived as in an earlier study (Schoemaker et al., 2005).

Participation rates were low: 51% for cases and 45% for controls. For 7% of the cases, and none of the controls, interviews were carried out with proxies. The relative risk of glioma was 0.94 (95% CI: 0.78–1.13) in relation to regular cellular phone use overall. There was no relation of risk to time since first use, cumulative years of use, or cumulative number of calls or hours of use, either for cellular phone use overall or for analogue and digital phone use separately. Risk of a tumor ipsilateral to reported side of phone use was significantly raised in relation to regular use (OR = 1.24, 95% CI: 1.02–1.52), whereas risk of a contralateral tumor was significantly reduced (OR = 0.75, 95% CI: 0.61–0.93). For 10 or more years of cumulative use, risk was nonsignificantly raised (OR = 1.60, 95% CI: 0.92–2.76) for an ipsilateral tumor and nonsignificantly reduced (OR = 0.78, 95% CI: 0.43–1.41) for a contralateral tumor.

The low participation rates in the study were a consequence of difficulty in conducting full population-based, rather than only hospital-based case recruitment, the lack of ethics agreement to approach close relatives as proxies for deceased or ill patients, (only relatives who approached the research team themselves were allowed to be interviewed as proxies) and general lack of willingness of subjects to take part in (unpaid) medical studies. Cellular phones were not emphasized in the study documentation as the main study hypothesis, however, which might have reduced the potential for selective participation of cellular phone users.

Germany, Brain Tumors and Acoustic Neuroma: (Schuz et al., 2006a; Schlehofer et al., 2007)

Schuz et al. (2006a) reported on data from the Interphone study in Germany. This population-based case–control study was carried out in three regions in Germany and included 366 glioma cases, 381 meningioma cases, and 1,494 controls post-hoc matched to cases on sex, region, and birth-year. Controls were identified from the German population register. Participation rates were 80% among glioma cases, 88% among meningioma cases, and 63% among controls. Proxy interviews were more common for glioma cases (11%) than for meningioma cases (1.3%), or controls (0.4%). Relative risks in relation to regular cellular phone use were 0.98 (95% CI: 0.74–1.29) for glioma and 0.84 (95% CI: 0.62–1.13) for meningioma. The study also reported that cordless phone use was not associated with risk of glioma or meningioma. For cellular phones, risk in subjects 10 or more years after start of use was nonsignificantly raised for glioma (OR=2.20, 95% CI: 0.94–5.11), and risks were somewhat raised for the top quartile of lifetime number of calls (OR=1.34, 95% CI: 0.86–2.07), and for daily use of 30 min or more (OR=1.54, 95% CI: 0.75–3.15), compared with never or nonregular use. Risk of meningioma was not substantially raised 10 years or more after start of phone use and was not appreciably raised for other indices of exposure. Subanalyses showed the relative risk after 10 or more years for glioma was due to a raised risk of high-grade glioma in females (OR=1.96, 95% CI: 1.10–3.50) but not males (OR=0.78, 95% CI: 0.53–1.14). On further inspection, the prevalence of phone use in the female high-grade glioma control group was considerably lower than in the rest of the female control group, whereas no such difference in exposure prevalence was observed in male controls. This sex and tumor type-specific finding might, therefore, be due to chance. Analyses restricted to tumors of the temporal lobes showed no significant excesses in risk in phone users either for glioma or meningioma.

It was noted that errors in recall were a potential problem because 18 out of the 37 long-term users reported subscribing to phone systems that were not in operation at the reported time of use. Control selection bias was also potentially present because participating controls were more likely to be a cellular phone user than nonparticipating controls who filled in a nonrespondent questionnaire; this was particularly true in the younger age groups.

A report on 97 cases of acoustic neuroma and 194 matched controls from the same German group showed a relative risk in relation to regular phone use of 0.67 (95% CI: 0.38–1.19) (Schlehofer et al., 2007). The relative risk was 0.53 (95% CI: 0.22–1.27) 5–9 years since first use, with no cases reporting first use 10 or more years ago. Relative risks in relation to the highest quartile of lifetime number of calls or duration of calls were (near) significantly reduced. Reported participation rates were 89% for cases and 55% for controls.

Nordic Countries and United Kingdom, Glioma: (Lahkola et al., 2007)

A pooled analysis of data on glioma risk and cellular phone use from five Interphone case–control studies in Denmark, Finland, Norway, Sweden, and Southeast England was published in 2007 (Lahkola et al., 2007). National data from the Danish, Swedish, and British studies had already been published earlier (Lonn et al., 2005; Christensen et al., 2005a; Hepworth et al., 2006). The analysis comprised 1,521 glioma cases, including 710 glioblastomas, and 3,301 controls. Sixty percent of all ascertained cases (37–81% between countries) and 50% of potential controls (range 42–69%) participated in the study. Twelve percent of case interviews were based on information provided by proxies, compared with less than 1% of control interviews. The relative risk in relation to regular cellular phone use was statistically significantly reduced for glioma overall (OR=0.78, 95% CI: 0.68–0.91) and for glioblastoma separately (OR=0.77, 95% CI: 0.64–0.93). No increase in risk of glioma was reported 10 or more years after start of phone use: the relative risk was 0.95 (95% CI: 0.74–1.23) for glioma and 0.86

(95% CI: 0.62–1.21) for glioblastoma, based on a total of 143 glioma cases and 220 controls, much the largest numbers of long-term users in publications so far. For cumulative number of calls or cumulative hours of use, risks were not raised for the highest quartile of exposure, compared with never or nonregular users. Results of trend tests for these variables depended on the method employed. Trend tests across categories of cumulative hours of use and number of calls, based on quartiles, showed no statistically significant trends of risk when including all subjects, and borderline significant positive trends for glioma in relation to cumulative number of calls ($p=0.05$), but not in relation to cumulative hours of use ($p=0.09$) after excluding never and nonregular users from the tests. Tests for trend across the actual values of exposure (i.e., continuous data rather than the categorized groups), showed a significant trend for cumulative hours of use (OR = 1.006, 95% CI: 1.002–1.010 per 100 h), but not for cumulative number of calls. No significant trends of risk were observed in relation to other exposure parameters such as time since first use.

With regard to type of phone, relative risks tended to be more reduced for digital than for analogue phone use across all indices of exposure, e.g., the odds ratio in relation to regular use was 0.75 (95% CI: 0.65–0.87) for digital and 0.85 (95% CI: 0.68–1.06) for analogue phone use.

Analyses of tumor laterality compared with reported side of phone use showed a relative risk for ipsilateral use of 1.13 (95% CI: 0.97–1.31) overall, and of 1.39 (95% CI: 1.01–1.92) for 10 or more years since first use, with a borderline statistically significant positive trend overall ($p=0.04$), which was not significant when the reference group was excluded ($p=0.18$). The relative risk of a contralateral tumor 10 or more years since first use was 0.98 (95% CI: 0.71–1.37). There were no significant trends of risk of either ipsilateral or contralateral tumors with lifetime years of use or cumulative hours of use, although for the latter, the relative risk of an ipsilateral tumor for the top quartile of cumulative hours of use was 1.24 (95% CI: 0.97–1.59).

Norway, Brain Tumors and Acoustic Neuroma: (Klaeboe et al., 2007)

Klaeboe et al. (2007) reported on risk of intracranial tumors in relation to cellular phone use in the Norwegian Interphone study, including 289 glioma, 207 meningioma and 45 acoustic neuroma patients and 358 frequency-matched controls. As for other Nordic studies, controls were identified through the population register. Cell phone use was evaluated up to 1 year prior to diagnosis for cases. For controls, the authors report that the censoring date was calculated as the mean of the diagnosis dates for the cases for each tumor site, suggesting that controls were interviewed substantially later than cases.

Reported participation rates were 74% for cases overall and 69% for controls. Data from the glioma and acoustic neuroma patients have also been included in pooled analyses elsewhere (Schoemaker et al., 2005; Lahkola et al., 2007). The presented odds ratios in relation to regular phone use were substantially below 1.0 for each of the tumor types, which the authors in part attribute to potential selection bias among controls, but could also in part be due to inadequate adjustment for controls having been interviewed on average later than cases, and the high proportion of proxies used for glioma cases (36%).

2.2. Case–Control Studies of Cellular and Cordless Phone use and Risk of Non-Intracranial Neoplasms

Several case–control studies have investigated a possible association between cellular and cordless phone use and risk of neoplasms other than the intracranial tumors described earlier. Such studies have focused on uveal melanoma, parotid gland tumors, testicular cancer, intratemporal facial nerve tumor, and non-Hodgkin's lymphoma. These studies are described here, with the exception of the study of parotid gland tumors by Auvinen et al. (2002) and that of intratemporal facial nerve tumor by Warren et al. (2003), which have already been discussed above because these studies also included intracranial tumors.

2.2.1. Uveal Melanoma

Germany, Uveal Melanoma: (Stang et al., 2001)

The possible association of uveal melanoma with cellular phone use was investigated in Germany (Stang et al., 2001). A total of 118 cases were included, 37 of which came from a population-based study (84% participation rate) and 81 from a single hospital (88% participation rate). Cases were diagnosed between 1995 and 1998. The study included 475 controls, matched on age, sex, and region of residence; 327 came from a population-based study (48% participation rate); and 148 controls were hospital-based (79% participation rate). Hospital-based controls had other ocular disease, excluding occupational eye injury. Assessment of cellular phone use was restricted to occupational use only. Subjects were asked about use of cellular phones "at your workplace for at least several hours per day." Ever use appeared to be defined as this level of use for at least 6 months or more, but this is not entirely clear from the paper. Cases were more likely to report that they were "certainly" or "probably" exposed to cellular phones at work to this degree than controls (5.1 vs. 3.2%), with a relative risk of 4.2 (95% CI: 1.2–14.5). This result derived largely from the hospital-based subjects (OR = 10.1, 95% CI: 1.1–484.4). The number of people reporting such a high level of use is somewhat surprising, as only occupational use was enquired about and in most office jobs at that time one would expect landlines to be used. Nonsignificantly raised risks were reported for use starting 5 or more years before diagnosis and cumulative use for 3 or more years. Cases were more often highly or lowly educated than controls, but analyses adjusting for education level did not affect the results.

2.2.2. Parotid Gland Tumors

Sweden, Parotid Gland Tumors: (Hardell et al., 2004a)

Hardell et al. (2004a) carried out a case–control study of parotid gland tumors in Sweden. In total, 267 cases diagnosed between 1994 and 2000 and identified from six regional cancer registries in Sweden were included, as well as 1,053 sex and age-matched controls. The study was carried out during the same period as Hardell's second study on brain tumors (Hardell et al., 2002a); for geographical areas that the two studies had in common, 815 controls were drawn from this brain tumor study. For cases for whom matching criteria were not fulfilled and for cases in nonoverlapping geographical areas, controls were selected from the population register. Reported participation rates were 91% for cases and 90% for controls, but one can derive from data in the paper that case participants actually represented 66% of all eligible cases. The study protocol was similar to that of the brain tumor study described earlier (Hardell et al., 2002a). The relative risk of parotid gland tumors was 1.02 (95% CI: 0.72–1.38) for use of any cellular or cordless telephone, and was not materially raised when considering analogue, digital, or cordless phones separately. Risk was not raised 10 or more years after first use, based on six cases.

Denmark and Sweden, Parotid Gland Tumors: (Lonn et al., 2006)

A pooled analysis of two population-based case–control studies on parotid gland tumors conducted within the Interphone collaboration was reported from Denmark and Sweden (Lonn et al., 2006), with methods as described earlier. The study included 60 malignant parotid gland tumors and 112 benign pleomorphic parotid tumors diagnosed during 2000–2002, and 681 controls. The reported participation rates were 85% for malignant cases, 88% for benign cases, and 70% for controls. The estimated relative risk in relation to regular cellular phone use was 0.7 (95% CI: 0.4–1.3) for malignant tumors and 0.9 (95% CI: 0.5–1.5) for benign tumors. Tumor risk was not associated with duration of use, time since first use, cumulative number of hours of use, or number of calls. Analyses in relation to laterality of phone use showed somewhat raised relative risks for ipsilateral tumors, but none of these associations were statistically significant, and contralateral risks were reduced.

2.2.3. Testicular Cancer

Germany, Testicular Cancer: (Baumgardt-Elms et al., 2002)

A case–control study was carried out in Germany to investigate risk of testicular cancer in relation to occupational RF exposures due to proximity to radiofrequency transmitters including radio sets and cellular phones, and radar (Baumgardt-Elms et al., 2002). In total 269 population-based cases of testicular cancer diagnosed between 1995 and 1997 were included and 797 age and region-matched controls selected from population registers. Reported participation rates were 76% for cases and 57% for controls, but it can be derived from data in the paper that the control participation rate was approximately 46% after taking into account people selected for the study who moved out of the area before first contact. Risk of testicular cancer was not increased in men who reported ever working in proximity to cellular phones or radios (OR=0.9, 95% CI: 0.6–1.2) or to radar (OR=1.0, 95% CI: 0.6–1.8), or in men who had a history of radar exposure as assessed by experts from their occupational history (OR=0.4, 95% CI: 0.1–1.2). There was no trend of increasing risk with increase in a score measuring duration of exposure and distance from the source.

Sweden, Testicular Cancer: (Hardell et al., 2007a)

Hardell et al. (2007a) investigated, as part of a study on exposure to PVC plastics, a possible association between testicular cancer and cellular and cordless phone use. Cases of testicular cancer in the Swedish male population incident during 1993–1997 and aged 20–75 years were retrospectively identified from the Swedish Cancer Registry. One control per case was sampled from the Swedish Population Registry, matched on 5-year age group of the case, irrespective of geographical area. The analysis included 888 cases, including 542 cases with seminoma and 346 cases with non-seminoma histologies, and 870 controls. The lower number of controls than cases suggests that no resampling was carried out for controls who refused to take part. The reported case participation rate was 91%, which we recalculate as representing 87% of all eligible cases after inclusion in the denominator of 37 deceased patients and 7 patients excluded for other reasons. As with other related case–control studies conducted by this research group, exposure information was collected by postal questionnaire and was supplemented by a telephone interview when considered necessary. Questions on cellular phone use were similar to those in previous studies and also included questions on the number of hours the phone was held on standby and the location of the phone when on standby. Analyses were carried out unmatched, with adjustment for age, history of cryptorchidism, and year of diagnosis. Relative risks of testicular cancer in relation to cellular phone use up to 1 year prior to diagnosis were around unity for analogue, digital, or cordless phones. Risks for the longest induction period available were somewhat (nonsignificantly) raised for analogue and digital phone use. Relative risks were somewhat higher for seminoma than for non-seminoma, separately. Tumor risk was also evaluated with regard to where the phone was held when on standby, during which only a contact signal is maintained with the base station. Keeping the phone in a pocket close to the testes did not appear to be associated with risk of testicular cancer.

2.2.4. Non-Hodgkin's Lymphoma

Sweden, Non-Hodgkin's Lymphoma: (Hardell et al., 2005d)

Parallel to studies of brain tumors and parotid gland tumors, Hardell et al. (2005d) conducted a case–control study of non-Hodgkin's lymphoma (NHL) in relation to cellular or cordless phone use. Cases were subjects diagnosed with NHL between December 1999 and April 2002 aged 18–74 years in four health service regions in

Sweden. Cases were identified through physicians treating lymphoma and through pathology departments. Controls were recruited from population registers in the same areas, frequency-matched to the cases. In total, 910 cases and 1,016 controls participated, with reported participation rates of 91% for cases (or 84% of all ascertained cases, as can be derived from data in the paper) and 92% for controls, respectively. No association of risk was observed with phone use overall or by phone type when all cases were considered. Analyses presented separately for B cell (819 cases), and T cell (53 cases) histological subtypes of NHL showed no association of risk for the B cell subtype, but for the T cell subtype, risks were nonsignificantly raised around 1.4–1.6 for use up to 1 year prior to diagnosis for each of the phone types, with 1.5- to 3.2-fold increased risks for use more than 10 years prior to diagnosis, highest for cordless phones (OR=3.2, 95% CI: 1.05–9.48). The authors presented greater relative risks when restricting the analyses to "certain T cell non-Hodgkin's lymphoma, e.g., of cutaneous and leukemia type," but this was based on 23 cases and it was unclear whether these subtypes were a priori expected to be related to phone use. Nevertheless, results based on T cell NHL overall and on this subset of 23 cases were emphasized in the abstract.

United States, Non-Hodgkin's Lymphoma: (Linet et al., 2006)

Linet et al. (2006) conducted a multicenter case–control study of non-Hodgkin's lymphoma in four geographical areas covered by the population-based National Cancer Institute (NCI) Surveillance, Epidemiology and End Results (SEER) program, including Iowa, Detroit, Los Angeles county, and Seattle. Cases were diagnosed between July 1998 and June 2000, at ages 20–74 years, and were resident within the study regions. Controls were selected within the same age range and from the same regions from the NCI SEER program. Controls who were younger than age 65 were identified through random digital dialing and older controls from Medicare eligibility files. Controls were frequency-matched to cases on age, sex, ethnic group, and geographical region. HIV-related cases and HIV-positive controls were excluded from the study. As part of a larger computer-assisted questionnaire, subjects were asked about lifetime use of cordless telephones, car telephones, and handheld cellular phones. Subjects who had ever used each type of phone were asked if lifetime use was less than 10 times, 10–100 times, or more than 100 times. Subjects who reported use more than 100 times were asked about the year of first and last use of each type of phone and the number of minutes that the phone was used during a typical week. The study comprised several components, such that questions on telephone use were given to none of the African-American participants and only half of the other participants in the study, whereas the rest received questions on medical history instead. A total of 551 NHL cases and 462 frequency-matched controls recruited to the study were asked the telephone questions, representing 79% of cases and 55% of cases whom the research team attempted to recruit. From the numbers provided in the paper, it can be calculated that the participants represented 61% of cases and 47% of controls who were eligible to receive the questionnaire containing questions on telephone use. Seventeen percent of cases and 15% of controls reported use of a handheld cellular phone more than 100 times in their lifetime. After adjustment for age, ethnic group, geographic area, and education, cellular phone use was not associated with NHL risk (OR = 0.9, 95% CI: 0.6–1.4) when comparing those with more than 100 calls in their lifetime to those who had never used a cellular phone. Twenty cases and 9 controls in the 100+ calls group had used a phone for more than 8 years, with a relative risk of 1.6 (95% CI: 0.7–3.8). Risks of NHL in the highest category of average number of minutes of use per week (>60 min) or cumulative hours of use (>209 h), compared with never use were close to unity. The authors reported that changing the reference category from never use to never use plus less than ten calls a lifetime, or to never plus less than 100 calls a lifetime revealed little difference in risk estimates. Analyses of risk by subtype of NHL showed little evidence for an association of cellular phone use with diffuse large B cell lymphoma or follicular NHL. Risk of NHL of "not otherwise specified" type was significantly raised in men for 6 or more years of use, and nonsignificantly raised in women, but these estimates were based on seven and two cases, respectively, with nonsignificant trend tests for dose–response. These findings could be due to chance, in particular in the light of the multiple tests carried out with analyses by sex as well as by NHL type. No substantial difference was reported in use of cordless or car phones between cases and controls.

2.3. Cohort Studies of Cancer Incidence or Mortality in Cellular Phone Subscribers

Two cohort studies of cellular phone subscribers have been published: one in the US and one in Denmark. They are described here.

US Cohort: (Rothman et al., 1996b; Dreyer et al., 1999)

Mortality was investigated in a cohort of 285,561 users of analogue cellular phones identified from two network operators in the US in 1993. Information on duration of use of phones was based on billing records from the network operator and mortality was ascertained through record linkage with the US National Death Index. Unfortunately, the cohort was only followed for 1 year, after which the study was terminated. Cause-specific mortality in 1994 was reported. Among 765 deaths that occurred in that year, 263 were from cancer, including 6 from brain cancer and 15 from leukemia. No relation was found between the use of hand-held vs. nonhandheld phones (the latter were considered as unexposed because the antenna is not part of the handset) and the rate of death from cancer overall, brain cancer or leukemia. There was also no relationship of these outcomes with the daily amount of time of use of handheld phones, or years of use of handheld phones, but usage was low with the maximal categories being >=2 min per day and >3 years, respectively.

Danish Cohort: (Johansen et al., 2001; Schuz et al., 2006b)

In 1996, researchers in Denmark identified all cellular phone subscribers in Denmark from the start of the service in that country, 1982, through to 1995, from operator records. These subjects were then linked to the Central Population Register for Denmark. The initial number of subscriptions was 723,421, but 200,507 had to be excluded because they were corporate subscriptions, and 102,819 because linkage was unsuccessful or for other reasons. A total of 420,095 remaining subscribers were followed for cancer incidence, death, and emigration by linkage to the Danish Cancer Registry. A first report was published in 2001, with follow-up to the end of 1996 (Johansen et al., 2001), and an update of follow-up was published in 2006, with follow-up until the end of 2002 (Schuz et al., 2006b). Because of the longer follow-up in the more recent report, we focus on that report here. Cancer incidence in the cohort was compared against that in the Danish population after taking into account the age and sex distribution of the cohort, the calendar period of incidence, and the fact that the cohort represented a sizeable proportion (15%) of the national population. The average length of follow-up was 8.5 years, with 15% of males and 5.5% of females having a cellular phone subscription for 10 or more years. A total of 14,249 cancers were incident during follow-up, statistically significantly fewer than expected based on general population rates (standardized incidence ratio (SIR) = 0.95, 95% CI: 0.93–0.97). This reduced risk was apparent in males (SIR = 0.93, 95% CI: 0.92–0.95) but not in females (SIR = 1.03, 95% CI: 0.99–1.07), and was attributed to a decreased risk of smoking-related cancers in males.

A total of 580 brain and nervous system neoplasms occurred during follow-up, with a SIR of 0.97 (no confidence interval reported). Brain and nervous system neoplasms included 257 gliomas, 68 meningiomas, and 32 cranial nerve tumors (31 confirmed as acoustic neuroma). No significantly raised risks were observed for glioma (SIR = 1.01, 95% CI: 0.89–1.14), meningioma (SIR = 0.86, 95% CI: 0.67–1.09), or tumors of the cranial nerves (SIR = 0.73, 95% CI: 0.50–1.03). When glioma risks were analyzed by lobe of the brain, risks were highest for temporal lobe tumors (SIR = 1.21, 95% CI: 0.91–1.58) and for tumors of other and unspecified lobes (SIR = 1.21, 95% CI: 0.88–1.64), and lowest for the parietal lobes (SIR = 0.58, 95% CI: 0.36–0.89). Risk of glioma in the temporal and parietal lobes combined was, however, not raised (SIR = 0.93, 95% CI: 0.73–1.17). Twenty-eight cases of brain and nervous system neoplasm occurred in subjects who started their subscription 10 or more years ago, corresponding to a statistically significantly reduced risk (SIR = 0.66, 95% CI: 0.44–0.95). The authors noted that this decreased risk is somewhat surprising because cohort members had a higher average annual income than the Danish general population, which was particularly pronounced in the early years, and brain tumors might be more common among people of higher socio-economic status (Inskip et al., 2003).

No increased risks were observed for tumors of the salivary gland (SIR = 0.77) and eye (SIR = 0.96), or leukemia (SIR = 1.00). For leukemia, the SIR 10 or more years after first subscription was 1.08 (95% CI: 0.74–1.52), based on 32 cases, with no statistical evidence for a trend with time since first subscription.

This study was based on subscription records, rather than cellular phone use, and has therefore the limitation of potential misclassification of exposure status. It is likely that a proportion of subscribers are not the actual or not the sole user of the phone for which they are contracted. Additionally, the exclusion of 200,507 corporate subscriptions might mean that heavy users with corporate subscriptions were excluded from the cohort (and included in the rest of the population which served as the comparison group). To obtain insight into the potential misclassification of exposure status, the researchers compared records for subjects who were also included as a control subject in their previously reported case–control study of brain tumors (Christensen et al., 2004, 2005a). On the basis of self-reported phone use in the case–control study, it was estimated that 61% of cohort members and 16% of the general population (i.e., noncohort members) might have been regular cellular phone users before 1996, when the cohort was constructed. Sensitivity analyses showed that this degree of misclassification would result in an attenuated estimate of risk if the true risk was raised, for example, from 1.5 to 1.2 or from 1.2 to 1.09. However, it would not lead to risk estimates below 1.0 if there was, in fact, an increased risk among cellular phone users.

3. STRENGTHS AND LIMITATIONS OF REPORTED STUDIES

Numerous studies have been conducted into cancer risks in relation to cellular phone use. Such studies have largely focused on brain and related intracranial tumors, with only a few addressing other types of cancer (Tables 1–4). Nearly all reported studies have been of case–control design, primarily because the tumors of interest are uncommon and large cohort studies would be needed to accrue enough cases. These studies have certain strengths and weaknesses, and it is important to take these into consideration in interpretation of the results.

3.1. Exposure Biases

Past studies have assessed cellular phone use as a (crude) measure of RF exposure. In case–control studies, such exposure assessment has been retrospective. The studies all relied on the subject's recall of cellular phone use, obtained through self-administered questionnaires or personal interview, with no measurement of exposures, except for the Finnish study by Auvinen et al. (2002), which was based on operator records. Past phone use is unlikely to be recalled accurately, because of complex and changing patterns of use, and there is therefore great potential for misclassification of exposure levels. Information on ever or regular use, and to some extent on number of years of use, might be expected to be relatively reliable, while information on intensity of use, such as time spent on the phone, or number of calls seems prone to inaccuracies. This is likely to be especially true for people with long-term use, a group of subjects that is of particular interest. The amount of phone use in terms of number of calls and call time has increased strongly in the last few years because of the reduction in costs of calls and introduction of subscriptions including free call time, and current usage patterns are therefore a poor reflection of frequency of use in early years. The validity of short-term recall of phone use was assessed in the Interphone study, showing large random error in recall of duration of use and number of calls (Vrijheid et al., 2006a). Such an error is likely to reduce the power of a study to detect an effect, if there is any. There was slight underestimation of call time in light users and overestimation in heavy users. Accuracy of recall in relation to long-term use has not been assessed.

The potential for errors and bias in recall is increased in studies of brain tumors, as the tumor itself may affect the ability to recall past use accurately. This is particularly likely for high-grade glioma, which is associated with memory loss, and cognitive and personality changes (Behin et al., 2003). Some evidence for this comes from the Danish case–control study, which found that performance in memory tests was poorer in cases than in controls, in particular in patients with high grade glioma (Christensen et al., 2005a). Such memory loss in cases could have an effect on risk estimates for cellular phone use, in an unknown direction.

Recall might also be biased due to the publicity that there has been in recent years about the possibility that cellular phones might cause brain tumors. Study participants might attribute their tumor to the phone, and over-report phone use as a result of this; such bias would result in spurious positive associations with phone use.

The use of operator records instead of self-reported phone use, as in the case–control study by Auvinen et al. (2002) and published cohort studies, has the advantage that problems with recall bias or quality of recall are avoided. A drawback, however, is that there is scope for misclassification of exposure, potentially leading to underestimation of effect estimates in relation to cellular phone use. The person who holds the personal subscription is not necessarily the user of the phone, i.e., those who appear to be exposed to cellular phones might not be, and, vice versa, people who might appear unexposed might have had use of other phones that are not listed under their name, e.g., from corporate subscriptions. Depending on the detail of information that the operators are willing or able to supply, there might be no information on frequency of use, or only information on outgoing, not incoming, calls. Furthermore, no information is available on use of hands-free devices or side of phone use.

3.2. Tumor Location

Analyses of tumor laterality in relation to side of phone use could in principle be informative about etiology, because of the strong exposure gradient in relation to side of phone use. Such analyses are, however, particularly prone to bias if information on reported side of phone use is collected from cases after development of their tumor, because subjects are aware of their tumor's laterality and might attribute it to their past phone use. This would lead to overestimation of risks of tumors on the same side as reported side of phone use, and underestimation of risks of tumors on the opposite side. Analyses restricted to tumors of specific lobes of the brain have been conducted to investigate whether risks are greater in lobes that might be expected to receive the highest exposure levels, but it is not clear at present which lobe(s) are most exposed (the only publication states parietal (Rothman et al., 1996a)), and analyses restricted to tumors in these lobes ipsilateral to side of phone use are susceptible to recall bias.

3.3. Cognition and Behavior

The preclinical stage of a tumor could also affect phone use, by its effects on cognition and personality or because illness may lead to absence from work. Such latent disease bias is particularly likely to occur for low-grade tumors, which may be present for a considerable time before diagnosis such that the phone use pattern could be in part a consequence of the tumor rather than the cause of it. Acoustic neuroma, in particular, is a slow growing tumor, and the tumor is likely to exist for several years before diagnosis, with symptoms, including unilateral hearing loss, for long periods before eventual diagnosis (Matthies and Samii, 1997); these symptoms might affect side of use of a phone, or indeed whether the phone is used at all. Reduced risks of acoustic neuroma as well as of glioma and meningioma have frequently been reported in relation to regular cellular phone use (Tables 1–3); latent disease bias could potentially explain at least part of these reduced risks.

3.4. Appropriateness of Controls

The choice of a control group gives potential for bias in case–control studies. The use of hospital-based controls in US case–control studies (Muscat et al., 2000, 2002; Inskip et al., 2001; Warren et al., 2003), as well as in a German study (Stang et al., 2001), gives increased potential for bias because the illnesses of controls might themselves be associated with phone use, positively or negatively, and because there may be different catchment populations for cases and controls. Controls with common conditions could, for example, be predominantly local residents whereas cases could have come from long distances for specialist treatment of their condition, leading to potential sociodemographic differences between cases and controls. Inskip et al. (2001) matched on residential proximity to the hospital in an effort to reduce the potential for this latter bias. The strength of hospital-based controls is, however, that participation rates have been high, as it is generally easier to recruit patients while they are at hospital than recruiting healthy volunteers from the general population. Random-digit dialing of landlines to select controls, as done in the Interphone study in Japan (Takebayashi et al., 2006) and in a US study of NHL (Linet et al., 2006) might also cause bias because the likelihood of being contactable by landline could be associated (perhaps inversely) with having a cellular phone, the main exposure of interest. Other main sources of population-based controls have been population registers, in the Nordic countries and Germany, and family doctor practices, in the United Kingdom; such sources are thought in principle to be less likely to give rise to bias.

Studies with population-based controls have shown a range of reported participation rates. This variation in rates in part reflects the willingness of different study populations to take part in (unpaid) medical studies, persistence of the research teams in contacting subjects, and other study-related aspects such as ethical agreements, but are also due to differences in how the participation rate was calculated. Rates greater than 90%, very high for population-based studies, have been reported in studies by Hardell et al. in Sweden, but with no details disclosed how these rates

Table 1. Summary of results from published studies on risk of malignant brain tumors or glioma in relation to cellular and cordless phone use

Study design, first author and year	Country, period	Total cases/ controls		Overall exposure: % exposed among controls	Overall exposure: Relative risk (95% CI)	Longest time since first exposure[a]: Time (years)	Longest time since first exposure[a]: No. exposed cases/controls	Longest time since first exposure[a]: Relative risk (95% CI)
Case–control								
Hardell et al., 1999[b]	Sweden, 1994–1996	136 brain cancer 425 controls		38%	1.0 (0.6–1.5)	–	–	–
Muscat et al., 2000	United States, 1994–1998	469 brain cancer 422 controls		18%[c]	0.8 (0.6–1.2)	≥4	17/22	0.7 (0.4–1.4)
Inskip et al., 2001	United States, 1994–1998	489 glioma 799 controls		22%	0.8 (0.6–1.2)	≥5	11/31	0.6 (0.3–1.4)
Auvinen et al., 2002	Finland, 1996	198 glioma 1,635 controls		11%	1.5 (1.0–2.4)	>2	11/37	1.7 (0.9–3.5)
Hardell et al., 2002a,b[b]	Sweden, 1997–2000	529 brain cancer 529 controls	Analogue[f]	19%	1.1 (0.8–1.6)	≥6	43/37[d]	1.2 (0.8–1.8)
			Digital[f]	33%	1.1 (0.9–1.5)	≥6	12/7[d]	1.7 (0.7–4.3)
			Cordless[f]	27%	1.1 (0.9–1.5)	≥6	36/23[d]	1.6 (0.9–2.6)
Christensen et al., 2005a	Denmark, 2000–2002	81 low-grade glioma		47%	1.1 (0.6–2.0)	≥10	6/9	1.6 (0.4–6.1)
		171 high-grade glioma 822 controls			0.6 (0.4–0.9)	≥10	8/22	0.5 (0.2–1.3)
Lonn et al., 2005	Sweden, 2000–2002	371 glioma 674 controls	Cellular	59%	0.8 (0.6–1.0)	≥10	25/38	0.9 (0.5–1.5)
			Cordless	–	0.8 (0.5–1.1)	–	–	–
Hardell et al., 2005c[b]	Sweden, 2000–2003	317 brain cancer 692 controls	Analogue[f]	11%	2.6 (1.5–4.3)	≥10	48/40	3.5 (2.0–6.4)
			Digital[f]	50%	1.9 (1.3–2.7)	≥10	19/18	3.6 (1.7–7.5)
			Cordless[f]	44%	2.1 (1.4–3.0)	≥10	30/35	2.9 (1.6–5.2)
Hardell et al., 2006b[e]	Sweden, 1997–2003	905 brain cancer 2,162 controls	Analogue[f]	14%	1.5 (1.1–1.9)	≥10	82/84	2.4 (1.6–3.4)
			Digital[f]	36%	1.3 (1.1–1.6)	≥10	19/18	2.8 (1.4–5.7)
			Cordless[f]	32%	1.3 (1.1–1.6)	≥10	33/45	1.8 (1.1–3.0)
Hepworth et al., 2006	United Kingdom, 2000–2004	966 glioma 1,716 controls		52%	0.9 (0.8–1.1)	≥10	66/112	0.9 (0.6–1.3)

Schuz et al., 2006a	Germany, 2000–2003	366 glioma 732 controls	Cellular	39%	1.0 (0.7–1.3)	≥10	12/11	2.2 (0.9–5.1)
			Cordless	24%	0.9 (0.7–1.3)	≥5	123/256	0.9 (0.7–1.2)
Klaeboe et al., 2007	Norway, 2001–2002	289 glioma 358 controls		63%	0.6 (0.4–0.9)	≥6	70/73	0.8 (0.5–1.2)
Lahkola et al., 2007	Denmark, Finland, Norway, Sweden, United Kingdom, 2000–2004	1,521 glioma 3,301 controls		59%	0.8 (0.7–0.9)	≥10	143/220	1.0 (0.7–1.2)
Cohort								
Schuz et al., 2006b	Denmark	257 glioma		Not applicable	1.0 (0.9–1.1)	≥10	28 cases of brain or nervous system tumor	0.7 (0.4–0.95)

[a]Time since first use, or otherwise duration of use if time since first use is not presented

[b]Results of studies by Hardell et al. were also published as several other reports. We have referred in this table to the first paper related to the respective study

[c]Ever having had a cellular phone subscription

[d]Discordant case–control pairs

[e]Pooled analyses of second and third study by Hardell et al. (2002a, 2005c, Hardell et al., 2005a)

[f]Odds ratios were only reported by type of phone, not for cellular phone use overall

Table 2. Summary of results from published studies on risk of meningioma in relation to cellular and cordless phone use

		Overall exposure				Longest time since first exposure[a]		
Study design, first author and year	Country, period	Total cases/ controls		% exposed among controls	Relative risk (95% CI)	Time (years)	No. exposed cases/controls	Relative risk (95% CI)
Case–control								
Hardell et al., 1999[b]	Sweden, 1994–1996	46/425		38%	1.1 (0.5–2.3)	–	–	–
Inskip et al., 2001	United States, 1994–1998	197/799		22%	0.8 (0.4–1.3)	≥5	6/31	0.9 (0.3–2.7)
Auvinen et al., 2002	Finland, 1996	129/1,635		11%	1.1 (0.5–2.4)	>2	2/12	0.8 (0.2–3.5)
Hardell et al., 2002a[b]	Sweden, 1997–2000	611/611[c]	Analogue[d]	15%[e]	1.1 (0.7–1.5)	–	–	–
			Digital[d]	30%[e]	0.8 (0.6–1.0)	–	–	–
			Cordless[d]	27%[e]	0.9 (0.6–1.1)	–	–	–
Christensen et al., 2005a	Denmark, 2000–2002	175/822		47%	0.8 (0.5–1.3)	≥10	6/8	1.0 (0.3–3.2)
Lonn et al., 2005	Sweden, 2000–2002	273/674	Cellular	59%	0.7 (0.5–0.9)	≥10	12/36	0.9 (0.4–1.9)
			Cordless	–	0.8 (0.5–1.2)	–	–	–
Hardell et al., 2005a[b]	Sweden, 2000–2003	305/692	Analogue[d]	11%	1.7 (1.0–3.0)	≥10	20/40	2.1 (1.1–4.3)
			Digital[d]	50%	1.3 (0.9–1.9)	≥10	8/18	1.5 (0.6–3.9)
			Cordless[d]	44%	1.3 (0.9–1.9)	≥10	19/35	1.9 (1.0–3.6)
Hardell et al., 2006a[f]	Sweden, 1997–2003	916/2,162	Analogue[d]	14%	1.3 (1.0–1.7)	≥10	34/84	1.6 (1.0–2.5)
			Digital[d]	36%	1.1 (0.9–1.3)	≥10	8/18	1.3 (0.5-3.2)
			Cordless[d]	32%	1.1 (0.9–1.4)	≥10	23/45	1.6 (0.9-2.8)
Schuz et al., 2006a	Germany, 2000–2003	381/762	Cellular	31%	0.8 (0.6–1.1)	≥10	5/9	1.1 (0.4–3.4)
			Cordless	23%	0.8 (0.6–1.0)	≥5	128/281	0.8 (0.6–1.1)

Study	Country	Cases	Participation	OR (95% CI)	Years	Exposed cases/controls	OR (95% CI)
Klaeboe et al., 2007	Norway, 2001–2002	207/358	63%	0.8 (0.5–1.1)	≥6	36/74	1.0 (0.6–1.8)
Cohort							
Schuz et al., 2006b	Denmark	68 incident cases	Not applicable	0.9 (0.7–1.1)	≥10	28 cases of brain or nervous system tumor	0.7 (0.4–0.95)

[a]Time since first use, or duration of use if time since first use is not presented

[b]Results of studies by Hardell et al. were also published as several other reports. We have referred in this table to the first paper related to the respective study

[c]Actual analyses are based on an undisclosed smaller number of subjects because incomplete case–control pairs were excluded from the analyses

[d]Odds ratios were only reported by type of phone, not for cellular phone use overall

[e]Among controls overall

[f]Pooled analyses of second and third study by Hardell et al. (2002a, 2005c, a)

Table 3. Summary of results from published studies on risk of acoustic neuroma in relation to cellular and cordless phone use

		Overall exposure				Longest time since first exposure[a]		
Study design, first author and year	Country, period	Total cases/ controls		% exposed among controls	Relative risk (95% CI)	Time (years)	No. exposed cases/ controls	Relative risk (95% CI)
Case–control								
Hardell et al., 1999[b]	Sweden, 1994–1996	13/13		38%	0.8 (0.1–4.2)	–	–	–
Inskip et al., 2001	United States, 1994–1998	96/799		22%	1.0 (0.5–1.9)	≥5	5/31	1.9 (0.6–5.9)
Muscat et al., 2002	United States, 1997–1999	90/86		27%	0.9 (–)	≥3	11/6	1.7 (0.5–5.1)
Hardell et al., 2002a[b]	Sweden, 1997–2000	159/159[c]	Analogue[d]	15%[e]	3.5 (1.8–6.8)	≥10	7/2[f]	3.5 (0.7–16.8)
			Digital[d]	30%[e]	1.2 (0.7–2.2)	≥5	2/1[f]	2.0 (0.2–22.1)
			Cordless[d]	27%[e]	1.0 (0.6–1.7)	≥10	2/1[f]	2.0 (0.2–22.1)
Warren et al., 2003	United States, 1995–2000	51/141		22%	1.0 (0.4–2.2)	–	–	–
Christensen et al., 2004	Denmark, 2000–2002	106/212		46%	0.9 (0.5–1.6)	≥10	2/15	0.2 (0.0–1.1)
Lonn et al., 2004a	Sweden, 1999–2002	148/604	Cellular	59%	1.0 (0.6–1.5)	≥10	14/29	1.9 (0.9–4.1)
			Cordless	–	0.7 (0.4–1.2)	–	–	–
Hardell et al., 2005a[b]	Sweden, 2000–2003	84/692	Analogue[d]	11%	4.2 (1.8–10)	≥10	7/40	2.6 (0.9–8.0)
			Digital[d]	50%	2.0 (1.1–3.8)	≥10	1/18	0.8 (0.1–6.7)
			Cordless[d]	44%	1.5 (0.8–2.9)	≥10	1/35	0.3 (0.0–2.2)
Schoemaker et al., 2005	Denmark, Finland, Norway, Sweden, United Kingdom, 1999–2004	678/3,553		54%	0.9 (0.7–1.1)	≥10	47/212	1.0 (0.7–1.5)

Takebayashi et al., 2006	Japan, 2000–2004	101/339		58%	0.7 (0.4–1.2)	≥8	4/12	0.8 (0.2–2.7)
Hardell et al., 2006a[g]	Sweden, 1997–2003	243/2,162	Analogue[d]	14%	2.9 (2.0–4.3)	≥10	19/84	3.1 (1.7–5.7)
			Digital[d]	36%	1.5 (1.1–2.1)	≥10	1/18	0.6 (0.1–5.0)
			Cordless[d]	32%	1.5 (1.0–2.0)	≥10	4/45	1.0 (0.3–2.9)
Klaeboe et al., 2007	Norway, 2001–2002	45/358		63%	0.5 (0.2–1.0)	≥6	8/67	0.5 (0.2–1.4)
Schlehofer et al., 2007	Germany, 2000–2003	97/194		38%	0.7 (0.4–1.2)	5–9	8/27	0.5 (0.2–1.3)
Cohort								
Schuz et al., 2006b	Denmark	32[h]		Not applicable	0.7 (0.5–1.03)	–	–	–

[a]Time since first use, or otherwise duration of use if time since first use is not presented

[b]Results of studies by Hardell et al. were also published as several other reports. We have referred in this table to the first paper related to the respective study

[c]Actual analyses are based on an undisclosed smaller number of subjects because incomplete case–control pairs were excluded from the analyses

[d]Odds ratios were only reported by type of phone, not for cellular phone use overall

[e]Among controls overall

[f]Discordant case–control pairs

[g]Pooled analyses of second and third study by Hardell et al. (2002a, 2005a)

[h]31 confirmed acoustic neuroma, 1 not known

Table 4. Summary of results from published studies on risk of non-intracranial neoplasms in relation to cellular and cordless phone use

			Overall exposure				Longest time since first exposure[a]		
Study design, first author and year	Country, period	Neoplasm	Total cases/ controls		% exposed among controls	Relative risk (95% CI)	Time (years)	No. exposed cases/con-trols	Relative risk (95% CI)
Case–control									
Stang et al., 2001	Germany, 1995–1998	Uveal melanoma	118/475		3%[b]	4.2 (1.2–14.5)	≥3	5/8	3.8 (0.8–19.7)
Auvinen et al., 2002	Finland, 1996	Parotid gland tumors	34/170		11%	1.3 (0.4–4.7)	>2	1/1	2.3 (0.2–25.3)
Hardell et al., 2004a	Sweden, 1994–2000	Parotid gland tumors	267/1,053		33%	1.0 (0.8–1.4)	≥10	6/38	0.7 (0.3–1.6)
Lonn et al., 2006	Denmark, Sweden, 2000–2002	Parotid gland tumors	172/681	Malignant	60%	0.7 (0.4–1.3)	≥10	2/36	0.4 (0.1–2.6)
				Benign		0.9 (0.5–1.5)	≥10	7/15	1.4 (0.5–3.9)
Baumgardt-Elms et al., 2002	Germany, 1995–1997	Testicular cancer	269/797		21%[c]	0.9 (0.6–1.2)	–	–	–
Hardell et al., 2007a	Sweden, 1993–1997	Testicular cancer	888/870	Analogue[d]	20%	1.0 (0.8–1.3)	≥10	14/19	1.5 (0.6–3.7)
				Digital[d]	16%	1.1 (0.8–1.5)	≥5	10/3	2.8 (0.8–11)
				Cordless[d]	19%	1.0 (0.8–1.4)	≥5	54/51	1.0 (0.7–1.6)
Warren et al., 2003	United States, 1995–2000	Intratemporal facial nerve tumor	18/141		22%	0.4 (0.1–2.1)	–	–	–
Hardell et al., 2005d	Sweden, 1999–2002	Non-Hodgkin's lymphoma	910/1,016		68%	1.1 (0.9–1.3)	≥10	116/125	1.0 (0.7–1.4)

Linet et al., 2006	United States, 1998–2000	Non-Hodgkin's lymphoma	551/462	15%	0.9 (0.6–1.4)	>8	20/9	1.6 (0.7–3.8)
Cohort								
Schuz et al., 2006b	Denmark	Parotid gland tumor	26	Not applicable	0.77 (–)[e]	–	32	1.1 (0.7–1.5)
		Eye tumors	44		0.96 (–)([e])	–		
		Leukemia	351		1.00 (–)([e])	≥10		
		Testicular cancer	522		1.1 (0.96–1.2)	–		

[a]Time since first use, or otherwise duration of use if time since first use is not presented
[b]Certain or probably exposure to cellular phones at the work place for several hours per day
[c]Working near radiofrequency transmitters such as radio sets and cellular phones
[d]Odds ratios were only reported by type of phone, not for cellular phone use overall
[e]Confidence interval not reported for both sexes combined

were calculated, in particular whether subjects who could not be contacted, who had moved out of the area, and those who had died were included in the denominator. Other population-based studies have reported control participation rates in the range of 45–70%. Low participation rates increase the potential for bias in that there is greater scope for participants to be unrepresentative of the source population, i.e., to be more, or less, likely to be phone users than the general, source population. Several studies found evidence suggesting that phone users were over-represented among control participants (Lonn et al., 2004a; Lahkola et al., 2005; Schuz et al., 2006a), but the validity of this conclusion is somewhat uncertain because information was only collected from a subset of nonparticipants.

3.5. Representativeness of the Study Population

Case participation rates have also shown large variation between studies, with on average higher rates in hospital-based than in population-based studies, as one might expect because it is easier to recruit cases from a limited number of cooperating hospitals, than it is to include all cases from a large geographical area, and because patients are often more likely to take part if approached while in hospital. There are difficulties in comparing case participation rates between population-based studies because of differences in the way these rates were calculated, as well as which patients were considered eligible for the study, and what systems were in place to identify all cases in the area. Participation rates can appear greater if cancer registration data are not used as a source of denominator data and as a consequence deceased cases, eligible cases treated outside the study region, and cases not treated at major centers are not included in the denominator of the rates. Rates can also appear greater if case ascertainment does not continue for a considerable time beyond the last diagnosis date of the study period, to identify retrospectively eligible cases that became known, for instance to cancer registries, well after their diagnosis. Unusually high participation rates have been reported for cases in studies by Hardell et al., but it is clear that the participation rates in these studies as conventionally calculated were considerably lower, due to the exclusion of patients who had died and other patients from the denominator of the published rates.

In studies generally, the highest rates of nonparticipation have usually been observed for patients with high-grade tumors. This would give bias if cellular phone use was related to prognosis of the tumor, but most studies found that relative risks in relation to phone use did not vary appreciably by tumor grade of glioma, arguing against this type of bias.

3.6. Data Collection Biases

In most studies, the interviewer could not be blinded to the case–control status of the interviewee, and the interviewer's preconceived ideas could have affected the way the questions were asked. Face-to-face interviews were used in the Interphone studies, but with a strongly structured computer-assisted interview, which may have reduced the potential for interviewer bias. The studies by Hardell et al. were based on self-administered questionnaires, reducing the likelihood of interviewer bias, but in some of their studies, selected respondents were telephoned because their questionnaires were "unclear" or of insufficient "quality." This leaves potential for bias if the questions asked by phone were influenced by case–control status, as this status is easily discovered on the telephone. Also, Hardell et al. (2002a) reported that cases were far more likely than controls to have been assisted by their relatives in completing the questionnaire, which might have affected the results in an uncertain direction. The use of proxies, e.g., close relatives of the case, for ill or deceased patients also makes the interpretation of some studies more difficult because in most studies proxies were used more often for cases than controls. The quality of recall of past phone use patterns by proxies has not been assessed.

Cohort studies have avoided the potential for recall bias and interviewer bias inherent in interview case–control studies by using network operator data recorded before cancer diagnosis. A further strength is that all tumors were included regardless of their aggressiveness, i.e., no subjects were lost to the study due to poor prognosis. Limitations were that exposure was based on subscription records, not direct evidence of use, and that there was no information on intensity of use, side of use or use of hands-free sets, nor on confounders. The omission of corporate users in the Danish cohort study (Schuz et al., 2006b) may have selected the cohort toward low exposure users, at least in the early years, due to the high cost of calls. Inclusion of these corporate users, who might on average have been the heaviest early users, in the reference group could have obscured a raised risk (if there was truly one) in comparisons with the reference group, although substantial risks should still have been detectable. Comparisons within the cohort, for instance according to time since first use, should not be subject to this problem.

Past studies have had the limitation that they had little power to examine risks in relation to prolonged use, long lag periods, or high intensities of use, in particular with regard to digital phone use. All studies have had problems with exposure assessment, and future studies would be improved if they could make use of operator records, ideally with more detailed information on frequency of use, as well as self-reported information to verify such exposures and obtain information on variables not available from operators, such as on use of hands-free devices. Furthermore, none of the published analyses have examined brain tumor risk in relation to distance of the tumor origin from the antenna, other than crudely by lobe of brain and laterality. Inclusion of tumors distant from the RF source could have diluted observed risks, if there were a real risk proportional to level of RF exposure. Information on cellular phone use is a crude measure of RF exposure, which is dependent on a range of factors such as geographical area and movement and location of the user (Erdreich et al., 2007) as well as the type of phone and the proximity of the antenna to the part of the brain under consideration. No studies have been published to date addressing cancer risks in relation to cellular phone use in childhood, despite concerns that children might be more susceptible to RF exposure (Independent Expert Group on Mobile Phones (IEGMP), 2000).

3.7. Bias Regarding Choice of Censoring Date

Cellular phone use as an exposure variable is unusual in that the prevalence of exposure in the population has increased very rapidly over time. For example, in the United Kingdom, the number of cellular phone subscriptions rose from 5 million in the year 1995 to 50 million in 2002 (Mobile Operators Association (MOA), 2008). Epidemiological case–control studies would need to consider this in the analysis. If, for example, exposure were evaluated, on average, up to a later date for controls than for cases in a particular study then, in the absence of a true association, the prevalence of phone use among controls will appear to be higher than among cases, leading to apparently reduced odds ratios. Conversely, artefactually raised risks could be observed if the exposure period for cases continued beyond the exposure period for controls. A study design in which controls are selected individually matched to cases often has the consequence that controls are interviewed later than cases, because the control can only be selected when the case has been identified. In such studies, this nonsynchronous recruitment can be controlled for by evaluating exposure for controls up to the same date, usually the case-diagnosis date, as the case it is matched to. This has the drawback, however, that the relevant exposure investigated, e.g., phone use, might be recalled after a longer time period for controls than cases, which might induce bias, of an unknown direction. In matched studies that do not state truncation on exact diagnosis date (e.g., Hardell et al., 2005c), there is scope for bias if the truncation has been cruder (e.g., based on year only or nonexistent). In studies in which controls are frequency-matched to cases, i.e., controls are selected based on, e.g., the age, and sex distribution of the entire set of cases, the choice of exposure censoring date for controls is less obvious. Adjustment for interview date in the US studies (Inskip et al., 2001; Muscat et al., 2000), with exposure evaluated up to interview, seems only appropriate if the lag time between diagnosis and interview for cases is very short. Several studies dealt with this issue by analytic correction for the cases lag time between diagnosis and interview as well as for nonsynchronous recruitment between cases and controls (e.g., Schoemaker et al., 2005; Lahkola et al., 2007).

4. INTERPRETATION OF STUDIES AND CONCLUSIONS

To draw a conclusion based on the published studies, one needs to evaluate the consistency in results between studies, the timing and magnitude of associations observed, and dose–response associations. The validity of individual studies and their results needs to be taken into account in this evaluation, which includes considering the size and quality of each study, and the likelihood of bias, confounding, and exposure misclassification. Furthermore, the epidemiological evidence needs to be evaluated against the biological plausibility of an effect. Overall, one needs to consider the classic "Bradford Hill criteria" of causality (Hill, 1965). We evaluate below the evidence for causality for each of the main types of neoplasms investigated, giving particular consideration to these factors.

4.1. Evidence for Causality by Tumor Type

4.1.1. Malignant Brain Tumors or Glioma

We have considered studies of malignant brain tumors or glioma together because glioma constitutes the majority of malignant brain tumors (CBTRUS, 2004). Studies of these types of tumor in relation to cellular phone use have shown inconsistent results. Increased risks of malignant brain tumors in cell phone users overall have been reported by Hardell et al. in Sweden and of glioma by Auvinen et al. in a record-based case–control study in Finland (Table 1). In contrast, findings from US studies, national or pooled Interphone studies, and the Danish cohort study did not show evidence of increased risks, although positive associations have been reported for certain parameters of phone use in some studies, albeit of no consistent pattern.

In terms of magnitude of effect estimates, the strongest increase in risk in phone users overall was reported in the third case–control study of malignant brain tumors by Hardell et al. (2005c), with significantly 1.9- to 2.6-fold increased risks in relation to ever use of analogue, digital, and cordless phones. There was evidence of increasing risk with increasing time since first use, with 2.9- to 3.6-fold increased risks after 10 years, but risks were already significantly raised after less than 5 years of first use (1.6- to 1.8-fold). These findings are unusual in that risks were significantly raised in subjects who had ever used a phone, regardless of elapsed time since first use, and raised risks were reported even for cordless phones, which have a considerably lower power output than cellular phones (AGNIR, 2003). Results of individual studies by Hardell et al. were published multiple times and appear to be based on selective analyses, with analyses and selection of subgroups varying between different papers by these authors, and this lack of internal consistency is of concern.

The record-based study by Auvinen et al. (2002) showed a 1.5-fold raised risk of glioma in cellular phone subscribers, which was 2.1-fold increased when restricted to analogue phones, lending some support to the findings by Hardell et al. It has the strength that it avoids problems of recall and selection bias, although exposure misclassification is a problem. Raised risks were observed after a relatively short duration of exposure; the average subscription time to an analogue phone was 2–3 years, and there was weak evidence for a trend in risk with duration of subscription. Confounding by socioeconomic status could potentially explain the results, as early phone users were likely to be affluent, and glioma tends to be more common among higher socioeconomic groups (Ohgaki and Kleihues, 2005), but it was reported that analyses adjusted for socioeconomic status did not affect the results.

Studies that did not show raised risks of malignant brain tumors or glioma in relation to cellular phone use overall include the early US studies, with low prevalence of phone use, the Nordic, British, and German studies that were part of Interphone, and the Danish cohort study. The cohort study is particularly important in assessing risks of cancer in phone users because it is very large, with relatively long follow-up and high quality cancer incidence data, and is based on unbiased recorded exposure information, albeit indirect. This study showed no raised risk of glioma in cellular phone subscribers overall, and no raised risk of brain and nervous system tumors 10 or more years after start of the subscription (Schuz et al., 2006b). Small increases in risk might not have been detectable in this cohort, however, because exposure assessment was indirect, based on subscriptions, not on phone use and because the "non-subscriber" group included corporate users who would have had exposure.

Some of the Interphone studies had low participation rates, a factor that has possibly contributed to the apparently reduced tumor risks in phone users observed in several studies, due to potential over-selection of exposed controls. The extent to which phone users were overrepresented among control participants, if at all, is currently unclear. An Interphone validation study has investigated this, but these results have not yet been published.

Most studies have not shown increased risks in relation to the longest duration of use or time since first use observed, including the largest studies to date: the British (Hepworth et al., 2006) and pooled Nordic-UK (Lahkola et al., 2007) studies. Significantly increased risks of glioma were observed 10 or more years after first phone use, however, in the third study by Hardell et al. (2005c) and the pooled analyses by Hardell et al. (2006b), and nonsignificantly increased risks in the second case–control study by Hardell et al. (2002b) and the German Interphone study (Schuz et al., 2006a). In a metaanalysis of pooled data from all previous case–control studies, Hardell et al. (2007b) recently reported a relative risk of 1.2 for glioma 10 or more years after first cellular phone use, with a wide confidence interval (0.8–1.9). The result is not statistically significant; however, it is of a magnitude that might have easily arisen from bias or error, and it is nearly entirely due to findings from one research group (i.e., Hardell et al).

The pooled Nordic-UK Interphone study by Lahkola et al. (2007), the largest study to date, showed no raised risks of glioma in relation to regular phone use or for long-term use overall, but reported a significantly positive trend of risk with cumulative hours of use, using one test (assessing cumulative hours of use as a continuous variable) but not another (assessing hours as a categorical variable). The dependence of the results on the statistical method employed, the lack of dose–response for other variables, and reports that duration of call time is very sensitive to errors in recall (Vrijheid et al., 2006a), make this finding weak as evidence on etiology. The increase in risk of glioma ipsilateral to reported side of phone use after 10 or more years of first use in this study could be due to reporting bias or, as it was borderline significant, to chance, so although compatible with etiology, it again gives weak evidence.

Thus, results of studies of glioma and cellular phones have been inconsistent. Evidence for an association has been observed in case–control studies by one research group in Sweden and a record-based study in Finland, but not in several other case–control studies elsewhere, or in a large cohort study of subscribers. There is no obvious methodological reason why the Hardell studies would be able to find true positives not detectable by other studies that do not appear to be methodologically inferior. One of the hallmarks of etiological findings is replicability by other investigators and other methods. The Hardell studies showed risks that were already raised after a relatively short period since first use and which were of a magnitude which, if true, would also be expected to have been observed in the Danish cohort study. Furthermore, if true, one would expect to observe an increase in recorded incidence rates linked to the introduction of cellular phones; such evidence has not been seen in studies to date of incidence (Lonn et al., 2004c; Muscat et al., 2006; Nelson et al., 2006; Roosli et al., 2007). Like the cohort data, this provides evidence against large effects within the time scale yet available.

4.2. Meningioma

Studies of *meningioma* risks have been fewer, and have included fewer subjects than for glioma. None of these studies have reported raised meningioma risks in relation to phone use overall, or in relation to long-term use, with the exception of the third study by Hardell et al. (2005a), in which a near-significantly raised risk was reported in relation to ever analogue phone use, which was significantly raised 10 years or more after first use (Table 2). Odds ratios for ever use of digital or cordless phones were somewhat increased (both OR = 1.3), and there was a modest increase in odds ratios with increasing induction time and with intensity of exposure, but no statistical trend tests were presented to exclude chance as a possible reason for this.

With regard to the longest induction time investigated, a metaanalysis by Hardell et al. (2007b) showed a nonsignificant pooled odds ratio of 1.3 from all previously published case–control studies for 10 or more years after first use, but the raised risk was heavily influenced by the results from Hardell's own study (Hardell et al., 2006a). These findings must also be compared with those from the Danish cohort study, which did not find a raised risk of meningioma (Schuz et al., 2006b).

Thus, the evidence overall does not suggest a raised risk of meningioma in relation to cellular phone use. The only increases in risk observed have been in the study by Hardell et al. (2005a), with a weak dose–response, and no support from any other studies, in particular from the Danish cohort study.

4.3. Acoustic Neuroma

Among studies of acoustic neuroma risks, Hardell et al. (2002a) reported a significantly raised risk in relation to ever analogue use in their second study and in relation to analogue and digital use in their third study (Hardell et al., 2005a) (Table 3). There was no trend of increasing risk with increasing induction time, however, as might be expected if the association was causal. Indeed, in the third study, the greatest risks were observed after the shortest induction time (1–5 years) (Hardell et al., 2005a). Acoustic neuroma is a slow growing tumor, and thus in general would be expected to have a long lag time between exposure and diagnosis. Although other studies did not report raised risks in phone users overall or in the longest term users, nonsignificantly raised risks in relation to long-term use were observed in several studies based on small numbers (Inskip et al., 2001; Muscat et al., 2002; Lonn et al., 2004a), but not in the Nordic-UK pooled analysis (Schoemaker et al., 2005), much the largest study to date, or the Danish cohort study (Schuz et al., 2006b). Hardell et al. (2007b) reported a pooled odds ratio of 1.3 (95% CI: 0.6–2.8) 10 or more years after first use based on all previous studies, similar to that for glioma and meningioma. The nonsignificant risk estimate with a wide confidence interval does not provide substantial support for etiology. Recorded incidence rates of cranial nerve tumors have been reported to be increasing, but, in a study in the UK, no evidence was found that the trend was linked to the introduction of cellular phones (Nelson et al., 2006).

Increased risks of an acoustic neuroma ipsilateral to side of phone use were observed after 10 years of use in the Swedish Interphone study (Lonn et al., 2004a), and in the Nordic-UK pooled data (Schoemaker et al., 2005). However, the interpretation of this finding is complicated because, additional to the problem of potential over-reporting of phone use on the same side as the tumor, tumor-induced hearing loss might cause the subject to hold the phone on the opposite side, and ipsilateral users might present to the doctor with hearing loss earlier than contralateral users. Such biases might well operate in opposite directions and potentially apply differently to short and long-term users, making the interpretation of laterality analyses of risk particularly uncertain for acoustic neuroma.

Thus, there is no convincing evidence that cellular phone use increases risk of acoustic neuroma. The only studies that observed raised risks overall came from one research group and have not been replicated in studies by other researchers. As acoustic neuroma is a slow-growing tumor, it is possible that a long induction time is required to develop a tumor. Several studies have suggested a raised risk after 10 years of use but much the largest one did not; evidence for this was not consistent between studies; and the scarcity of data make these findings very uncertain. Increases in risk ipsilateral to side of phone use have been reported, but all studies conducted to date are susceptible to bias with regard to such laterality analyses, and are particularly difficult to interpret.

4.3.1. Non-Intracranial Tumors

Studies on non-intracranial tumors have been relatively few (Table 4), and have collectively not shown evidence for associations of risk of parotid gland tumors, non-Hodgkin's lymphoma, testicular cancer, or other investigated neoplasms with phone use. The study on uveal melanoma in Germany (Stang et al., 2001) found a fourfold increased risk in relation to cellular phone use, but the small size of the study, unsatisfactory assessment of exposure, and combination of data from two different study designs do not give confidence in the results. Furthermore, the Danish cohort study did not show raised risk of uveal melanoma (Schuz et al., 2006b), and no marked secular increase in national incidence rates of this tumor has been observed in Denmark (Johansen et al., 2002), which would be expected if there were a truly raised risk of that magnitude. A raised risk of T cell, but not B cell, non-Hodgkin's lymphoma in relation to cellular phone use was reported by Hardell et al. (2005d), but with no a priori justification for such subset analysis and no raised risk for non-Hodgkin's lymphoma overall. A raised risk of non-Hodgkin's lymphoma in relation to cellular phone use was not found in a US-based case–control study (Linet et al., 2006), or in the Danish cohort study (Schuz et al., 2006b); therefore, there is no substantial evidence of an association of cellular phone use with non-Hodgkin's lymphoma.

Other Epidemiological and Biological Evidence on Cancer Risk in Relation to RF Exposure

Information on cancer risks in relation to long-term exposures to radiofrequency fields comes also from studies of occupational groups who have been exposed to radiofrequency fields for several decades, such as radar technicians and radio and telegraph operators. Risks in such groups have been reviewed previously (AGNIR, 2003). Although increased risks of brain tumors and/or lymphatic and hematopoietic neoplasms have been reported in a few studies, e.g., in Polish military personnel (Szmigielski, 1996), in men who had worked in a job with likely RF field exposure (Thomas et al., 1987), and in Norwegian electrical workers (Tynes et al., 1992), several other studies did not support this (Milham, 1985; Tynes et al., 1996; Morgan et al., 2000; Groves et al., 2002), and overall there has been no consistent evidence of a raised risk of cancer of any site, although most studies had methodological weaknesses, in particular a lack of individual exposure measurement (AGNIR, 2003; Ahlbom et al., 2004).

Biological Plausibility

Potential biological mechanisms by which RF fields could cause neoplasms have been subject to considerable research efforts. The general scientific consensus is that there is no established biological mechanism by which RF exposure could induce or promote cancer (Independent Expert Group on Mobile Phones (IEGMP), 2000; Health Council of the Netherlands, 2002; AGNIR, 2003; The Royal Society of Canada, 2004; Moulder et al., 2005; SCENIHR (2001)).

5. SUMMARY

In summary, epidemiological studies published to date show, on balance, no convincing or consistent evidence for a raised risk of cancer in relation to cellular phone use, and this is consistent with the background epidemiological and biological literature on the potential relation of RF to cancer risk. The overall evidence from case–control and cohort studies of phone use suggests that it is unlikely that there are large increased risks of intracranial tumors, or of other cancers, in relation to cellular phone use in the lag period yet assessable. The trends in recorded incidence rates of these cancers reinforce this conclusion, given the enormous uptake of this technology by the population.

However, past studies have had limitations, in particular that exposure measurement has been crude, and data on risk after long lag periods, prolonged use, high intensities of use, and for childhood exposures are still limited, and the possibility of risk in relation to these remains open.

ACKNOWLEDGMENT. The authors thank the reviewers for their helpful comments with regard to this chapter.

REFERENCE

AGNIR, 2003, Health effects from radiofrequency electromagnetic fields: Report of an independent Advisory Group of Non-Ionising Radiation. *Doc NRPB* 14: 1–177.

Ahlbom, A., and Feychting, M., 1999, Re: Use of cellular phones and the risk of brain tumours: A case-control study. *Int J Oncol* 15: 1045–1047.

Ahlbom, A., Green, A., Kheifets, L., Savitz, D., and Swerdlow, A., 2004, Epidemiology of health effects of radiofrequency exposure. *Environ Health Perspect* 112: 1741–1754.

Auvinen, A., Hietanen, M., Luukkonen, R., and Koskela, R.S., 2002, Brain tumors and salivary gland cancers among cellular telephone users. *Epidemiology* 13: 356–359.

Baumgardt-Elms, C., Ahrens, W., Bromen, K., Boikat, U., Stang, A., Jahn, I., Stegmaier, C., and Jockel, K.H., 2002, Testicular cancer and electromagnetic fields (EMF) in the workplace: Results of a population-based case-control study in Germany. *Cancer Causes Control* 13: 895–902.

Behin, A., Hoang-Xuan, K., Carpentier, A.F., and Delattre, J.Y., 2003, Primary brain tumours in adults. *Lancet* 361: 323–331.

Cardis, E., Richardson, L., Deltour, I., Armstrong, B., Feychting, M., Johansen, C., Kilkenny, M., McKinney, P., Modan, B., Sadetzki, S., Schuz, J., Swerdlow, A., Vrijheid, M., Auvinen, A., Berg, G., Blettner, M., Bowman, J., Brown, J., Chetrit, A., Christensen, H.C., Cook, A., Hepworth, S., Giles, G., Hours, M., Iavarone, I., Jarus-Hakak, A., Klaeboe, L., Krewski, D., Lagorio, S., Lonn, S., Mann, S., McBride, M., Muir, K., Nadon, L., Parent, M.E., Pearce, N., Salminen, T., Schoemaker, M., Schlehofer, B., Siemiatycki, J., Taki, M., Takebayashi, T., Tynes, T., van Tongeren, M., Vecchia, P., Wiart, J., Woodward, A., and Yamaguchi, N., 2007, The INTERPHONE study: Design, epidemiological methods, and description of the study population. *Eur J Epidemiol* 22: 647–664.

CBTRUS, 2004, Statistical report: Primary Brain Tumors in the United States 1997–2001. Central Brain Tumor Registry of the United States, Chicago, USA.

Christensen, H.C., Schuz, J., Kosteljanetz, M., Poulsen, H.S., Thomsen, J., and Johansen, C., 2004, Cellular telephone use and risk of acoustic neuroma. *Am J Epidemiol* 159: 277–283.

Christensen, H.C., Schuz, J., Kosteljanetz, M., Poulsen, H.S., Boice, J.D., Jr., McLaughlin, J.K., and Johansen, C., 2005a, Cellular telephones and risk for brain tumors: A population-based, incident case-control study. *Neurology* 64: 1189–1195.

Christensen, H.C., Schuz, J., Kosteljanetz, M., Poulsen, H.S., Boice, J.D., Jr., McLaughlin, J.K., and Johansen, C., 2005b, Erratum re: Cellular telephones and risk for brain tumors: A population-based, incident case-control study. *Neurology* 65: 1324.

Dimbylow, P., and Mann, S.J., 1999, Characterisation of energy deposition in the head from cellular phones. *Radiat Prot Dosimetry* 83: 139–141.

Dreyer, N.A., Loughlin, J.E., and Rothman, K.J., 1999, Cause-specific mortality in cellular telephone users. *JAMA* 282: 1814–1816.

Erdreich, L.S., Van Kerkhove, M.D., Scrafford, C.G., Barraj, L., McNeely, M., Shum, M., Sheppard, A.R., and Kelsh, M., 2007, Factors that influence the radiofrequency power output of GSM mobile phones. *Radiat Res* 168: 253–261.

Groves, F.D., Page, W.F., Gridley, G., Lisimaque, L., Stewart, P.A., Tarone, R.E., Gail, M.H., Boice, J.D., Jr., and Beebe, G.W., 2002, Cancer in Korean war navy technicians: Mortality survey after 40 years. *Am J Epidemiol* 155: 810–818.

Hardell, L., Nasman, A., Pahlson, A., Hallquist, A., and Hansson, M.K., 1999, Use of cellular telephones and the risk for brain tumours: A case- control study. *Int J Oncol* 15: 113–116.

Hardell, L., Nasman, A., Pahlson, A., and Hallquist, A., 2000, Case-control study on radiology work, medical X-ray investigations, and use of cellular telephones as risk factors for brain tumors. *MedGenMed* 2: E2.

Hardell, L., Mild, K.H., Pahlson, A., and Hallquist, A., 2001, Ionizing radiation, cellular telephones and the risk for brain tumours. *Eur J Cancer Prev* 10: 523–529.

Hardell, L., Hallquist, A., Mild, K.H., Carlberg, M., Pahlson, A., and Lilja, A., 2002a, Cellular and cordless telephones and the risk for brain tumours. *Eur J Cancer Prev* 11: 377–386.

Hardell, L., Mild, K.H., and Carlberg, M., 2002b, Case-control study on the use of cellular and cordless phones and the risk for malignant brain tumours. *Int J Radiat Biol* 78: 931–936.

Hardell, L., Hansson, M.K., Sandstrom, M., Carlberg, M., Hallquist, A., and Pahlson, A., 2003a, Vestibular schwannoma, tinnitus and cellular telephones. *Neuroepidemiology* 22: 124–129.

Hardell, L., Mild, K.H., and Carlberg, M., 2003b, Further aspects on cellular and cordless telephones and brain tumours. *Int J Oncol* 22: 399–407.

Hardell, L., Hallquist, A., Hansson, M.K., Carlberg, M., Gertzen, H., Schildt, E.B., and Dahlqvist, A., 2004a, No association between the use of cellular or cordless telephones and salivary gland tumours. *Occup Environ Med* 61: 675–679.

Hardell, L., Mild, K.H., Carlberg, M., and Hallquist, A., 2004b, Cellular and cordless telephone use and the association with brain tumors in different age groups. *Arch Environ Health* 59: 132–137.

Hardell, L., Carlberg, M., and Hansson, M.K., 2005a, Case-control study on cellular and cordless telephones and the risk for acoustic neuroma or meningioma in patients diagnosed 2000–2003. *Neuroepidemiology* 25: 120–128.

Hardell, L., Carlberg, M., and Hansson, M.K., 2005b, Use of cellular telephones and brain tumour risk in urban and rural areas. *Occup Environ Med* 62: 390–394.

Hardell, L., Carlberg, M., and Mild, K.H., 2005c, Case-control study of the association between the use of cellular and cordless telephones and malignant brain tumors diagnosed during 2000–2003. *Environ Res* 100: 232–241.

Hardell, L., Eriksson, M., Carlberg, M., Sundstrom, C., and Mild, K.H., 2005d, Use of cellular or cordless telephones and the risk for non-Hodgkin's lymphoma. *Int Arch Occup Environ Health* DOI 10.1007/s00420-005-0003-5.

Hardell, L., Carlberg, M., and Hansson, M.K., 2006a, Pooled analysis of two case-control studies on the use of cellular and cordless telephones and the risk of benign brain tumours diagnosed during 1997–2003. *Int J Oncol* 28: 509–518.

Hardell, L., Carlberg, M., and Hansson, M.K., 2006b, Pooled analysis of two case-control studies on use of cellular and cordless telephones and the risk for malignant brain tumours diagnosed in 1997-2003. *Int Arch Occup Environ Health* 79: 630–639.

Hardell, L., Mild, K.H., Carlberg, M., and Soderqvist, F., 2006c, Tumour risk associated with use of cellular telephones or cordless desktop telephones. *World J Surg Oncol* 4: 74.

Hardell, L., Carlberg, M., Ohlson, C.G., Westberg, H., Eriksson, M., and Hansson, M.K., 2007a, Use of cellular and cordless telephones and risk of testicular cancer. *Int J Androl* 30: 115–122.

Hardell, L., Carlberg, M., Soderqvist, F., Hansson, M.K., Morgan, L.L., 2007b, Long-term use of cellular phones and brain tumours - Increased risk associated with use for > 10 years. *Occup Environ Med* 64: 626–632.

Health Council of the Netherlands, 2002, *Mobile telephones: An evaluation of health effects*. Health Council of the Netherlands, The Hague, Netherlands.

Hepworth, S.J., Schoemaker, M.J., Muir, K.R., Swerdlow, A.J., van Tongeren, M.J., and McKinney, P.A., 2006, Mobile phone use and risk of glioma in adults: Case-control study. *BMJ* 332: 883–887.

Hill, A.B., 1965, The environment and disease: Association or causation? *Proc R Soc Med* **58**: 295–300.

Independent Expert Group on Mobile Phones (IEGMP), 2000, *Mobile Phones and Health*. National Radiological Protection Board, Chilton, Didcot, UK.

Inskip, P.D., Tarone, R.E., Hatch, E.E., Wilcosky, T.C., Shapiro, W.R., Selker, R.G., Fine, H.A., Black, P.M., Loeffler, J.S., and Linet, M.S., 2001, Cellular-telephone use and brain tumors. *N Engl J Med* 344: 79–86.

Inskip, P.D., Tarone, R.E., Hatch, E.E., Wilcosky, T.C., Fine, H.A., Black, P.M., Loeffler, J.S., Shapiro, W.R., Selker, R.G., and Linet, M.S., 2003, Sociodemographic indicators and risk of brain tumours. *Int J Epidemiol* 32: 225–233.

Johansen, C., Boice, J., Jr., McLaughlin, J., and Olsen, J., 2001, Cellular telephones and cancer – A nationwide cohort study in Denmark. *J Natl Cancer Inst* 93: 203–207.

Johansen, C., Boice, J.D., Jr., McLaughlin, J.K., Christensen, H.C., and Olsen, J.H., 2002, Mobile phones and malignant melanoma of the eye. *Br J Cancer* 86: 348–349.

Klaeboe, L., Blaasaas, K.G., and Tynes, T., 2007, Use of mobile phones in Norway and risk of intracranial tumours. *Eur J Cancer Prev* 16: 158–164.

Lahkola, A., Salminen, T., and Auvinen, A., 2005, Selection bias due to differential participation in a case-control study of mobile phone use and brain tumors. *Ann Epidem* 15: 321–325.

Lahkola, A., Auvinen, A., Raitanen, J., Schoemaker, M.J., Christensen, H.C., Feychting, M., Johansen, C., Klaeboe, L., Lonn, S., Swerdlow, A.J., Tynes, T., and Salminen, T., 2007, Mobile phone use and risk of glioma in 5 North European countries. *Int J Cancer* 120: 1769–1775.

Linet, M.S., Taggart, T., Severson, R.K., Cerhan, J.R., Cozen, W., Hartge, P., and Colt, J., 2006, Cellular telephones and non-Hodgkin lymphoma. *Int J Cancer* 119: 2382–2388.

Lonn, S., Ahlbom, A., Hall, P., and Feychting, M., 2004a, Mobile phone use and the risk of acoustic neuroma. *Epidemiology* 15: 653–659.

Lonn, S., Forssen, U., Vecchia, P., Ahlbom, A., and Feychting, M., 2004b, Output power levels from mobile phones in different geographical areas; implications for exposure assessment. *Occup Environ Med* 61: 769–772.

Lonn, S., Klaeboe, L., Hall, P., Mathiesen, T., Auvinen, A., Christensen, H.C., Johansen, C., Salminen, T., Tynes, T., and Feychting, M., 2004c, Incidence trends of adult primary intracerebral tumors in four Nordic countries. *Int J Cancer* 108: 450–455.

Lonn, S., Ahlbom, A., Hall, P., and Feychting, M., 2005, Long-term mobile phone use and brain tumor risk. *Am J Epidemiol* 161: 526–535.

Lonn, S., Ahlbom, A., Christensen, H.C., Johansen, C., Schuz, J., Edstrom, S., Henriksson, G., Lundgren, J., Wennerberg, J., and Feychting, M., 2006, Mobile phone use and risk of parotid gland tumor. *Am J Epidemiol* 164: 637–643.

Matthies, C., and Samii, M., 1997, Management of 1000 vestibular schwannomas (acoustic neuromas): Clinical presentation. *Neurosurgery* 40: 1–9.

Milham, S., Jr., 1985, Silent keys: Leukaemia mortality in amateur radio operators. *Lancet* 1: 812.

Mobile Operators Association (MOA), 2008, History of cellular mobile communications. http://www.mobilemastinfo.com/information/history.htm

MobileTracker, 2006, 2 billion GSM subscribers by the end of the month. http://www.mobiletracker.net/archives/2006/06/13/2-billion-gsm . 13-6-2006.

Morgan, R.W., Kelsh, M.A., Zhao, K., Exuzides, K.A., Heringer, S., and Negrete, W., 2000, Radiofrequency exposure and mortality from cancer of the brain and lymphatic/hematopoietic systems. *Epidemiology* 11: 118–127.

Moulder, J.E., Foster, K.R., Erdreich, L.S., and McNamee, J.P., 2005, Mobile phones, mobile phone base stations and cancer: A review. *Int J Radiat Biol* 81: 189–203.

Muscat, J.E., Malkin, M.G., Thompson, S., Shore, R.E., Stellman, S.D., McRee, D., Neugut, A.I., and Wynder, E.L., 2000, Handheld cellular telephone use and risk of brain cancer. *JAMA* 284: 3001–3007.

Muscat, J.E., Malkin, M.G., Shore, R.E., Thompson, S., Neugut, A.I., Stellman, S.D., and Bruce, J., 2002, Handheld cellular telephones and risk of acoustic neuroma. *Neurology* 58: 1304–1306.

Muscat, J.E., Hinsvark, M., and Malkin, M., 2006, Mobile telephones and rates of brain cancer. *Neuroepidemiology* 27: 55–56.

Nelson, P.D., Toledano, M.B., McConville, J., Quinn, M.J., Cooper, N., and Elliott, P., 2006, Trends in acoustic neuroma and cellular phones: Is there a link? *Neurology* 66: 284–285.

Ohgaki, H., and Kleihues, P., 2005, Epidemiology and etiology of gliomas. *Acta Neuropathol (Berl)* 109: 93–108.

OPCS, 1995, *Morbidity Statistics from General Practice; Fourth national study 1991–1992*. HMSO Report. London, UK.

Roosli, M., Michel, G., Kuehni, C.E., and Spoerri, A., 2007, Cellular telephone use and time trends in brain tumour mortality in Switzerland from 1969 to 2002. *Eur J Cancer Prev* 16: 77–82.

Rothman, K.J., 2000, Epidemiological evidence on health risks of cellular telephones. *Lancet* 356: 1837–1840.

Rothman, K.J., Chou, C.K., Morgan, R., Balzano, Q., Guy, A.W., Funch, D.P., Preston-Martin, S., Mandel, J., Steffens, R., and Carlo, G., 1996a, Assessment of cellular telephone and other radio frequency exposure for epidemiologic research. *Epidemiology* 7: 291–298.

Rothman, K.J., Loughlin, J.E., Funch, D.P., and Dreyer, N.A., 1996b, Overall mortality of cellular telephone customers. *Epidemiology* 7: 303–305.

Schlehofer, B., Schlaefer, K., Blettner, M., Berg, G., Bohler, E., Hettinger, I., Kunna-Grass, K., Wahrendorf, J., and Schuz, J., 2007, Environmental risk factors for sporadic acoustic neuroma (Interphone Study Group, Germany). *Eur J Cancer* 43: 1741–1747.

Schoemaker, M.J., 2007, An epidemiological study of risk factors for intracranial tumours in adults. Thesis/Dissertation, University of London, London, UK.

Schoemaker, M.J., Swerdlow, A.J., Ahlbom, A., Auvinen, A., Blaasaas, K.G., Cardis, E., Christensen, H.C., Feychting, M., Hepworth, S.J., Johansen, C., Klaeboe, L., Lonn, S., McKinney, P.A., Muir, K., Raitanen, J., Salminen, T., Thomsen, J., and Tynes, T., 2005, Mobile phone use and risk of acoustic neuroma: results of the Interphone case-control study in five North European countries. *Br J Cancer* 93: 842–848.

Schuz, J., Bohler, E., Berg, G., Schlehofer, B., Hettinger, I., Schlaefer, K., Wahrendorf, J., Kunna-Grass, K., and Blettner, M., 2006a, Cellular phones, cordless phones, and the risks of glioma and meningioma (Interphone Study Group, Germany). *Am J Epidemiol* 163: 512–520.

Schuz, J., Jacobsen, R., Olsen, J.H., Boice, J.D., Jr., McLaughlin, J.K., and Johansen, C., 2006b, Cellular telephone use and cancer risk: Update of a nationwide Danish cohort. *J Natl Cancer Inst* 98: 1707–1713.

Scientific Committee on Emerging and Newly identified Health Risks (SCENIHR). Possible effects of Electromagnetic Fields (EMF) on Human Health. 21-7-2001.

Slattery, M.L., Edwards, S.L., Caan, B.J., Kerber, R.A., and Potter, J.D., 1995, Response rates among control subjects in case-control studies. *Ann Epidemiol* 5: 245–249.

Stang, A., Anastassiou, G., Ahrens, W., Bromen, K., Bornfeld, N., and Jockel, K.H., 2001, The possible role of radiofrequency radiation in the development of uveal melanoma. *Epidemiology* 12: 7–12.

Szmigielski, S., 1996, Cancer morbidity in subjects occupationally exposed to high frequency (radiofrequency and microwave) electromagnetic radiation. *Sci Total Environ* 180: 9–17.

Takebayashi, T., Akiba, S., Kikuchi, Y., Taki, M., Wake, K., Watanabe, S., and Yamaguchi, N., 2006, Mobile phone use and acoustic neuroma risk in Japan. *Occup Environ Med* 63: 802–807.

The Royal Society of Canada, 2004, *Recent Advances in Research on Radiofrequency Fields and Health: 2001–2003, Report.* Ottawa, Canada.

Thomas, T.L., Stolley, P.D., Stemhagen, A., Fontham, E.T., Bleecker, M.L., Stewart, P.A., and Hoover, R.N., 1987, Brain tumor mortality risk among men with electrical and electronics jobs: A case-control study. *J Natl Cancer Inst* 79: 233–238.

Tynes, T., Andersen, A., Langmark, F., 1992, Incidence of cancer in Norwegian workers potentially exposed to electromagnetic fields. *Am J Epidemiol* 136: 81–88.

Tynes, T., Hannevik, M., Andersen, A., Vistnes, A.I., Haldorsen, T., 1996, Incidence of breast cancer in Norwegian female radio and telegraph operators. *Cancer Causes Control* 7: 197–204.

Vrijheid, M., Cardis, E., Armstrong, B.K., Auvinen, A., Berg, G., Blaasaas, K.G., Brown, J., Carroll, M., Chetrit, A., Christensen, H.C., Deltour, I., Feychting, M., Giles, G.G., Hepworth, S.J., Hours, M., Iavarone, I., Johansen, C., Klaeboe, L., Kurttio, P., Lagorio, S., Lonn, S., McKinney, P.A., Montestrucq, L., Parslow, R.C., Richardson, L., Sadetzki, S., Salminen, T., Schuz, J., Tynes, T., and Woodward, A., 2006a, Validation of short term recall of mobile phone use for the Interphone study. *Occup Environ Med* 63: 237–243.

Vrijheid, M., Deltour, I., Krewski, D., Sanchez, M., and Cardis, E., 2006b, The effects of recall errors and of selection bias in epidemiologic studies of mobile phone use and cancer risk. *J Expo Sci Environ Epidemiol* 16: 371–384.

Warren, H.G., Prevatt, A.A., Daly, K.A., and Antonelli, P.J., 2003, Cellular telephone use and risk of intratemporal facial nerve tumor. *Laryngoscope* 113: 663–667.

Chapter 4

Cognitive Effects of Electromagnetic Fields in Humans

Alan W. Preece

ABSTRACT

Electromagnetic fields interact with human tissue in a number of ways, depending on power level and frequency, and have been long suspected by some to give rise to harmful effects. In particular, the use of a mobile phone against the head has aroused suspicions of various cognitive effects. Accordingly, there have been a large number of studies of behavioural effects from ELF, RF and microwave exposure, mostly as provocation experiments. This chapter discusses the ways and means of doing this, the confusion of physiological effects variously observed and the problem of inconsistency of results. In animal studies, exposure levels have been sufficient to elicit responses, which largely appear to be related to heating effects, whereas similar studies are not ethically easy in humans. Accordingly, many of the studies have sought to explore non-thermal responses, mediated either through induced currents in neurological tissue or biochemical responses directly in cells. Effects reported for ELF tend to be dissimilar to those for RF and this may be more to do with the physical interaction with tissue than anything more fundamental. It is the possible existence of non-thermal effects that has largely been considered here and also the problem of studies in children who have been urged to restrain

A.W. Preece Bristol Oncology Centre, Horfield Road, Bristol, BS2 8ED, UK, e-mail: A.W.Preece@bristol.ac.uk

J.C. Lin (ed.), *Advances in Electromagnetic Fields in Living Systems, Health Effects of Cell Phone Radiation, Volume 5*,
DOI 10.1007/978-0-387-92736-7_4, © Springer Science+Business Media, LLC 2009

mobile phone use. Some important consequences have been observed such as the effect of phone use on driving, but largely the conclusion of detailed provocation studies is that short term or acute exposure to ELF or RF within the established guidelines may not be a hazard to humans, which suggests that the increasing use of RF technology may not itself be harmful.

1. INTRODUCTION

Considerable interest in the effects of RF on human cognitive behaviour has arisen from the dramatically growing use of mobile phone technology that uses both radiofrequency energy transmission to convey intelligence and also involves holding the source of energy close to the head. This is dictated obviously by the presence of the speech and hearing organs in the head and the need for a compact device that can convey speech to the ear and detect sound from the vocal system of the user, whilst communicating with other devices or with relay systems (base-stations). The technology as far as the public is concerned is new in that it has rapidly evolved to personal use over the last two decades. Furthermore, the physical energy is only detectable with instruments and is therefore to the public an unknown quantity which is difficult to deliberately avoid. This is exactly as the situation with ionising radiation, which has unavoidable natural levels and avoidable levels associated with industry and medical uses. Unfortunately, radiofrequency 'radiation' has a synonymy with ionising radiation, which is known to be harmful and therefore radiofrequency also has come under suspicion as a possible aetiologic agent. There has been some reinforcement of this suspicion because of reports of headaches, stress effects and general sensitivity to the radiated field, certainly when the complainant is aware of the operating source of radiofrequency. The concept of possible harm caused by holding a small powerful source of radiofrequency energy against the head has reawakened research interest in the interaction of electromagnetic fields with the brain and with neurological systems in general. The impact of mobile phone technology on economic growth, commercial profit and political development is so enormous that there is also a suspicion by some of the public that politics, profit and economy may rate higher than health and safety, and that commercial pressures may bear unfairly and unethically (Carlo and Schram, 2001) on research effort. Such general fears and a degree of mistrust of government and institutions have been highlighted by another author (Burgess, 2003) who feels that as a consequence there is over-reaction by certain sectors of the government in a move to allay public anxiety and mistrust.

This chapter is intended to look at the current research that is examining some of the possible interactions of electromagnetic fields with human cognition. Although much is known about neurophysiology, cognition itself is such a complex topic that understanding neurophysiology and its effects on behaviour and cognition as well as understanding the physics of electromagnetic field propagation and tissue interaction, and engineering technology, is a daunting if not impossible task. Nevertheless, there is sufficient research to be able to consider whether there is a plausible link between physiological effects and cognitive changes and even measurable electrophysiological changes resulting from electromagnetic exposure. There have been a number of good reviews of the evidence, particularly regarding mobile phones, indicating that there is research of variable quality and equally variable results. The tendency has been for quality to improve with time as the need for tight quality control, for good design and careful execution has been imposed by bodies supplying funding and by referees reviewing potential publications. In undertaking, reviewing or even understanding the outcomes of this research, there are key aspects involved in the design of studies such as level of exposure from either handset or cell-phone mast, the details of and uniformity of subjects, as well as consideration of important aspects such as statistics and experimental integrity, i.e. whether double-blind or open. There has been no uniformity of project design and there are no specific rules or guidelines for undertaking human research other than those imposed by research ethics committees. As a result there is a hodgepodge of conclusions to be drawn. In UK, a review in 2000 by the Independent Expert Group on Mobile Phones (IEGMP, 2000) led recently to the launch of human volunteer studies under the guidance of the Mobile Telephones Health Research programme (MTHR, 2004). In this programme a series of targeted studies were designed and commissioned using standard hand-sets calibrated and checked by the National Physical Laboratory, Teddington, UK, and using standard exposure protocols, but aimed at a range of outcome measures designed to replicate previous findings. The protocols were agreed by contract between the researcher and the MTHR committee and the conduct monitored by a sub-group carrying out monitoring visits. The provocation studies were accompanied by international epidemiological studies, and by base station research, largely targeted at the question of hypersensitivity. This research is only recently concluded and not all reports are finalised, but some of the specific findings are discussed later.

2. DESIGN ASPECTS

2.1. Exposure Levels

The human body operates in a relatively narrow physiological range of body temperature, and levels of electrolytes and nutrients in the blood, and it has powerful feedback mechanisms via the hypothalamus to maintain the homeostatic levels, or the *milieu intérieur*. This includes core temperature, which is normally stabilised within very narrow limits and adjusted by control of heat loss from the skin together with brain blood flow. Therefore, it is possible to detect alterations of blood flow to use as a surrogate for temperature effect. Certainly radiofrequency, if powerful enough, can affect temperature by overwhelming the stabilizing mechanism, but there remains a question of whether there are non-thermal interactions as well as the thermal response. At microwave frequencies, the main mechanism of energy deposition is by induced relaxation losses; at lower frequencies below about 200 MHz, conduction losses are also present but always resulting in heating, although at low field strengths this may be undetectable. The question is therefore whether the heating is sufficient to induce a physiological response, and then if not, whether a non-thermal mechanism is present.

2.2. Exposure System

Most recent research has used, largely because it is convenient and easy to produce, the maximum exposure produced by a particular model of phone where the power saving and power control mechanisms are bypassed to give maximum output. This is normally well within guidelines and rarely are experiments designed to produce the maximum recommended public or occupational levels required by the respective legislations. One of the problems with the former method is that two different phones from the same manufacturer can generate SAR levels in the head that differ by a factor of 15, for example, the Nokia 9300 at 0.07 W/kg compared with the Nokia 6086 at 0.15 W/kg (Mobile Phones, 2007), even though the higher source is still well within guidelines (2 W/kg averaged over 10 g in Europe and 1.6 W/kg averaged over 1 g in USA and most other countries). To study the maximum level recommended by the guidelines requires a specially designed exposure antenna and power source, since by definition a commercially available handset should be designed to give exposures lower than the guidelines. To make the cognitive testing as realistic as possible, it is desirable to make the use of the phone as similar as possible to normal usage. This is difficult when the hands need to be free and involved in the cognitive response such as using a mouse, tracking device or a keyboard. The use of a hands-free kit is also undesirable since the exposure to the head will not be realistic, so this usually needs a harness or headset of some kind. At a simple level it is quite possible to simulate a phone operating at the recommended maximum level by constructing an electrical analogue of the phone (see Fig. 1) and supplying power from an external source. This method of use is illustrated in Fig. 2 and shows a simple design easily adapted for different phone designs. The MTHR studies used a more consistent and accurate arrangement (Fig. 3), since it is obvious from Fig. 2 that the head to antenna spacing will be a variable depending on exact head harness placement.

The design of an exposure system is extremely important since this affects not only the absolute level of exposure, but also the spatial distribution. Unfortunately, there is no standard design of phone and no simple way to compare a situation designed by one laboratory compared with another. Additionally, small differences in positioning of the source with respect to the head also alter both the level of heating and the spatial distribution. It is in this context that the unsuitability of animal models shows itself except for studying whole organ or whole body exposure effects. This is because at high frequencies the level of penetration of radiofrequency into lossy (highly conductive) animal tissue is small. For example, the plane wave penetration ($1/e^2$) into muscle (or brain) at 900 MHz is of the order of 30 mm, reducing to about 20 mm for near field sources (Preece et al., 1987). The simplest exposure system is a real phone mounted in the normal use arrangement in a plastic cradle attached to a non-metallic headband. Most phones can be computer controlled to

Figure 1. Model of an analogue 900 MHz using a sheet metal ground plane and simple quarter wave antenna pragmatically trimmed to resonance in proximity to the head. The soft ear muff locates the phone over the pinna. Since there is no battery or power module, heating from other than radiated power is minimal.

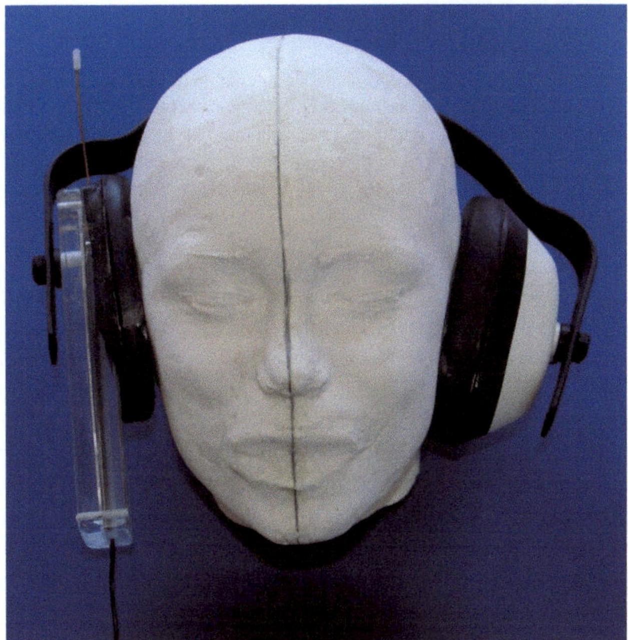

Figure 2. The simulated phone in position on a non-metallic ear defender. The plaster cast is that of a 10-year old used to model the acrylic shell for a realistic head phantom of a child (cast and shell produced by the mould room of the Oncology Centre, Bristol, UK).

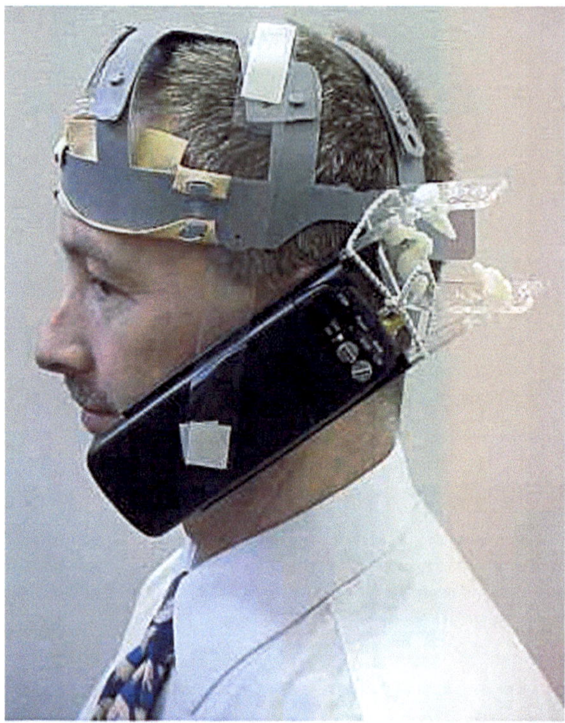

Figure 3. The MTHR setup used for a GSM phone simulation. In this case the phone is more controlled in position, and is a self contained unit transmitting GSM, or constant wave, or transmitting into an internal dummy load. This means that the total power consumption will be constant, although there may be slightly more power dissipated in the phone as a result of the power deposited in the load as opposed to that radiated in the active state (Photo courtesy of Phil Chadwick).

determine power-level and frequency. The shortfall in this approach is that it is not easy to conceal the operating state from the subject and experimenter and would normally require an independent researcher to follow a randomisation table out of sight of the person managing the subject. This approach was refined in the MTHR studies where a conventional phone was modified to produce GSM, constant wave or sham output under the control of an encoded binary switch with sufficient code selections to allow the complete 'blinding' of the phone status. In this case the major problems in the design were battery life (some cognitive sessions can last more than one hour), battery pack audible pulsing in time to the 217 Hz modulation, and interference with the computer monitor as raster modulation or audio modulation. The approach from the Zurich group is to use slot antennas at about 11 cm from the head (Achermann, 2000) for some of the sleep and cognitive studies. This will give a higher SAR and more uniform exposure (Schmid et al., 2007) but it could be argued that such exposure does not represent the true state of phone use with its highly localised SAR and rapid fall-off in energy.

2.3. Statistical Power

There are intrinsic and natural variations in any biological parameter and therefore a study needs to be powerful enough and sufficiently protected against operator or subject bias. A scattergun approach to testing, or 'data trawling', is very likely to throw up a chance association, and therefore good design requires an a priori hypothesis, sound randomisation and careful choice of subjects to reduce the intrinsic variation in the subject material. With any cognitive testing, a single set of measurements such as of choice reaction time in response to a stimulus can yield multiple outcomes. In this example, there is mean reaction time and accuracy. With higher cognitive loads such as the verbal memory, there is response time, accuracy and 'intruders' that can be tested from a single trial. To deal with this situation, it is normally necessary to use a Bonferroni correction for multiple outcomes. The Bonferroni correction is applied with the objective of excluding Type 1 errors, that is to say, to reduce the likelihood of falsely attributing significance to effects that are not biological but have occurred by chance. If numerous statistical tests are carried out on a data set, a number of them will yield calculated probabilities less than 0.05 by chance alone. To counter this, the Bonferroni correction adjusts the calculated significance in proportion to the number of comparisons. However the procedure has the disadvantage of increasing the likelihood of Type 2 errors, that is to say, dismissing effects as non-significant, which are biologically genuine. It is possibly better, particularly in a replication study, to consider an a priori hypothesis and focus on a particular outcome or alternatively, where there are, for example, several reaction time tests to consider grouping these. If the significance appears to increase, then it is likely that a genuine phenomenon has been observed, if it decreases then the probability was that the single significant result is truly chance.

2.4. Corroborative Evidence

Any cognitive process will have an associated neurological phenomenon and ideally the association of these effects will lend support to the demonstration of cognitive effects – for example, reaction involves electroencephalographic (EEG) effects that can be detected by electrophysiology, such as the P60 and P300 cortical signals. Sleep effects can be characterised by a range of physiological and EEG measurements, and stress effects by cardiac responses. In addition, cerebral blood flow is a physical measurement that can demonstrate neurological responses.

2.5. Acute vs. Continuous Exposure Effects

Neurological responses are normally quite fast, so it has been usual to design protocols for acute exposure, and only to wait 10 min or so to reach a steady state of exposure. Consideration has to be given also to the effects of chronic exposure, where adaptation of cumulative effects could occur. This also includes consideration of whether subjects are regular or heavy phone users where adaptation or chronic changes could have happened. This is a particular concern for studies in electro-hypersensitivity where the subject feels it may take some time for effects to occur and also a considerable time for effects to abate after stimulus withdrawal. At least one provocation study has arranged for exposure of real or sham to occur on different days in the subjects' own homes. This works well for mood and emotion effects, which involve self reporting, usually by self-completion questionnaire.

2.6. Ethical Aspects

Apart from the need to maintain an ethical approach to integrity of research, the question of exposing volunteers to a physical source of energy needs to be considered. By definition, a provocation study involves a challenge with an agent that is under research because it is unknown whether it is harmful or not. It is necessary to inform the subject what is known about the risks and at what level these might occur. This is to allow fully informed consent so that the subjects can assess whether the risks of the procedure are worthy of the interest in an outcome. The subject must also be made aware of exactly what will be done and what the subject has to contribute in terms of time and commitment. Some cognitive studies can take hours to complete and require several sessions at precise time intervals, as well as requiring abstinence from drugs, alcohol, stimulants and perhaps mobile phone use. This is a major commitment by the volunteer so that the question of payment arises for loss of time, freedom and inconvenience as well as 'out of pocket' expenses. Such payments are a problem for ethics committees who have to ensure that it is at a level of genuine compensation and does not become an inducement. In our own case, with student volunteers, payment was not permitted so by way of compensation for time given up, a voucher for a meal for two at a local restaurant was offered after completion of the trial as a way of saying 'thank you'. For children, a book token was offered, again after completion by all subjects. The form and level of such payment is a matter of negotiation between the researcher and the ethics committee and varies enormously. Technically, such research needs altruism by the volunteer. Certainly volunteers are a problem particularly with research weary students and with hypersensitive individuals who anticipate real harm.

Consent becomes exceedingly difficult in children where it is assumed that they may not be competent to give fully informed consent below the age of 16, or possibly 18 years depending on the country. It is important to distinguish between therapeutic and non-therapeutic research. In the former there is intent to benefit the patient and the distinction is seen by research ethics committees as critical. Not all human research ethics committees will accept that exposure of a child for research purposes (non-therapeutic research) at levels below the relevant exposure guidelines is justified or acceptable, even though the results may be important to the population of children as a whole. There is absolutely no doubt that the mobile phone is a great aid to social development of children and an extremely powerful security tool. Thus it is important to know whether there is potential for detriment in the use by children (IEGMP, 2000). The reluctance of research ethics committees to consider non-therapeutic research in children can be a stumbling block to designing a research protocol. However, it is possible to argue that children now voluntarily expose themselves to mobile phone radiation and provided they are phone users, the levels are no greater than the child is normally exposed to, so that if they wish to take part, there should be no moral or ethical objection.

2.7. Test System

With few exceptions each of the cognitive studies used different test systems, although commonly these are computer based. For example, the Finnish and Swedish groups have in the past used in-house software that requires the tester to manually load the test sequences. This makes double-blind conditions difficult if not impossible. There are, however, fully automated systems that require completion of one task before presenting the next automatically in a pre-arranged time sequence. Such designs are important for consistency between subjects. In our own case, we were provided with commercial cognitive testing software (Wesnes, 2006) designed for studying drug effects in healthy and ill patients and which had been extensively validated. This required no intervention by the tester, and the software would only allow tests in the right sequence and after the appropriate training sessions at the right intervals. Such standardization potentially could lead to better inter-comparison of results.

The assessment of any changes in cognitive function depends on the ability to measure it, and since many aspects of cognitive function are difficult to observe, then it is difficult to identify changes without the application of appropriate tests. Furthermore, such tests should be relevant to everyday activities. These are attention, decision making, memory, reasoning and co-ordination. Cognitive function is assessed by asking individuals to complete tasks involving the function under investigation. Thus, if memory is assessed, the test involves memorizing information and the outcome measure reflects how well the information was memorized. There have been thousands of cognitive tasks developed in the past, but because they are so diverse it is difficult to generalise results. Automated testing using well-designed commercial software allows easy test administration by non-specialists such as engineers or even self-administered. This improves sensitivity and allows difficult parameters such as reaction time or retrieval time from memory to be accurately quantified. In addition, each step of the procedure can be interlocked so that each test has to be completed before the subject can move onto the next. The one particular cognitive condition not amenable to testing is electrical hypersensitivity. This involves mood and emotion and can only be adequately assessed by self-reporting questionnaires.

2.8. Subjects

These are the greatest source of variation in any testing. Cognitive function declines with age. On the one hand, typically the composite score (time to respond) of attention, episodic and working memory increases by 50% from age 18–25 to age 80 (Williams et al., 2007). On the other hand, motor control such as finger tapping rates at age 6–8 were 920 ms, age 10–12 were 720 ms and adults 460 ms (Wolff et al., 1998), which

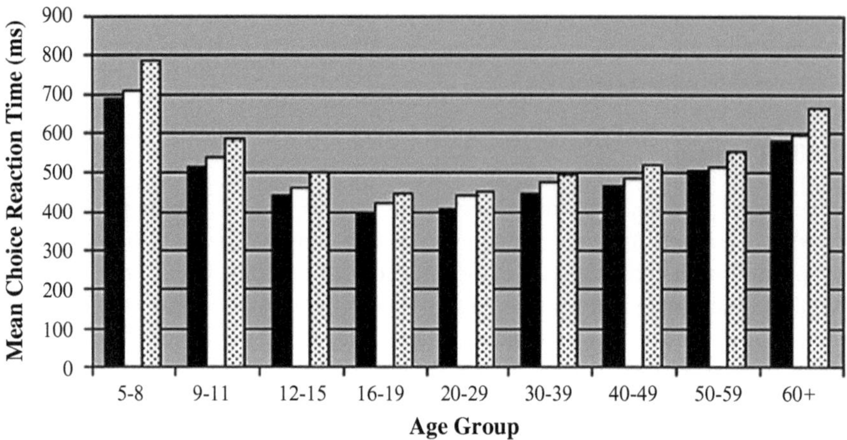

Figure 4. Mean choice reaction time – change with age (Williams et al., 2007).

is a 50% decrease with increasing age. Similarly the health status of an individual has a marked effect and this may vary from one test day to the next. The main effect is a large increase in the variability of response particularly with untreated neurological illness. Gender, it appears from the literature, has little effect on performance, but in any case this would be automatically controlled for by use of repeated measure testing (Fig. 4).

It is, therefore, possible to set out to design a study to test the effects of exposure on cognitive functioning in one or more of a myriad of ways. It is also possible to use similar or identical protocols to study ELF, RF and microwave exposure, although the interpretation has to take into account the interaction with the body and the degree of absorption.

3. COGNITIVE EFFECTS (COGNITION)

The strict definition of cognition involves knowing or perceiving, in contrast to emotion and volition, and therefore involves perception and sensation. Cognition describes the mental processes involved in attending to, acquiring, storing, retrieving and manipulating information. Cognitive function underpins all of our everyday activities, and intact cognitive processes are essential to physical, mental and emotional well being. These can all be disturbed by environmental and chemical influences to produce an alteration in function that could, depending on the context, be a problem or even a hazard. Consideration of cognitive effects is often extended and refined to be a study of cerebral, neurological and physiological processes such as memory, learning, reaction and sleep as well as the electrophysiological processes involved in heart rate, reaction speed, sensation and hearing.

There are four quite distinct elements:

1. Motor control
   (a) Ability to co-ordinate eye and hand movements, and can be tested by *tracking tests*
   (b) Overall control of body sway, checked by examining *postural stability*
   (c) Capacity to rapidly execute simple repetitive manual operations, which can be measured as *tapping rates*
2. Attention
   (a) Alertness or power of concentration are the primary stage of information processing, and can be measured as *simple reaction time*
   (b) Alertness and power of concentration added to stimulus discrimination and response organisation is examined *by choice reaction time*
   (c) Sustained attention and intensive vigilance combined with the ability to ignore distraction is measured by *digit vigilance*
3. Memory
   (a) Sub-vocal rehearsal of digit sequences as a component of working *memory* can be shown by *numeric memory* tests
   (b) Visuo-spatial sub-loop of working memory is examined as *spatial memory* using patterns or shapes
4. Episodic secondary memory
   (a) Ability to store and recall verbal information, capacity for uncued retrieval of words shown by episodic secondary verbal recall is testable by *word recall*
   (b) Ability (speed and sensitivity) to discriminate novel from previously presented words is episodic secondary recognition requires *word recognition*
   (c) Ability to discriminate novel from previously presented pictorial information is episodic secondary non-verbal visual recognition and can be tested with *picture recognition*
   (d) Ability to discriminate novel from previously presented faces shown by *face recognition*

In addition, the term 'mood' has been used in the context of cognitive effects and is reflected by any relatively short-lived emotional state. However, mood and emotions are not aspects of cognitive function but are subjective experiences, which can only be assessed by self-report. Mood and emotion can influence cognitive function, and thus need to be assessed particularly in relation to hypersensitivity states.

Some or all of these features can be involved in the very difficult field of electromagnetic hypersensitivity and the consequent ills of allergy, headaches, migraine, tension and depression. All the above effects have been considered in the power frequency (50–60 Hz) and the radiofrequency (1 MHz–5 GHz) ranges, and exposure guidelines have been designed to control the various effects. In its simplest form, at low frequencies, the guidelines are designed to protect against the physiological effects of induced currents in the body (NRPB, 2003). These are based on using a safety factor applied to the minimum detectable level, and over time the safety factor has tended to be increased, particularly for uncontrolled public exposure. The minimum detectable levels have also been lowered as more studies contribute effects, always maintaining the belief that the minimum detectable level is both harmless and reversible, and therefore the exposure level is even more so.

For radiofrequency exposure, the assumption is that RF is not electro-physiologically active, and it is only the thermal effects that have to be considered. The heat distribution depends on the frequency since the human body is varyingly conductive at different frequencies, affecting the penetration, and therefore heat distribution. At low RF frequencies, it is usually whole body heating with consideration of body constrictions that is important. Various studies suggest that 1°C rise can be detectable, and exposure limits look to limiting to a safety factor on this. Frequently, this is a factor of 10 or 50. At higher RF frequencies, localized effects are important because the penetration of RF is very low and heating of superficial structures is the limiting factor. Additional problems such as lack of heat clearance in the eye and its size lead to the assumption that particular frequencies such as the 2.45 GHz from microwave cookers are a special hazard, notwithstanding the animal experiments that suggest the guidelines are adequate. The NRPB document sets out the evidence from research that now leads to an endorsement of the ICNIRP guidelines (ICNIRP, 1998), thereby lowering the UK recommended levels for occupational and public exposure to about one quarter of the previously accepted limits, and for the first time including an additional factor of five for public restrictions. The effect of this is that before about 2003, studies were carried out with power frequency exposures up to 1.8 mT. After 2003, this has not been ethically acceptable and limits for human volunteers are now normally restricted by research ethics committees to 100 μT. For power frequencies, the adoption of these recommended public exposure limits has not acknowledged any possible associations with childhood leukaemia. The International Agency for Cancer Research has classified power-line frequency exposure above 0.3–0.4 μT as a possible carcinogen in children, and this has been discussed in the context of the Health Protection Agency (UK). Not all bodies or scientists accept these observations but, nevertheless, the onset of cognitive effects are therefore associated with levels considerably above these and can be considered by some to represent a cancer risk, albeit in a restricted group of humans. The possibility that exposure to mobile phone base stations as sources of RF inducing cognitive change has been researched, particularly with respect to behavioural effects and hypersensitivity. In this case, except for occupational exposure of riggers and others working 'live' on antennas, the effective tissue levels will be very small. Nevertheless, the background levels found in towns (usually ~1 V/m or that order) have been tested in the laboratory because the only problem is the possibility of challenging 'sensitive' subjects with something they would regard as 'noxious'.

4. POWER FREQUENCY

Cognitive effect has under its heading memory, reaction, spatial awareness, vision, hearing, speech, behaviour and sleep. These are all in some way involving electrical processes in neurological tissue and cannot be separated from biological electrical effects in the EEG, ECG and EMG. A party trick from a century ago was the demonstration of magneto-phosphenes, which are seen as a flashing light when the head is exposed to 10–40 mT magnetic fields. Threshold values for magneto-phosphenes are a function of magnetic field frequency as well as colour and intensity of the background illumination. A typical sensitivity maximum demonstrated by Lövsund et al. (1979) is 20–25 Hz. In vivo electrode recordings show electrical responses in the frog retina exposed to the same type of fields. Retinal activity, induced by the fields, was recorded from the ganglion cell layer by means of microelectrode technique. Interest in this phenomenon has been awakened recently because of its relevance to functional MRI and has been reviewed by Taki et al. (2003). Their estimates of the field at threshold (about 5 mT at 25 Hz) varied from about 5 to 25 mV/m, and current density of ~18 mA/cm^2 depending on the model used and the estimate of retina conductivity.

There is thus much confusion in the results and controversy whether alternating or pulsed magnetic fields have an effect and relevance in real life, i.e., in environmental levels below the millitesla range. Weaver and Astumian (1990, 1992) produced a mathematical model that showed that the possibility of meeting the requirement for detection of electrical signals could be met if the signal to background plus noise (Johnson or thermal) was ≥1. Their argument was that this was not likely to be achieved with environmental levels of magnetic field. In contrast, Kaune (2002) argued that the thermal noise in a specific power band (e.g. 50 or 60 Hz) may be insufficient to mask the effect of external applied fields thus allowing the small internal fields that result from the relatively high conductivity of human tissue to be sufficient just across membranes. There have been both supporters (Vincze et al., 2005; Bier, 2005) of this observation albeit with a different interpretation of the data, and critics such as Adair (2003), who felt the case was still a speculation and an oversimplification. The importance of these arguments is that when extended to radiofrequency and microwaves, since the higher frequencies (above a few kilohertz) are not 'biologically' active, an additional mechanism becomes essential. This is that a non-linear element in tissue is needed to decode any pulsatile elements (including amplitude modulation) to derive frequencies that are 'biologically active'. Such a non-linear system is the cell membrane. However, at high frequencies the high capacity of the membrane presents a low impedance barrier, so that developed voltages are small, making it increasingly implausible that direct electrical effects would be significant at high frequencies.

It remains, however, that cognitive effects, depending on the frequency, are likely to come from one of the two possible mechanisms:

1. Direct electrical action alters or interferes with normal electrophysiology, for example by adding in electrical noise to a normal resting potential,

thereby exceeding a threshold where self-triggering will occur. This would inevitably require rectification at high frequencies but, in the power frequency range, the response time of some neurological systems is quite adequate to respond to the alternating component (e.g. electromyographic (EMG) voltages are in the kilohertz range). Conductivity of tissue increases with frequency, particularly across the lipid cell membrane, thus providing a larger potential difference at low frequencies. Heating effects that occur largely at high frequencies may lead to increased metabolic rates, which also in turn could increase blood flow. In addition, there are mechanisms whereby even small increases in temperature will cause heat shock protein (HSP) generation in vessel walls. HSPs in vessel walls can induce vasodilatation (Beall et al., 1997; Jerius et al., 1999), further amplifying the effect, and moreover such effects would be seen to persist, even for some hours.

2. Any cognitive effects not due to heating should either show simultaneous or consequential alterations in the electrophysiology. This is the basis of much research into cognitive effects simply because electrophysiology can be directly measured, whereas the cognitive outcomes (executive effects) require controlled observation of more complex tasks such as response time, memory or accuracy. These are inevitably subjected to individual variation through training, age or health. For example, reaction times should be demonstrable in the P100 (positive peak at 100 ms after an auditory or visual stimulus) and P300 (similarly the positive peak at 300 ms) responses, which show small variation between subjects. Event-related processes and evoked potentials may be used to measure the timing of neuronal activity underlying sensory and cognitive processes, or direct observation of intrinsic activity may be used instead. Evidence for electrophysiological effects comes from heart rate variability (HRV) studies. Baldi et al. (2007) in a pilot study showed that HRV could be affected by the transition from exposed to non-exposed condition generated by a 3 Hz square wave modulation of a 50 Hz signal. Here the field level was small (~5 μT) at the level of the heart, but coming from a source of 1.6 mT at floor level. Similar effects have been reported in occupational exposure of RF welders (Wilen et al., 2007) who were also exposed to pulsed low frequency fields. A problem here is that occupational exposure will tend to be non-uniform and highly variable and such phenomena are better analysed in controlled exposure environments.

Many studies of cognitive and physiological effects have been carried out over the last century, with systematic studies in the last two decades reviewed by Paneth (1993). Universally, these studies were neither double blind nor critically analysed, and tended to be at relatively high fields, or were epidemiological studies with limited dosimetry. Paneth noted no particular outcomes other than at very high fields of the order of 130 kV/m, although there were apparent effects on circadian rhythms. This led to the postulate of entrainment of cyclical patterns in the pineal (mediated by melatonin) by electromagnetic fields. The involvement of such complex mechanisms denies the Occam principle (basically that the simplest explanation is usually the best and probably the correct one), but in particular Wever (1979) reported lengthening of the circadian pattern by remarkably small (2.5 V/m) electric fields.

In one of a series of studies by Graham (1987), an effect of slowing of heart rate was noted, but a later replication study (Graham et al., 1994), using a between-subject design in 18 males, tested whether the exposure levels at which the greatest effects occur differ for different endpoints. Three matched groups had two 6 h exposure test sessions with different mixtures of field (low group: 6 kV/m, 10 μT; medium group: 9 kV/m, 20 μT; and high group: 12 kV/m, 30 μT). The study was performed double-blind, with exposure order counterbalanced. Significant slowing of heart rate, as well as alternations in the latency and amplitude of event-related brain potential measures derived from the EEG, occurred in the group exposed to the 9 kV/m, 20 μT combined field (medium group). Significant decrements in reaction time and performance accuracy on a time estimation task were observed only in the low group. This rather confusing result suggested that it is a particular combination of E and H that affects particular end points. The purpose of this study was to reproduce and extend an earlier investigation of the effects of human exposure to combined, 60-Hz electric and magnetic fields representing a more normal occupational exposure for electricity workers, i.e. a high electric field and moderate magnetic field. The authors concluded that these data indicate that changes in exposure level (rate of change parameter) may be more important than duration of exposure for producing effects in human beings. The outcomes were again slowing of heart rate and changes in reaction time, mainly related to accuracy.

More recently, the studies have been constrained by ethics approval and the quality of review. It has become usual to keep strictly within public exposure guidelines and use very much more reproducible exposure conditions together with double blind randomized protocols. These recent studies were reviewed by Cook et al. (2002) and updated by Cook et al. (2006). With reference only to the ELF studies, they discussed the event-related potentials (ERP) separately from the executive effects, although logic would suggest that these should be inextricably linked. In particular, the P300 and the N100 (negative peak at 100 ms) responses should be direct measures of processing speed. Unfortunately, the highest magnetic fields used were restricted to 100 μT or less and results were conflicting. Other effects on EEG waveform such as increased alpha and decreased delta and theta have been reported, but these are harder to understand than ERPs since the roles of alpha and theta waves are more poorly understood. One of the possible caveats for these kinds of studies is that collection of waveforms from the head, where the retrieved signal is of 2–10 μV, in the presence of 50/60 Hz magnetic fields of the order of 1 mT, is exceedingly difficult and subject to artefact. Contact potentials and polarization effects are a constant problem, as is amplifier saturation. In general, there is some evidence of the existence of magnetic field effects, which are ephemeral at low intensities, but become increasingly certain at higher intensities, finally being reproducible and overtly an electric field effect at 1 mT or more as a result of induced currents passing through relatively high-resistance membranes.

The debate is, therefore, whether similar electric phenomena (non-thermal) can occur at communication frequencies or whether observed effects are purely a thermal phenomenon.

5. MOBILE PHONES

5.1. Cognitive Effects and EEG

The effect of radiofrequency on behaviour has been extensively studied in animals and reviewed (D'Andrea et al., 2003). This mainly examined the effects in rats and mice, non-human primates, which had been extensively researched but also touched on some of the effects in humans. In doing this, the authors looked at six human studies and a recent review. These can to some extent be compared with the much larger body of animal data referenced and described in this work. Note that since these animal studies were carried out at 2.5–6.7 W/kg, there clearly may be thermal effects induced at these power levels. The conclusion from animal work is that a temperature rise of 1°C is able to disrupt learning behaviour. Of the human studies available (the first six in Table 1) at the time, the conclusion was that the evidence that mobile phone irradiation can affect human behaviour is weak.

Table 1. Some of the recent studies on human cognitive function

Study	Year	Significant finding
Preece et al., 1999	1999	Reduced reaction time
Koivisto et al., 2000a	2000	Reduced reaction time
Krause et al., 2000	2000	Alteration of EEG during cognitive load
Koivisto et al., (2000b)	2001	No subjective symptoms
Lee et al., 2001	2001	Improvement of trail-making task
Edelstyn and Oldershaw, 2002	2002	Facilitation of digital and spatial cognition
Haarala et al., 2003	2003	Changes in cerebral blood flow
Haarala et al., 2003b	2003	Unable to reproduce cognitive effects
Smythe and Costall, 2003	2003	Improved memory in males not in females
D'Costa et al., 2003	2003	Alterations of alpha and beta EEG bands
Curcio et al., 2004	2004	Improvement of simple and choice reaction
Haarala et al., 2004	2004	No effect on short term memory
Maier et al., 2004	2004	Reduction in auditory cognitive performance
Haarala et al., 2005	2005	No effect on children's cognition
Preece et al., 2005	2005	No effect on children's cognition
Besset et al., 2005	2005	No effect of daily phone use
Wilen et al., 2006	2006	Subjective symptoms above 0.5 W/kg
Russo et al., 2006	2006	No effect on attention
Papageorgiou et al., 2006	2006	Decreased EEG in males, increased in females
Keetley et al., 2006	2006	Slowing of simple and choice reaction times
Vecchio et al., 2006	2006	Alters coherence of EEG
Krause et al., 2006	2006	Affects EEG in children
Ferreri et al., 2006	2006	Changes in brain excitability
Krause et al., 2007	2007	Affects EEG but not behaviour
Haarala et al., 2007	2007	No effects on behavioural or cognitive tasks

Interest in effects on the brain is a direct result of provocation experiments on humans. In the earliest study of effects on humans, Eulitz et al. (1998) looked at the electrophysiological response. When the signal was present, the phone signal (GSM) modulated mainly the hemisphere directly exposed to the electromagnetic radiation, but only in combination with task-relevant stimuli detected by changes in the P300 response shown as an increase in spectral power. The first outcome study (Preece et al., 1999) used a battery of computerised tests designed to look for drug-induced effects on cognition during exposure to a simulated GSM and analogue phone signal. Sessions lasted 30 min were balanced for test order, and at least 48 h washout time was allowed between sessions. In that study it was only the un-modulated signal, which was about six times stronger than a comparative modulated signal that decreased choice reaction time. Retrospective measurement of the analogue signal on a head phantom indicated that the SAR of 1.6 W/kg was within the later 2 W/kg standard, although at the time, the UK used the NRPB standard of 10 W/kg head exposure limits. This particular exposure corresponded to a 900 MHz band analogue phone signal, which was still in UK use at the time. That study was rapidly followed by a larger study from Finland (Koivisto et al., 2000a, b), which also showed effects of speeded up response times in the simple reaction time and vigilance tasks. Additionally, the cognitive time needed in a mental arithmetic task was decreased and accuracy was also enhanced. This was using a genuine GSM phone rather than an externally powered dummy phone and twice the power of that in the previous study. A study of memory in the same group suggested that, particularly in the higher task loads, mobile phone exposure speeded up that response also. Functionally, this was supported by the electrophysiologists who showed that RF significantly increased EEG power in the 8–10 Hz frequency band only and altered the ERD/ERS responses in other frequency bands as a function of time and memory task (encoding vs. retrieval). This suggested that the exposure to EMF did not alter the resting EEG per se but modified the brain responses significantly during a memory task. Also, giving support to these first findings, Lee et al. (2001) examined the cognitive abilities of young students, half of which used mobile phones for total times in excess of 175 min. The phone user group appeared in particular to be better at the trail making test, which examines visual and spatial ability. They concluded that exposure to mobile phone RF may have a mild facilitating effect on attention functions, which they considered consistent with previous observations of a facilitating effect on cognitive processing.

Rather than assuming that this is a radiofrequency effect, the possibility occurs that mobile phone users may have benefited from learning to multitask with phone use, or possibly the overall phone use reflects other social or intelligence characteristics. In the same year, Lass et al. (2002) showed a similar effect of improved performance on higher task loads, but to a 450 MHz, 7 Hz modulated signal. Edelstyn and Oldershaw (2002) showed that volunteers exposed to a mobile phone EMF for 15 min showed improved immediate verbal memory capacity, immediate visuospatial working memory capacity and sustained attention. This was evident from the subject's performance on forwards digit span, backwards spatial span and serial subtraction tasks. They concluded that the biological effects of EMF are fast acting rather than a slow chemical or thermal effect, thus further suggesting a direct electrical effect rather than a self-selection or training effect by mobile phone users. A major criticism of this last study using these particular tests is the way the controls were different subjects – a 'between subjects' study that possibly is influenced by inherent biological variability (see Table 2 for example) and therefore difficult to randomise. In addition, the time scale of the effects was different in that the former study concerned chronic effects of habitual phone use and the latter observed acute responses. However, in general principle, these last six studies all seemed to support each other's conclusions, namely that there was an apparent speeding up of reaction times involving more than a simple reflex response, i.e., involving some degree of cortical function. Since they were either single or, in some cases, double-blind, this seemed to indicate that there were effects detectable within the exposure standards. Such effects were not necessarily seen as deleterious, or indeed persistent, but they did attract media attention and also considerable government concern. This was notwithstanding that all the results reported to date had suggested a positive effect of enhanced functioning – it was simply disturbing that any effects had been reported. It was largely as a result of the earliest of these concordant studies that the Independent Expert Group on Mobile Phones was set up late in 1999 under the chair of Sir William Stewart. It reported in May 2000, having taken evidence from a number of sources, to include work in progress and in press. The result was the setting up of the Mobile Telephones Health Research Programme in 2001, the aim of which was to promote targeted research using very controlled design and quality assurance with adequate technical backup.

By 2003, the first of the replication studies (Haarala et al., 2003a, b) in which the earlier study was repeated in two centres, both using double blind techniques, reported. The tests were essentially the same as their earlier studies but this time double-blind, and they concluded that there was no effect on reaction. A replication of the Lee study on phone use by Besset et al. (2005) also failed to demonstrate any changes. This was followed by a study of the effect of a GSM signal from a real phone on maximum power under computer control in children (Preece et al., 2005), and like their earlier study found no effect of GSM on very similar tests to those undertaken by adults. The use of the much more powerful signal from an analogue phone was not permitted, so it is difficult to know if this was a non-replication or simply an underpowered (literally) study. A direct comparison with adults was not possible because children show very different reaction time rates and different responses to accuracy tests compared with adults. Because analogue systems were

Table 2. Difference between adult and children's reaction times (mean of control and exposure conditions) (from Preece et al., 1999, 2005)

	Simple reaction time (ms)	Choice reaction time (ms)
Adults	227	381
Children (age 10–12 years)	301 (33% slower)	485 (27% slower)

by then discontinued, the ethics committee would not permit studies at greater exposure. Coincidentally, Haarala's group (Haarala et al., 2005) had also carried out a study in a wider age group of children, with identical results to the other children's study, of no detectable effect of phone exposure. The response time data could almost be overlaid on each other. These are to date the only children's studies and they badly need investigating further, either with higher powers within the guidelines or with a much larger study.

Adult studies sponsored by the MTHR included one from Russo et al., (2006) using a large (168) sample of volunteers undertaking a series of cognitive tasks previously reported sensitive to RF exposure (a simple reaction task, a vigilance task and a subtraction task). Participants performed those tasks twice, in two different sessions, thereby acting as their own controls. In one session, they were exposed to RFs, either GSM signals or CW signals, while in the other session they were exposed to sham signals. No significant effects of RF exposure on performance for either GSM or CW were found, independent of whether the phone was positioned on the left or on the right side. A different outcome was obtained by Keetley et al. (2006) in comparing the performance of 120 volunteers undertaking 8 neuropsychological tests during real or sham exposure to a real phone set to maximum available radiofrequency power output. Several parameters showed alterations at significance levels of $p<0.05$, and of these, simple and choice reaction times (CRT) showed strong evidence of impairment in direct contrast to studies at other centres. However, performance on a trail-making-task improved, supporting the hypothesis that modulated radiofrequency emissions improve the speed of processing of information held in working memory. This study was almost as large as the Russo study and also restricted the number of variables – a possible requirement to reduce the risk of a type 1 error – and yet showed some characteristics (enhanced reaction times) of the earlier less rigorous studies.

In summary, the early studies are sometimes supported by later and better defined studies, but also include non-replications (for example Curcio et al., 2008), which are difficult to explain in view of the very definite early findings.

5.2. Brain Blood Flow

Human studies diverge from animal ones because the power levels and resultant SAR levels have to be lower to comply with human exposure guidelines, and therefore the behavioural changes noted in animals because of general or local thermal stress do not occur in humans. It is assumed that the impact on the human head is non-thermal. Whether this is strictly true is uncertain. Van Leeuwen et al. (1999) calculated the thermal change in the cortex of the brain from mobile phone exposure. The temperature rise from indefinite exposure (reaching a peak in 6 min) was a surprising 0.11°C, which would normally be sensed by the brain and removed by a generalised increased blood flow. However, thermal control is a function of the hypothalamus, and in humans that area of the brain is relatively unaffected by RF exposure, unlike in small animals such as the rat, because of absorption in the brain tissue. Most of the radiation is absorbed in the outer 20 mm of the cortex. The calculations of temperature rise are supported by direct studies of blood flow by PET scanning (Huber et al., 2005). However, the authors of this study were convinced that it was the complex nature of the mobile phone signal that resulted in the blood flow changes and that modulation was an essential component. Haarala et al. (2003a, b) also observed an increase in regional cerebral blood flow, but these authors ascribed the effect to a response to the audible signal by stimulation of the auditory cortex because the location of maximum relative blood flow did not coincide with maximum SAR. The same group repeated these studies coincidentally with cognitive testing (N-back tests) and again showed increases in relative blood flow, but in more than one area, again ascribing this to neuronal activity not physiological heating.

5.3. Associated Electrophysiology

Mechanistically, the effects on electrophysiology were also confirmed by Krause et al. (2006) in children showing as event-related desynchronization/synchronization (ERD/ERS) responses in the approximately 4–8 Hz EEG frequency band, which was similar in character to the adult responses. Phone exposure transformed these brain oscillatory responses in the approximately 4–8 Hz and also at approximately 15 Hz. Haarala et al. (2007) extended their work with children into a more rigorous study on 52 adults and further examined the effect of a simulated phone exposure (thereby overcoming criticism of clues from power module buzz and battery heating clues) in three sessions for each subject with pulsed, CW and sham exposure. Unlike Keetley, but like Russo, this study yielded no significant effects of mobile phone exposure, whether exposed on right or left hemisphere. Similarly, the same group used an auditory threshold task, which used signals only able to induce a response 60 or 70% of the time to test for interference in this threshold by GSM signals (Cinel et al., 2007). Again no significant changes that could be ascribed to interference with the electrophysiology were found in a quite large study group.

These results seem, therefore, to be exceedingly confusing. Some appear to replicate the earlier studies, with enhanced cognitive effects functioning, a few showing impairment, whilst other well controlled and double-blind studies show no effect. The electrophysiological studies similarly show different or no effects of RF exposure whether pulsed or CW. This suggests that the earlier studies may have been poorly performed, a result of perhaps a Type 1 error, or just chance. There is, however, a small amount of evidence of real physiological effects on the brain that can be demonstrated directly. Following the calculations of Van Leeuwen et al. (1999), which showed that a 0.1°C rise was feasible, other visualisations have been made. Huber et al. (2005) used PET scanning to show that pulse-modulated EMF exposure increased relative cerebral blood flow in the dorsolateral prefrontal cortex on the same side as the exposure. This is forward of the exposure area. In contrast, Haarala et al. (2003a, b) found a decrease in blood flow on exposure on the side irradiated, whereas Aalto et al. (2006) showed both an increase and a decrease in different areas. The main changes were in the left fusiform gyrus in the posterior inferior temporal cortex below the antenna (the suggested source of increased performance in choice reaction time), while an increase in relative cerebral blood flow was seen also bilaterally in the superior and medial frontal gyri. However, in *this last study* the authors concluded – 'The EMF had no effects on reaction times or the accuracy of responses Also, the reaction times and the accuracy of responses behaved fairly similarly as a function of time during both conditions'.

The authors suggest that these results cannot be compared with those of Huber because the source patterns were different. The Huber study used post-exposure PET scanning after exposure to a planar antenna, whereas the Aalto study used a real phone with small antenna and PET scanning contemporaneously with exposure. Both these studies conclude that the blood flow changes are a consequence of alterations of neuronal activity, which will be associated with cognitive effects, but that these particular studies were not interrogating the particular responses. Some indication of possible differences between studies is highlighted in a recent study by Boutry et al. (2007), who examined the resultant SAR from the Zurich (Huber et al., 2005), the Turku (Haarala et al., 2003a, b and associated studies) and the Swinbourne (Keetley et al., 2006) systems and showed that the maximum SARs were 1.02, 0.31 and 0.19 W/kg, respectively, but with different spatial patterns. The surprising result is that it is the Turku study that detected no cognitive effects. This again suggests that the presence or absence of an effect of exposure will be affected by the antenna pattern, i.e. the area of the brain exposed, and by the power. Indeed the earlier study by Preece et al. only showed an effect with the analogue signal that generated a maximum brain SAR of 1.6 W/kg and, in unpublished work, alterations in blood flow could be detected ultrasonically in the middle cerebral artery only during antenna powers of 2 W or more where the SAR would have been nearer 5 W/kg. The relative confusion of results, either showing small effects or no effect, or not capable of being replicated strongly, suggests that in the real world the effect of phone use is marginal, and particularly with more modern efficient GSM or 3G phones, the brain exposure is rarely sufficient to produce a change that is not buried in normal biological noise or normal physiological variation. If the Van Leeuwen results are indeed correct and the maximum change in brain temperature is only of the order of 0.1°C then it is not surprising that test results are ephemeral. This would support the argument that tests should be carried out not with real life phones, but with sources that generate SARs at the ICNIRP or IEEE recommended maximum public levels.

A careful study of 24 patients (Valentini et al., 2007) showed in general little effect of the GSM signal on simple finger tapping tasks, but they were unable to exclude evidence for a trend to facilitation in simple reaction time tasks, which they felt mirrored results form foregoing studies. In considering all available studies in a 'comprehensive review' of neurophysiological effects, the same authors (Valentini et al., 2007) decided that possibly the 0.25 W GSM systems were too low powered for direct RF effects. In particular, the evidence appeared to point to effects on sleep, both immediate and persistent, direct effects on EEG particularly in the theta and alpha bands (also relevant to sleep), and increased susceptibility to transcranial magnetic excitation. They also concluded that the results, even restricting the analysis to those that they described as 'adequate' studies, firmly pointed to the occurrence of signal demodulation. If this is true, then the possibility of membrane depolarisation as a result of RF exposure definitely could exist. The effects were seen at even very high frequencies. Since different phone systems use different modulation patterns, different phone designs affect the SAR distribution, and different researchers use different head mounts for the phones, this may explain the rather ephemeral results of mobile phone studies. They further concluded that simple SAR related changes in cerebral blood flow were an unlikely mechanism because of the lack of direct relationship between the locations of SAR and blood flow changes. There seems to be a need to examine this carefully to determine whether the mechanism is heat induced on a 'micro' scale, or, is a result direct electrical stimulation. This could be done by use of CW signals at maximum permitted SAR (as defined by ICNIRP, HPA-RPD and IEEE) in comparison with lesser power signals with an impressed physiologically relevant modulation rather than the complex waveforms and modulation methods in use to maximise channel use and minimise interference between subscribers.

However, such suggestions are not generally accepted by research ethics committees who take the view that if there is evidence of a 'possible health effect' then research activity should be limited to a controlled study of the exposure that volunteers would voluntarily expose themselves to in using phones. This is particularly so in the case of children. The reasons for this were highlighted in the IEGMP report – namely that there may be physical differences in the amounts of energy absorbed in a child's head, children have longer time to express detriment and longer time to be exposed, and children's tissues may be more sensitive (as in ionising radiation).

5.4. Children and Mobile Phones

5.4.1. The Problem

The question arises whether children should be considered a special case and this has been deliberated by the Stewart report and by a number of other bodies. Most reviews up until and including the Expert Panel report from the Royal Society of Canada considered only the cancer risk to children and held no views on any other possible health effects. The Stewart report acknowledged that there was some evidence of cognitive effects within the exposure standards, and that this might be greater in children by virtue of the different physical size affecting SAR distribution, similarly the physical thickness and properties of head tissues, and the relatively undeveloped state of neurological tissues compared with adults. The WHO comments (2000) did not include any reference to hand-held devices with respect to children, only the observations of possible sensitivity about siting base stations – 'Siting base stations near kindergartens, schools and playgrounds may need special consideration. Open communication and discussion between the mobile phone operator, local council and the public during the planning stages for a new antenna can help create public understanding and greater acceptance of a new facility'. This clearly does not provide advice about children and phone use. The extent of children's phone use has increased dramatically (Schuz, 2005). In 2002, the use by 15- to

19-year-old children varied from 40% in the US to 91% in Sweden. Over the same period, the phone use and ownership in Italy by children showed the startling figure of 8% use by 5-year olds up to 68% ownership by 13-year olds. These figures have undoubtedly increased since 2002, particularly since the cost of both phone ownership and use have decreased. Thus the question of whether children constitute a special case for consideration is important.

Even before the Stewart report, Schonborn et al. (1998) modelled the differences in absorption between adults and children because of their differing anatomies based on MRI scans of an adult and two children, using the same voxel size ($2 \times 2 \times 1.1$ mm^3) of the ages of 3 and 7 years. In addition, they incorporated ten different tissue types. The differences in absorption were investigated for 900 MHz and 1,800 MHz using 0.45 lambda dipoles. The results revealed no significant differences in the absorption of electromagnetic radiation in the near field of sources between adults and children. The same conclusion holds when children are approximated as scaled adults. This contrasts with the results of Martinez-Burdalo et al. (2004) who studied similar frequencies using patch or quarter lambda antennas in adults and children and concluded absorption rates up to 60% higher in children than in adults. A very variable result was obtained by Keshvari and Lang (2005) at Nokia laboratories, who showed large variations in SAR between males, females and two ages of children, which also varied with frequency. The conclusions of this study is that it is the geometry and anatomy that affects the absorption and has the dominant effect, not the gross size of the head (5.7 kg largest to 3.3 kg, smallest studied). Nevertheless, there were clearly large differences in relative SAR distribution, which were difficult to predict. The absolute values were not quoted in the study, which is, therefore, difficult to compare with other studies using absolute measurements.

Detailed differences for children were highlighted by Gandhi et al. (1996), who claimed that the peak absorbed power could be 50–55% higher in children. Indeed as a result of this study, in a later study, Gandhi and Kang (2002) pointed out the possible effect of a change from the existing 1999 IEEE guidelines to the proposed IEEE SCC 28.4 with a relative increase of about 4 in radiated power. This contrasts strongly with Bit-Babik et al. (2005), who, again using modelling, concluded that head size was irrelevant and absorption levels were similar for adult and children's heads. The elevated exposure findings of Gandhi are replicated also by modelling by de Salles et al. (2006), who also found that elevations of up to 60% occurred in small heads. Since ownership of mobile phones ranges from 97% of 9+ year-old children in Italy (Dimonte and Ricchiuto, 2006), 76% in Hungary (Mezei et al., 2007) and 36% in Germany (Bohler and Schuz, 2004), then consideration of exposure effects is particularly important. There was also a particular call from the IEGMP that precaution should be exercised by children, and a note that research was needed to study cognitive and other effects.

5.4.2. Studies on Children

There are not many studies on children. The first, using similar tests to that used in adults was carried out on the Isle of Man with sixteen 10- to 12-year olds, participants from a longitudinal healthy study (Preece et al., 2005). Care was taken to double-blind the tests, to have very consistent conduct, to balance the study and to create a reproducible exposure system. A real (at the insistence of the Research Ethics committee) computer controlled phone mounted on a plastic headset was used as the source. The a priori hypothesis was that there would be effects on reaction time, particularly choice reaction time, similar to that reported in adults by various studies. The only deviation from the original adult computer-based tests was the substitution of joystick-operated double attention task for the adult sound and vision mixed test. However, no effects were detected. Differences from adult performance showed in a much greater inter-subject variation in children and very much slower reaction times. This latter result was unexpected, although at that age some of the motor responses are not fully developed. For example, simple tasks such as finger tapping are much slower in young children (Garvey et al., 2003) and change rapidly during the ages studied. This causes considerable problems since a large inter-subject variation will affect results, even with a within subject design. Considering that differences in reaction time under exposed or non-exposed states are only 4–5% then the possibility of detecting a change is small. In our study on children, the standard deviation of individual performance was much larger for children than for adults. Since in children the mean values showed a trend to shorter reaction time (by 6%) in exposed conditions, even more than in adults, then the case for children being unaffected by mobile phones is still unproven.

In that study it was calculated that the number needed to compensate for the variation was in excess of 70 subjects. This, considering the need to match ages, is still appearing to be an insurmountable task. Do we need to be concerned? Probably not, since a change of 6% is well within normal physiological range and is far less than the effect that aging has on reaction times. Haarala et al. (2005) were faced with a similar problem in their study. In this case the age range was wide in developmental terms and covered 10–14 years. In addition the N-back tasks used up to a 3-back condition, which required a considerable load on memory and was therefore subject to even greater variation in neurological development and attention span. Haarala et al. also noticed the slowing of reaction time of about 125 ms compared with adults, but a speeding during the demanding memory tasks of 30–90 ms, once again emphasising the importance of cognitive development as well as age matching where possible. The other comparison deriving from the two children's studies is that even in a simple task such as Simple Reaction Time, there are differences inherent in the computer-based tests used as shown by the very different control values for similar age groups.

The conclusion for children's use of mobile phones has not changed. It is possible that there are cognitive or physiological effects within the existing exposure standards, but that the likely consequences for real exposure with standard phones using power reduction technology are probably not detectable. This implies that any such effects would be well within normal physiological limits and of no practical consequence and certainly not a health effect. This has nothing to contribute to the debate about non-thermal sub-cellular effects that might have health consequences for cancer or development or teratology.

6. DRIVING

Of all the practical consequences of cognitive effects of mobile phone use, none can compare with that of the perceived effect on driving. It has become illegal in most countries to use a mobile phone, whilst driving a vehicle unless a suitable 'hands-free kit' has been installed. However, it is not merely the hands-on-the-wheel status that may be important. Evidence comes from many sources. Although the possible effect of phone use on safe driving was questioned in 1997, the first definitive study by Lamble et al. (1999) was an intervention (provocation) study in real cars using either a visual attention task to simulate dialling, or a memory and addition task to simulate non-visual attention. Both conditions resulted in a 1.0 s increase in response time to an emergency. Unfortunately, the study also indicated that the use of 'hands-free' systems was unlikely to remove the effect.

These findings were similar in two studies by Alm and Nilsson (1994, 1995) using a simulator, where the use of a phone had a negative effect on road position, response time and speed. In particular, it was noted that the driver did not make allowance for the phone task by leaving a larger gap between vehicles, and where choice was involved, the apparent delay induced by the phone was of the order of several seconds. This situation would have led to a collision in a situation where the vehicle in front undertook hard braking.

These tasks were undertaken with cognitive load – conversation or response to memory and calculation tasks using hands-free devices so as not to impair physical control of the vehicle, but without the presence of RF. All the studies have indicated major impairment of performance. In one recent study, it has been estimated that in the US up to 330,000 accidents per annum are directly related to phone use, leading to 2,600 deaths (Cohen and Graham, 2003).

Personal note. If such studies were undertaken with real phones emitting RF energy, what might be the effect? Some years ago we constructed a study with medical students where the subject used a computer-based driving simulator with simple vehicle controls. Three conditions were tested – a sham exposure, 1 W at 915 MHz CW from a simulated phone mounted on the left of the head, and 1 W on the right in randomised but balanced order. This was equivalent to an old analogue phone but without conversation and without a phone-derived cognitive load. When the results began to indicate improvement in accuracy and reaction time for the exposed situation the series was abandoned. In comparison with the effect of the cognitive load in all the other studies, any apparent beneficial effect would have sent the wrong message – it was basically an unrealistic experiment and scientifically unethical because it would have, through the media, been at risk of providing the wrong message to society.

7. HYPERSENSITIVITY

Although technically a cognitive effect, hypersensitivity (or electrosensitivity) is not easily measured by physical or objective means, although the reports of this phenomenon attach physiological, well being and mood changes to the effects. As stated in Sect. 3, 'mood' has been used in the context of cognitive effects and is related to a relatively short-lived emotional state. These are subjective experiences, which can only be assessed by self-report. Such mood and emotion can influence other cognitive functions and thereby may be involved in hypersensitivity states. The first reports of electrosensitivity reported skin changes (dermatitis) evidenced in VDU operators (Linden and Rolfsen, 1981), and later a number of reports also in Scandinavia of symptoms of fatigue, headaches and migraine associated with mobile phone use (Oftedal et al., 2000). Attempts to demonstrate cognitive-behavioural effects by means of provocation studies was undertaken by Andersson et al. (1996). In these, subjects were exposed to EMFs and tested for hormonal changes related to stress, such as prolactin and cortisol, at the same time as completing self-evaluation questionnaires. These were all 'sensitive' subjects but showed no response to the electromagnetic challenge. In a later study (Lonne-Rahm et al., 2000), the same group deliberately subjected sensitive participants together with age and sex-matched controls to deliberate stress, to fields where the subjects were aware of the exposure and to fields when they were not aware. Quite clearly, sensitive subjects showed symptomatic response to the fields only when aware, but quite randomly when not. Since the topic of the study was VDU use, it was mainly skin symptoms that had been reported in the sensitive group. It would be expected that in genuine symptoms, the response would have involved invasion of mast cells mediating the inflammatory response. In spite of including skin biopsies as part of the assessment procedure, no evidence was found of mast cell invasion even in the stress provocation studies.

One of the problems associated with this kind of experiment is that the subjects are studied in a totally strange environment (they are usually laboratory conditions), the subjects are out of routine and the tests or stimuli are applied briefly and in relatively rapid sequences. Subjects protest that the symptoms take a while to develop and to abate, thus blurring the awareness of stimulus and reported symptom. Cognitive-emotional response can only be measured by self-reported measures such as questionnaires or interview which by themselves, for most subjects, also act as a stressor. To overcome this criticism, another Swedish group carried out a home-based study (Flodin et al., 2000) with exposure to electric equipment as a field source. There were separate days for the different conditions and the estimate of response was taken 24 h after exposure. These were hypersensitive patients as before, but even under these conditions no field discrimination could be detected.

Again in Sweden but with the additional neurological support from Russia (Lyskov et al., 2001) another study of volunteers used a battery of electrophysiological measures including blood pressure, heart rate, EEG and VEP as well as flicker fusion rates and electrodermal activity. Subjects were either electrosensitive or controls, all of which were subjected to the stress of undertaking mathematical calculation. Clearly this approach is fraught with the risks of showing a type 1 error simply by the large number of potential outcome measures. Nevertheless, the results were entirely negative both for the ability to detect electromagnetic fields and non-subjective responses to field exposure. There were, however, clear differences between those who claimed hypersensitivity and controls, again suggesting that the susceptibility is an expression of inherent physiological differences. This is supported by Hillert et al. (2002) from a cross-sectional study in Sweden of 15,000 individuals who completed a questionnaire. A proportion (1.5%) reported electro-sensitivity, which was associated with a sensitive response to a number of other conditions, but it did indicate that there may be a general concern in the population about the risks of all electromagnetic exposures. Whether this is a true observation may be compromised by the study design where the application of RF exposure was not done in a balanced design, so that the order of administration could have affected the results.

The possibility of such perceptions applying to mobile phones was investigated by Hietanen et al. (2002) in a provocation study of 20 volunteers in a remote forest area to eliminate the confounding effect of ELF, strong TV or cell phone signals and even any other electrical equipment. The results of the study indicated that the number of reported symptoms was higher during sham exposure than during real exposure. Additionally, no subject could distinguish real RF exposure from sham exposure. The authors concluded that adverse subjective symptoms or sensations, though unquestionably perceived by the test subjects, were not produced by cellular phones. These symptoms mirrored the reports of symptoms in the general public, namely an increase in headaches, dizziness and nervousness. Like many other cognitive studies, the question here is whether 20 subjects, even under these remarkably well-controlled conditions, are sufficient to detect what is a moderately rare condition in the general public.

Within the UK, ahead of studies commissioned by the MTHR, a review was undertaken by Rubin et al. (2005) where they reviewed some 31 hypersensitivity studies involving 725 subjects with claimed hypersensitivity. Most (77%) of these studies were negative and of the remaining 7 studies, 5 were either unable to be repeated or were statistically flawed. Only two studies showed a positive association but the results were in some disagreement with each other. Their conclusion to a meta-analysis was that although the symptoms were severe and sometimes disabling, there was no evidence that the symptoms were related to EM field exposure. This echoes the earlier studies which all suggest there is a phenomenon that is ascribed to EM fields, but which may be related to other personality differences. A very similar outcome came from our own study of exposure to military radar in Cyprus (Preece et al., 2007). This was an interview-conducted study of three communities, two of which were close to a military radar source. The third, a control village, was in a remote area with poor radio and TV reception and no cell phone mast. It was additionally chosen to be off the antenna radiation pattern of the military source and at 15 km distance. All three communities showed elevated perception of risk compared with the UK and Sweden, but the two 'exposed' communities in addition showed very definite elevation of incidence of headaches, dizziness and depression. In symptoms, this mirrored the two-country Scandinavian study of cell phone exposure and use (Oftedal et al., 2000). This was a cross-sectional epidemiological study of 17,000 people using a mobile phone in their job. In Norway, 31% were heavy phone users experiencing symptoms, and in Sweden 13% had experienced at least one symptom in connection with MP use. In Cyprus, concurrent measurement of exposure to the military source and the other sources of RF, namely a cell phone base station and a medium wave relay antenna, indicated that the average exposure was no different than in the centre of Bristol, Gothenburg or even Nicosia. In all three of these locations, and in the two study communities, it was the cell phone base station that provided the highest field as measured simply as volts per metre and was also the most persistent.

The media continues to express an interest in the phenomenon. This is fuelled by the large number of permutations of cell phone modes and growing use of WiFi with the introduction of new technologies and the suspicion that more sources means ever rising field levels, whereas in practice the new sources are very low power and therefore effectively short range. A novel approach to this was undertaken in Austria (Leitgeb et al., 2005) using a questionnaire survey of general practitioners about their beliefs, and the kind of enquiry received from the public. Over two-thirds of the doctors had been consulted about problems believed to be derived from EM fields and up to 96% of the doctors believed to some extent in the relevance of EM fields in health effects. However, only a very small percentage was aware of public exposure limits. The interesting thing is that there exists a dichotomy between physicians' opinions and the international statements on health risk guidelines. It is such a condition that probably leads to the continued propagation of the beliefs about the harmful effects of low level exposure and concerns about hypersensitivity in spite of the consistently negative results of studies involving provocation of hypersensitivity or even simple perception of fields.

Attempts to demonstrate induced effects in human physiology, particularly those associated with the immune system, in hypersensitive subjects have been carried out by a number of groups. One such study (Markova et al., 2005) compared the in vitro lymphocyte responses of hypersensitive and control subjects to GSM signals and detected response similar to heat shock. Although there were no differences in cellular responses between the two groups, this experiment demonstrated a response to all subjects of very low levels of RF, thought not to be harmful, and taking the form of a stress response. This would be significant if it were not for the final outcome of the very similar finding on transgenic *Coenorhabditis elegans* with human stress genes that showed a response to non-thermal levels of microwaves (de Pomerai et al., 2000). This study was retracted by the journal Nature (2006) when it was discovered that the very marginal temperature rise was shown to be equally able to induce heat stress protein expression in the absence of microwaves. The main part of the error derived from the skin effect in the microwave transmission line perturbed the heat distribution, something only detected by independent and improved modelling. Direct measurement of very small temperature changes is difficult even in the absence of relative large RF fields, and in their presence, errors can be introduced at any number of different points because of the presence of metal, non-linear elements and sensitive amplifier inputs.

Other studies continue to demonstrate the existence of an illness described as electrical hypersensitivity and to investigate methods for 'curing' the condition (Rubin et al., 2006) and to investigate the characteristics (Leitgeb et al., 2007) or prevalence (Mortazavi et al., 2007) in the general public or in special subgroups such as students. Always there are findings of symptoms that can be allayed or managed by information or treatment, but no-one has found a definitive physiological effect specific to the electro-hypersensitive subject. Pressure from public groups, possibly a minority, requires that research continues possibly supported by the general belief in an adverse effect by a significant number of the population.

8. OBSERVATIONS AND CONCLUSIONS

1. Reported effects of RF have been quite dissimilar from ELF both in the response to low levels and in the effects controlled by the guidelines.
2. The ability to draw conclusions whether cognitive effects of phones are real is seriously hampered by the great variation in technique, experimental design and the enormous variation in phone design resulting in very different exposure level and distribution.
3. Some of this should be addressed by recent co-ordinated studies through the EU and bodies such as the UK MTHR programmes.
4. As techniques have improved and the focus is more on replication studies, the earlier 'positive' effects have not been upheld and many of the later studies have been 'negative'.
5. The disappearance of effects in successive studies may be the result of the reduction in exposure caused by phone development and the use of GSM systems, rather than use of the early high exposure analogue systems for the experiments (many of the early phones had higher SAR levels than the later and current designs and not likely to be a result of habituation to ubiquitous RF use).
6. There are consistent reports of effects on blood flow alterations even at the low exposure levels induced by GSM real phones or simulations, similarly with the reported effects on EEG and some other electrophysiological functions.
7. Overall, there is little evidence that cognitive function in humans is upset or detrimentally altered by the standard modern GSM phone, and that any effects, if they occur, are very subtle and extremely difficult to detect. There are indications that current exposure levels are on the margin for detectable effects.

REFERENCES

Aalto, S., Haarala, C., Bruck, A., Sipila, H., Hamalainen, H., and Rinne, J.O., 2006, Mobile phone affects cerebral blood flow in humans. *J Cereb Blood Flow Metab* 26: 885–890.

Achermann, P., 2000, http://www.pharma.unizh.ch/sleep/handy/index.htm

Adair, R.K., 2003, Biophysical limits on athermal effects of RF and microwave radiation. *Bioelectromagnetics* 24: 39–48.

Alm, H., and Nilsson, L., 1994, Changes in driver behaviour as a function of handsfree mobile phones–a simulator study. *Accid Anal Prev* 26: 441–451.

Alm, H., and Nilsson, L., 1995, The effects of a mobile telephone task on driver behaviour in a car following situation. *Accid Anal Prev* 27: 707–715.

Andersson, B., Berg, M., Arnetz, B.B., Melin, L., Langlet, I., and Liden, S, 1996, A cognitive-behavioral treatment of patients suffering from "electric hypersensitivity". Subjective effects and reactions in a double-blind provocation study. *J Occup Environ Med* 38: 752–758.

Baldi, E., Baldi, C., and Lithgow, B.J., 2007, A pilot investigation of the effect of extremely low frequency pulsed electromagnetic fields on humans' heart rate variability. *Bioelectromagnetics* 28: 76–79.

Beall, A.C., Kato, K., Goldenring, J.R., Rasmussen, H., Brophy, C.M., 1997, Cyclic nucleotide-dependent vasorelaxation is associated with the phosphorylation of a small heat shock-related protein. *J Biol Chem* 272: 11283–11287.

Besset, A., Espa, F., Dauvilliers, Y., Billiard, M., and de Seze, R., 2005, No effect on cognitive function from daily mobile phone use. *Bioelectromagnetics* 26: 102–108.

Bier, M., 2005, Gauging the strength of power frequency fields against membrane electrical noise. *Bioelectromagnetics* 26: 595–609.

Bit-Babik, G., Guy, A.W., Chou, C.K., Faraone, A., Kanda, M., Gessner, A., Wang, J., and Fujiwara, O., 2005, Simulation of exposure and SAR estimation for adult and child heads exposed to radiofrequency energy from portable communication devices. *Radiat Res* 163: 580–590.

Bohler, E., and Schuz, J., 2004, Cellular telephone use among primary school children in Germany. *Eur J Epidemiol* 19: 1043–1050.

Boutry, C.M., Kuehn, S., Achermann, P., Romann, A., Keshvari, J., and Kuster, N., 2008, Dosimetric evaluation of different exposure apparatuses used in human provocation studies. *Bioelectromagnetics* 29: 11–19.

Burgess, A., 2003, *Cellular Phones, Public Fears and a Culture of Precaution.* Cambridge University Press, Cambridge.

Carlo, G.L., and Schram, M., 2001, *Cell Phones: Invisible Hazards in the Wireless Age.* Carroll and Graf, New York.

Cinel, C., Boldini, A., Russo, R., and Fox, E., 2007, Effects of mobile phone electromagnetic fields on an auditory order threshold task. *Bioelectromagnetics* 28: 493–496.

Cohen, J.T., and Graham, J.D., 2003, A revised economic analysis of restrictions on the use of cell phones while driving. *Risk Anal* 23: 5–17.

Cook, C.M., Thomas, A.W., and Prato, F.S., 2002, Human electrophysiological and cognitive effects of exposure to ELF magnetic and ELF modulated RF and microwave fields: A review of recent studies. *Bioelectromagnetics* 23: 144–157.

Cook, C.M., Saucier, D.M., Thomas, A.W., and Prato, F.S., 2006, Exposure to ELF magnetic and ELF-modulated radiofrequency fields: The time course of physiological and cognitive effects observed in recent studies (2001–2005). *Bioelectromagnetics* 27: 613–627.

Curcio, G., Ferrara, M., De Gennaro, L., Christrani, R., D'Inzeo, G., Bertini, M., 2004, Tune course of electromagnetic field effect on human performance and tympanic temperature. *Neuro report* 15: 161–164.

Curcio, G., Valentini, E., Moroni, F., Ferrara, M., De Gennaro, L., and Bertini, M., 2007, Psychomotor performance is not influenced by brief repeated exposures to mobile phones. *Bioelectromagnetics* 29: 237–241.

D'Andrea, J.A., Chou, C.K., Johnston, S.A., and Adair, E.R., 2003, Microwave effects on the nervous system. *Bioelectromagnetics* Suppl 6: S107–S147.

de Pomerai, D., Daniells, C., David, H., Allan, J., Duce, I., Mutwakil, M., Thomas, D., Sewell, P., Tattersall, J., Jones, D., and Candido, P., 2000, Non-thermal heat-shock response to microwaves. *Nature* 405: 417–418.

de Salles, A.A., Bulla, G., and Rodriguez, C.E., 2006, Electromagnetic absorption in the head of adults and children due to mobile phone operation close to the head. *Electromagn Biol Med* 25: 349–360.

Dimonte, M., and Ricchiuto, G., 2006, Mobile phone and young people. A survey pilot study to explore the controversial aspects of a new social phenomenon. *Minerva Pediatr* 58: 357–363.

Edelstyn, N., and Oldershaw, A., 2002, The acute effects of exposure to the electromagnetic field emitted by mobile phones on human attention. *Neuroreport* 13: 119–121.

Eulitz, C., Ullsperger, P., Freude, G., and Elbert, T., 1998, Mobile phones modulate response patterns of human brain activity. *Neuroreport* 9: 3229–3232.

Flodin, U., Seneby, A., and Tegenfeldt, C., 2000, Provocation of electric hypersensitivity under everyday conditions. *Scand J Work Environ Health* 26: 93–98.

Gandhi, O.P., and Kang, G., 2002, Some present problems and a proposed experimental phantom for SAR compliance testing of cellular telephones at 835 and 1900 MHz. *Phys Med Biol* 47: 1501–1518.

Gandhi, O.P., Lazzi, G., and Furse, C.M., 1996, Electromagnetic absorption in the human head and neck for mobile telephones at 835 and 1900 MHz. *IEEE Trans Microw Theory Tech* 44: 1884–1897.

Garvey, M.A., Ziemann, U., Bartko, J.J., Denckla, M.B., Barker, C.A., and Wassermann, E.M., 2003, Cortical correlates of neuromotor development in healthy children. *Clin Neurophysiol* 114: 1662–1670.

Graham, C., Cohen, H.D., Cook, M.R., Phelps, J., Gerkovich, M.M., and Fotopoulous, S.S., 1987, A double band evaluation of 60-H_2 field effects on human performance physiology and subjective state. In: Anderson, L.E., et al. (eds.) "Interaction of Biological Systems with Static and ELF Electric and Magnetic Fields" CONF-841041 Spring field VA: NTIS, pp 471–486.

Graham, C., Cook, M.R., Cohen, H.D., and Gerkovich, M.M., 1994, Dose response study of human exposure to 60 Hz electric and magnetic fields. *Bioelectromagnetics* 15: 447–463.

Haarala, C., Bjornberg, L., Ek, M., Laine, M., Revonsuo, A., Koivisto, M., and Hamalainen, H., 2003, Effect of a 902 MHz electromagnetic field emitted by mobile phones on human cognitive function: A replication study. *Bioelectromagnetics* 24(4): 283–288.

Haarala, C., Aalto, S., Hautze, H., Julkunen, L., Rinne, J.O., Laine, M., Krause, B., and Hamalainen, H., 2003a, Effects of a 902 MHz mobile phone on cerebral blood flow in humans: a PET study. *Neuroreport* 14: 2019–2023.

Haarala, C., Bjornberg, L., Ek, M., Laine, M., Revonsuo, A., Koivisto, M., and Hamalainen, H., 2003b, Effect of a 902 MHz electromagnetic field emitted by mobile phones on human cognitive function: A replication study. *Bioelectromagnetics* 24: 283–288.

Haarala, C., Bergman, M., Laine, M., Revonsuo, A., Koivisto, M., and Hamalainen, H., 2005, Electromagnetic field emitted by 902 MHz mobile phones shows no effects on children's cognitive function. *Bioelectromagnetics* Suppl 7: S144–S150.

Haarala, C., Takio, F., Rintee, T., Laine, M., Koivisto, M., Revonsuo, A., and Hamalainen, H., 2007, Pulsed and continuous wave mobile phone exposure over left versus right hemisphere: Effects on human cognitive function. *Bioelectromagnetics* 28: 289–295.

Hietanen, M., Hamalainen, A.M., and Husman, T., 2002, Hypersensitivity symptoms associated with exposure to cellular telephones: No causal link. *Bioelectromagnetics* 23: 264–270.

Hillert, L., Berglind, N., Arnetz, B.B., and Bellander, T., 2002, Prevalence of self-reported hypersensitivity to electric or magnetic fields in a population-based questionnaire survey. *Scand J Work Environ Health* 28: 33–41.

Huber, R., Treyer, V., Schuderer, J., Berthold, T., Buck, A., Kuster, N., Landolt, H.P., and Achermann, P., 2005, Exposure to pulse-modulated radio frequency electromagnetic fields affects regional cerebral blood flow. *Eur J Neurosci* 21: 1000–1006.

ICNIRP, 1998, Guidelines for limiting exposure to time-varying electric, magnetic and electromagnetic fields (up to 300 GHz). *Health Phys* 74: 494–522.

IEGMP, 2000, Mobile phones and health. http://www.iegmp.org.uk/report/text.htm

Jerius, H., Karolyi, D.R., Mondy, J.S., Beall, A., Wootton, D., Ku, D., Cable, S., and Brophy, C.M., 1999, Endothelial-dependent vasodilation is associated with increases in the phosphorylation of a small heat shock protein (HSP20). *J Vasc Surg* 29: 678–684.

Kaune, W.T., 2002, Thermal noise limit on the sensitivity of cellular membranes to power frequency electric and magnetic fields. *Bioelectromagnetics* 23: 622–628.

Keetley, V., Wood, A.W., Spong, J., and Stough, C., 2006, Neuropsychological sequelae of digital mobile phone exposure in humans. *Neuropsychologia* 44: 1843–1848.

Keshvari, J., and Lang, S., 2005, Comparison of radio frequency energy absorption in ear and eye region of children and adults at 900, 1800 and 2450 MHz. *Phys Med Biol* 50: 4355–4369.

Koivisto, M., Krause, C.M., Revonsuo, A., Laine, M., and Hamalainen, H., 2000a, The effects of electromagnetic field emitted by GSM phones on working memory. *Neuroreport* 11: 1641–1643.

Koivisto, M., Revonsuo, A., Krause, C., Haarala, C., Sillanmaki, L., Laine, M., and Hamalainen, H., 2000b, Effects of 902 MHz electromagnetic field emitted by cellular telephones on response times in humans. *Neuroreport* 11: 413–415.

Krause, C.M., Sillanmaki, L., Koivisto, M., Haggqvist, A., Saarela, C., Revonsuo, A., Laine, M., and Hamalainen, H., 2000, Effects of electromagnetic field emitted by cellular phones on the EEG during a memory task. *Neuroreport* 11: 761–764.

Krause, C.M., Bjornberg, C.H., Pesonen, M., Hulten, A., Liesivuori, T., Koivisto, M., Revonsuo, A., Laine, M., and Hamalainen, H., 2006, Mobile phone effects on children's event-related oscillatory EEG during an auditory memory task. *Int J Radiat Biol* 82: 443–450.

Lamble, D., Kauranen, T., Laakso, M., and Summala, H., 1999, Cognitive load and detection thresholds in car following situations: Safety implications for using mobile (cellular) telephones while driving. *Accid Anal Prev* 31: 617–623.

Lass, J., Tuulik, V., Ferenets, R., Riisalo, R., and Hinrikus, H., 2002, Effects of 7 Hz-modulated 450 MHz electromagnetic radiation on human performance in visual memory tasks. *Int J Radiat Biol* 78: 937–944.

Lee, T.M., Ho, S.M., Tsang, L.Y., Yang, S.H., Li, L.S., Chan, C.C., and Yang, S.Y., 2001, Effect on human attention of exposure to the electromagnetic field emitted by mobile phones. *Neuroreport* 12: 729–731.

Leitgeb, N., Schrottner, J., and Bohm, M., 2005, Does "electromagnetic pollution" cause illness? An inquiry among Austrian general practitioners. *Wien Med Wochenschr* 155: 237–241.

Leitgeb, N., Schrottner, J., and Cech, R., 2007, Perception of ELF electromagnetic fields: Excitation thresholds and inter-individual variability. *Health Phys* 92: 591–595.

Linden, V., and Rolfsen, S., 1981, Video computer terminals and occupational dermatitis. *Scand J Work Environ Health* 7: 62–64.

Lonne-Rahm, S., Andersson, B., Melin, L., Schultzberg, M., Arnetz, B., and Berg, M., 2000, Provocation with stress and electricity of patients with "sensitivity to electricity". *J Occup Environ Med* 42: 512–516.

Lövsund, P., Oberg, P.A., and Nilsson, S.E., 1979, Influence on vision of extremely low frequence electromagnetic fields. Industrial measurements, magnetophosphene studies volunteers and intraretinal studies in animals. *Acta Ophthalmol* 57: 812–821.

Lyskov, E., Sandstrom, M., and Mild, K.H., 2001, Provocation study of persons with perceived electrical hypersensitivity and controls using magnetic field exposure and recording of electrophysiological characteristics. *Bioelectromagnetics* 22: 457–462.

Markova, E., Hillert, L., Malmgren, L., Persson, B.R., and Belyaev, I.Y., 2005, Microwaves from GSM mobile telephones affect 53BP1 and gamma-H2AX foci in human lymphocytes from hypersensitive and healthy persons. *Environ Health Perspect* 113: 1172–1177.

Martinez-Burdalo, M., Martin, A., Anguiano, M., and Villar, R., 2004, Comparison of FDTD-calculated specific absorption rate in adults and children when using a mobile phone at 900 and 1800 MHz. *Phys Med Biol* 49: 345–354.

Mezei, G., Benyi, M., and Muller, A., 2007, Mobile phone ownership and use among school children in three Hungarian cities. *Bioelectromagnetics* 28: 309–315.

Mobile Phones UK, 2007, SAR values and mobile phone health. http://www.mobile-phones-uk.org.uk/sar.htm

Mortazavi, S.M., Ahmadi, J., and Shariati, M., 2007, Prevalence of subjective poor health symptoms associated with exposure to electromagnetic fields among university students. *Bioelectromagnetics* 28: 326–330.

MTHR, 2004, http://www.mthr.org.uk/index.htm

Nature, 2006, Retraction of De Pomerai D., et al. *Nature* 405: 417–418.

NRPB, 2003, *Health Effects from Radiofrequency Electromagnetic Fields: Report of an Independent Advisory Group on Non-Ionising Radiation*, 14, No. 2. NRPB, UK.

Oftedal, G., Wilen, J., Sandstrom, M., and Mild, K.H., 2000, Symptoms experienced in connection with mobile phone use. *Occup Med* 50: 237–245.

Paneth, N., 1993, Neurobehavioral effects of power-frequency electromagnetic fields. *Environ Health Perspect* 101 Suppl 4: 101–106.

Papageorgion, C.C., Nanou, E.D., Isiafukis, Y.G., Capsalis, C.N., and Rabavilas, A.D., 2004, Gender related differences on the EEG during a simulated mobile phone signal. *Neuro report* 15: 2557–2560.

Preece, A.W., Murfin, J.L., and Johnson, R.H., 1987, RF field penetration from electrically small hyperthermia applicators. *Phys Med Biol* 32: 1595–1601.

Preece, A.W., Iwi, G., Davies-Smith, A., Wesnes, K., Butler, S., Lim, E., and Varey, A., 1999, Effect of a 915-MHz simulated mobile phone signal on cognitive function in man. *Int J Radiat Biol* 75: 447–456.

Preece, A.W., Goodfellow, S., Wright, M.G., Butler, S.R., Dunn, E.J., Johnson, Y., Manktelow, T.C., and Wesnes, K., 2005, Effect of 902 MHz mobile phone transmission on cognitive function in children. *Bioelectromagnetics* Suppl 7: S138–S143.

Preece, A.W., Georgiou, A.G., Dunn, E.J., and Farrow, S.C., 2007, Health response of two communities to military antennae in Cyprus. *Occup Environ Med* 64: 402–408.

Rubin, G.J., Das Munshi, J., and Wessely, S., 2005, Electromagnetic hypersensitivity: A systematic review of provocation studies. *Psychosom Med* 67: 224–232.

Rubin, G.J., Hahn, G., Everitt, B.S., Cleare, A.J., and Wessely, S., 2006, Are some people sensitive to mobile phone signals? Within participants double blind randomised provocation study. *BMJ* 332: 886–891.

Russo, R., Fox, E., Cinel, C., Boldini, A., Defeyter, M.A., Mirshekar-Syahkal, D., and Mehta, A., 2006, Does acute exposure to mobile phones affect human attention? *Bioelectromagnetics* 27: 215–220.

Schmid, G., Cecil, S., Goger, C., Trimmel, M., Kuster, N., and Molla-Djafari, H., 2007, New head exposure system for use in human provocation studies with EEG recording during GSM900 and UMTS-like exposure. *Bioelectromagnetics* 28: 636–647.

Schonborn, F., Burkhardt, M., and Kuster, N., 1998, Differences in energy absorption between heads of adults and children in the near field of sources. *Health Phys* 74: 160–168.

Schuz, J., 2005, Mobile phone use and exposures in children. *Bioelectromagnetics* Suppl 7: S45–S50.

Taki, M., Suzuki, Y., and Wake, K., 2003, Dosimetry considerations in the head and retina for extremely low frequency electric fields. *Radiat Prot Dosimetry* 106: 349–356.

Valentini, E., Curcio, G., Moroni, F., Ferrara, M., De Gennaro, L., and Bertini, M., 2007b, Neurophysiological effects of mobile phone electromagnetic fields on humans: A comprehensive review. *Bioelectromagnetics* 28: 415–432.

Van Leeuwen, G.M., Lagendijk, J.J., Van Leersum, B.J., Zwamborn, A.P., Hornsleth, S.N., and Kotte, A.N., 1999, Calculation of change in brain temperatures due to exposure to a mobile phone. *Phys Med Biol* 44: 2367–2379.

Vincze, G., Szasz, N., and Szasz, A., 2005, On the thermal noise limit of cellular membranes. *Bioelectromagnetics* 26: 28–35.

Weaver, J.C., and Astumian, R.D., 1990, The response of living cells to very weak electric fields: the thermal noise limit. *Science* 247: 459–462.

Weaver, J.C., and Astumian, R.D., 1992, Estimates for ELF effects: Noise-based thresholds and the number of experimental conditions required for empirical searches. *Bioelectromagnetics* Suppl 1: 119–138.

Wesnes, K.A., 2006, Cognitive function testing: The case for standardization and automation. *J Br Menopause Soc* 12: 158–163.

Wever, R., 1979, *The Circadian Rhythm in Man.* Springer-Verlag, New York.

WHO, 2000, Electromagnetic fields and public health: mobile telephones and their base stations. Fact street 193, http://www.who.int/mediacentre/factsheets/fs193/en/

Wilen, J., Wiklund, U., Hornsten, R., and Sandstrom, M., 2007, Changes in heart rate variability among RF plastic sealer operators. *Bioelectromagnetics* 28: 76–79.

Williams, B.R., Strauss, E.H., Hultsch, D.F., and Hunter, M.A., 2007, Reaction time inconsistency in a spatial Stroop task: Age-related differences through childhood and adulthood. *Neuropsychol Dev Cogn B Aging Neuropsychol Cogn* 14: 417–439.

Wolff, P.H., Kotwica, K., and Obregon, M., 1998, The development of interlimb coordination during bimanual finger tapping. *Int J Neurosci* 93: 7–27.

Chapter 5

Electromagnetic Hypersensitivity

Norbert Leitgeb

ABSTRACT

Electromagnetic hypersensitive persons (EHS) attribute their nonspecific health symptoms to environmental electromagnetic fields (EMF) of different sources in or outside their homes. In general, causal attribution is not restricted to specific EMF frequencies but involves a wide range from extremely low frequencies (ELF) up to radio frequencies (RF) including mobile telecommunication microwaves and radar. EHS argue that existing exposure limits were not low enough to account for their increased sensitivities. Results of measurement campaigns are summarized. They demonstrate that environmental fields in the ELF and RF range are usually orders of magnitudes below exposure limits. The rational and biological background of recommended exposure limits are described. The existing scientific studies are reviewed, including investigations on the prevalence of EHS among the general population, ability of EHS to perceive and/or react to exposures to weak EMF (assessed in laboratory provocational studies or to the vicinity of EMF sources studied by epidemiologic approaches), and the existence of a specific symptom cluster, which could characterize a suspected EHS syndrome, or individual EHS-specific factors such as electric perception thresholds, neurophysiologic parameters, and cognitive performance and behavior. However, in spite

N. Leitgeb Institute of Health Care Engineering, Graz University of Technology, Inffeldgasse 18, A-8010, Graz, Austria, e-mail: norbert.leitgeb@tugraz.at

J.C. Lin (ed.), *Advances in Electromagnetic Fields in Living Systems, Health Effects of Cell Phone Radiation, Volume 5*,
DOI 10.1007/978-0-387-92736-7_5, © Springer Science+Business Media, LLC 2009

of the variety of scientific attempts, a causal role of EMF remains yet unproven. This does not mean that the suffering could be ignored. It is recognized that EHS cases deserve help. Therapeutic approaches are described and the conclusion of the World Health Organisation (WHO) is summarized.

1. INTRODUCTION

An increasing number of people suffering from sometimes severe nonspecific health symptoms of unclear origin attribute their health problems to external sources such as various environmental multiple chemical or physical factors, among them environmental EMF. Frequently, affected people explain the fact that most others do not exhibit symptoms due to suspected factors at levels well below existing exposure limits by postulating being hypersensitive to such influences.

Electromagnetic hypersensitive (EHS) persons attribute their health symptoms to environmental EMF to different sources in or outside their rooms emitting EMF. In general, attribution is not restricted to specific frequencies but involves a large range of frequencies from extremely low frequencies (ELF) up to radio frequencies (RF), mobile telecommunication microwaves, and radar. Suspected electromagnetic hypersensitivity was argued to challenge EMF exposure limits. Petitions were brought forward to lower existing EMF exposure limits by several orders of magnitude. EHS has already become a social issue. In many countries, EHS self-aid groups have been established. For example, in Sweden, the association for EHS is recognized as a handicap organization. An overwhelming majority of general practitioners do not exclude or are even convinced environmental EMF could be causally related to nonspecific health symptoms and multiply their opinions during their contacts with patients and related diagnostic conclusions.

Scientific attempts to investigate and substantiate personal convictions on hypersensitivity and electromagnetic allergy began two decades ago. Since then, a body of scientific studies has been published on EHS issues. To demonstrate a causal link between environmental EMF and the development of health symptoms on the basis of the hypothesis of electromagnetic hypersensitivity the following questions were investigated:

Is there an EHS subgroup within the population characterized by a sensitivity to electromagnetic fields which is increased beyond the normal range?

Is increased sensitivity to EMF causally linked with the development of health symptoms?

If it exists, what is the prevalence of EHS within the general population?

Are the reduction factors implemented in the derivation of EMF exposure limits sufficient to account for EHS groups?

This chapter reviews the existing literature and provides answers to these questions.

2. EXPOSURE LIMITS

To protect from known adverse health effects, exposure limits for ELF and RF electromagnetic fields have been proposed and already implemented in numerous countries worldwide. Protection strategy is based on "basic restrictions" limiting intracorporal quantities relevant for biologic interactions derived from the first health-relevant interaction level, which is lowered by tenfold accounting for uncertainties of knowledge to determine the basic restriction for occupational exposure. To account for potential higher sensitivities in certain population groups such as frail and/or elderly, infants and young children, and people with diseases or taking medications, which may compromise their perception ability and/or thermal tolerance, to limit exposure of the general population, a factor of 5 had been introduced to further reduce electric current density in the ELF range and specific absorption rate (SAR) in the RF range, respectively.

In the ELF range electric and magnetic fields interact with the body by inducing intracorporal electric field strengths and current densities, although governed by different laws and, hence, with different pathways. Consequently, basic restrictions limit intracorporal current densities or intracorporal electric field strengths within a region of interest, namely, the central nervous system (CNS), which is composed of the brain and spinal cord. Starting from the excitation threshold 100 mA/m² of central nervous tissue, the basic restriction has been set to 10 mA/m² for occupationally exposed and 2 mA/m² for general population (ICNIRP, 1998).

The main biologic interaction mechanism of RF electromagnetic fields is heating due to absorption of RF EMF energy. Consequently, basic restrictions limit the SAR, which is absorbed power related to tissue mass. SAR limits are defined for whole body and for local exposure by relating the absorbed power either to the whole body mass (SAR_{WB}) or to any 10 g tissue (SAR_{10g}), respectively (ICNIRP, 1998). Starting from initiation of thermal regulation at 1 C temperature rise which is caused by 4 W/kg SAR_{WB}, the basic restriction has been set to 0.4 W/kg for occupational exposure, and 0.08 W/kg for the general population. This means that the maximum permitted heating by RF EMF absorption is considerably lower than that of the human metabolic rate, which is about 1.2 W/kg at rest and can increase up to 12 W/kg during heavy exercise.

Because in daily life testing compliance with basic restrictions is difficult, for practical reasons, "reference levels" of easily measurable external field quantities such as electric or magnetic field strength were derived, linking worst case homogeneous field whole body exposures to basic restriction levels. If reference levels are met, compliance with basic restrictions can be assumed. However, at more favourable exposure conditions, reference levels could be exceeded without violating basic restrictions.

Nonionising radiation is characterized by the fact that amplitudes have to exceed biological thresholds to cause health relevant effects. Such threshold effects are stimulation of nerve and muscle cells by induced electric current densities or electric field strengths in the ELF range, and heat-triggered onset of thermoregulation due to absorbed RF EMF radiation energy (ICNIRP, 1998; IEEE, 2002, 2005).

The existence of biological thresholds allows excluding these effects rather than just reducing their probability of occurrence.

3. TERMINOLOGY AND SYMPTOMS

Although widely used in public media and scientific literature, electromagnetic hypersensitivity is associated with different meanings. There is a need to separate different aspects of this term (Leitgeb and Schröttner, 2003; WHO, 2005). In general,

> "Sensibility" addresses the ability to perceive exposures without necessarily developing health symptoms
>
> "Sensitivity" addresses the development of health symptoms as a causal reaction to exposures
>
> "Hypersensitivity" addresses the development of health symptoms as a causal reaction to exposures at much lower levels than required for the general population

Attributing nonspecific health symptoms to EMF seems to be neither a problem of the rich nor the poor, nor does it depend on education. It seems to be a problem of adults; however, there is no linear dependence on age. Females and persons with high tendency to somatisation report more frequent and more severe EMF-associated symptoms than others (Frick et al., 2002). An early attempt to identify a specific symptom cluster characterizing a syndrome based on an inquiry and involving 11 European countries failed (Bergqvist et al., 1997). Both symptoms and attributions varied among individuals. Throughout Europe a north-south gradient has been found with decreasing prevalence towards the south. Until now, reported EMF-associated symptoms (Table 1) include neurasthenic, vegetative and dermatological symptoms. However, the collection of symptoms is not part of any recognized syndrome (Bergqvist et al., 1997; Frick et al., 2002; Hillert et al., 2002; ICNIRP, 2003; WHO, 2005; Mild et al., 2006; Schreier et al., 2006; Schüz et al., 2006).

The World Health Organisation (WHO) concluded that EHS resembles multiple chemical sensitivities (MCS), another disorder associated with low-level environmental exposures to chemicals. Therefore, it proposed a preference for the more general term "idiopathic environmental intolerance" (IEI) already used for sensitivities to environmental factors. This term would not insinuate unproven causation or physiological mechanisms and does not already imply chemical etiology, immunological sensitivity or EMF susceptibility (WHO, 2005). Consequently, WHO recommended replacing the term EHS with "idiopathic environmental intolerance related to EMF" (IEI-EMF). This addresses an acquired disorder with multiple recurrent symptoms without forming a characteristic symptom cluster, associated with environmental factors or situations which are tolerated by the majority of people and cannot (yet) be explained by any known medical or psychological mechanism. However, this recommendation was rarely followed, and the common use of the term EHS persists.

Table 1. Reported symptoms associated with exposures to electric, magnetic, and electromagnetic fields (in alphabetical order)

Abdominal pain	Headache	Numb limbs
Anxiety	Head pressure	Phosphenes (flickering)
Appetite loss	Heart beat irregularity	Rash
Arousal decreased	Heart palpitation	Restlessness
Blood pressure increase	Hormonal disorder	Skin burning
Breathlessness	Hypersensitivity to medication	Skin redness
Chest pain	Hypersensitivity to noise	Skin tingling
Concentration difficulties	Intestinal trouble	Sleep disturbance
Crankiness	Irregular bowl movement	Stress
Daytime sleepiness	Irritation	Sweating
Digestive problem	Itching skin	Swollen eyes
Dizziness	Limb pain	Swollen joints
Dry skin	Metabolic disorder	Tachycardia
Exhaustion	Mood changes	Tenseness
Faintness	Mood depression	Tiredness
Fatigue	Muscle cramps	Toothache
Fear	Muscle pain	Trembling
Feebleness	Nausea	Unfeelingness
Feeling hot	Neck pain	Vision blurring
Forgetfulness	Neuralgia	Vomiting
Hair loss	Neurasthenia	Weariness

Physicians are already used to the term electromagnetic hypersensitivity, and many of them are deeply convinced that environmental EMF can play a causal role in the development of nonspecific health symptoms. A survey among Austria's general practitioners (Leitgeb et al., 2005a, b) with a response rate of 49% found an overwhelming majority of 96% not excluding, and 33% deeply convinced, that EMF could cause adverse health effects. Almost two thirds of the practitioners (61%) were making such a diagnosis. In Switzerland, based on a response rate of only 28%, the majority of general practitioners (54%) judged the association between EMF and health symptoms to be plausible. Physicians practising complementary medicine were much more convinced of this hypothesis. Overall, 14% had considered EMFs as a potential cause for symptoms they had experienced themselves (Huss and Röösli, 2006).

4. PREVALENCE

Despite the lack of scientific evidence of a causal relation, EHS cases in terms of people suffering from health symptoms which they attribute to EMF do exist. Some of them are suffering severely. In extreme cases, individuals can become disabled and even unable to pursue normal work or social life. Estimates on the prevalence of EHS within the general population differ widely. Initially, mainly case descriptions were published based on self-reported nonspecific symptoms such as eye discomfort, headache, muscular pain and skin disorders, frequently associated with work at video display units (VDU) (Knave et al., 1985; Bergdahl, 1995). An early prevalence study (Leitgeb, 1994, 1995) was based on an inquiry among a random sample of 200 men and women of the Austrian population. The results were dependent on the kind of assessment. On the basis of the questionnaire and self-definition, 10% declared themselves to be very sensitive to electricity without actually suffering from health symptoms. On the basis of the measurements of perception thresholds for directly applied electric currents on a randomly selected sample of 200 persons of the general population, it could be estimated that less than 2% of the general population are EHS. This was confirmed by an enlarged measurement campaign of electric current perception involving 708 adults (349 men and 359 women) aged between 16 and 60 years (Leitgeb et al., 2005a, b). A Swedish postal questionnaire survey among 10,670 adults with a response rate of 75% identified 1.5% individuals reporting to be hypersensitive or very allergic to electricity (Hillert et al., 2002). A Californian telephone interview-based study among 2,072 adults found 3.2% allergic or very sensitive to being near electric appliances, computers or power lines (Levallois et al., 2002). A Swiss telephone interview survey among 2,048 persons older than 14 years resulted in 5% EHS (Schreier et al., 2006).

A German telephone interview-based survey (Ulmer and Bruse, 2006) of a sample of 2,406 inhabitants identified 6% attributing repetitively experienced health symptoms to EMF. However, only about 1% reported themselves to be hypersensitive to EMF. EHS did not differ with regard to any socio-demographic parameter except education. EHS persons were more highly educated: 26% of EHS had a university-entrance diploma compared to 15% of the general population. Symptoms were attributed to RF-EMF sources (mobile phones and mobile phone basestations) as well as to ELF-EMF sources (TV set, alarm clock).

Apart from regional and cultural differences and prevalence-driving parameters such as public and media attention, different estimates can be explained by the weak definition of the term electromagnetic hypersensitivity as such: prevalence numbers might refer to a percentage of individuals suffering from health symptoms and attributing them to EMF or to persons just believing themselves to be hypersensitive without suffering from health symptoms. Lacking confirmation by specific EMF-related experience or perception, individual's beliefs are mostly based on their general sensitivity and/or experiencing sensitivities to other influences such as weather changes or temperature. Further, the wording of questions asking about electromagnetic hypersensitivity strongly influences the assessed prevalence numbers. In addition to that, in Germany investigations of a random sample of the general population comprising 340 individuals (177 female, 163 men, mean age 43.6 ± 13.0 years) demonstrated that the frequency of health complaints considerably depends on factors influencing perception of risks such as media attention and the context in relation to other risks (Frick et al., 2002).

5. ENVIRONMENTAL FIELDS

Environmental levels of ELF electric and magnetic fields and RF electromagnetic fields are usually several orders of magnitude below existing limits. However, this does not necessarily apply to electric devices. Under nominal load condition and in proximity, emissions of electric appliances can approach or even exceed reference field levels; those for electric fields up to 11-fold (Leitgeb et al., 2008b) and those for magnetic fields up to 80-fold (Leitgeb et al., 2008a). However, field levels rapidly decrease with distance (Preece et al., 1997; Kaune et al., 2002; ICNIRP, 2003; Leitgeb et al., 2008a; WHO, 2007). Figure 1 shows the results measured at 1,146 devices of 166 different categories comparing root mean square (rms) B_{rms} values with frequency-weighted sums of identified spectral peaks with amplitudes larger than twice the signal to noise ratio (SNR). The summation formula (ICNIRP, 1998) was slightly modified to generate an equivalent induction $B_{equ,ICNIRP}$ as follows (Leitgeb et al., 2008a):

$$B_{equ,ICNIRP} = B_{RL,50} \cdot \sum_{i=1}^{N} \frac{B_{peak,i}}{B_{RL,i}} \tag{1}$$

with $B_{peak,i} > 2SNR$.

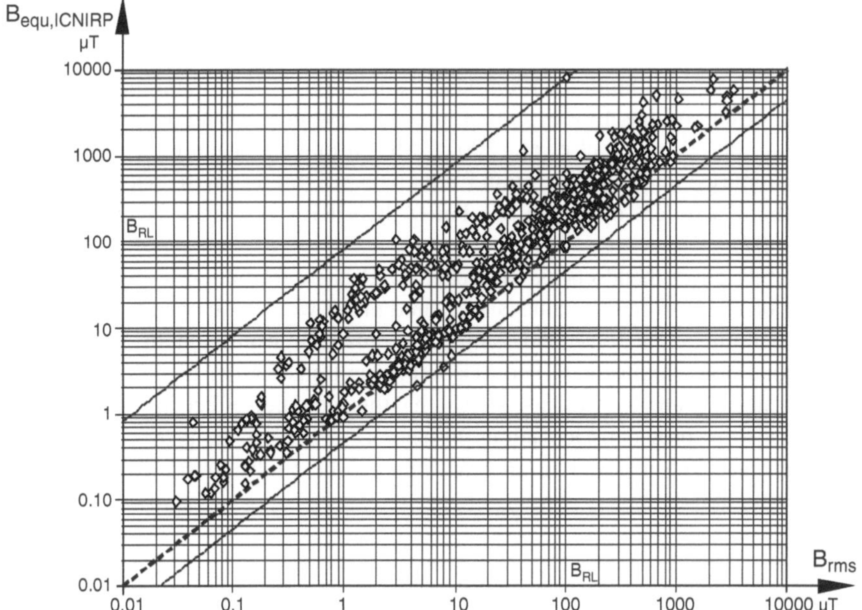

Figure 1. 50 Hz-equivalent magnetic induction $B_{equ,ICNIRP}$ emitted by 1,146 devices of 166 different categories in dependence on the B_{rms} value calculated in the frequency range 4 Hz–2 kHz (Leitgeb et al., 2008a). B_{RL}, 50 Hz magnetic field reference level; *fat dashed line*, direct proportionality; *dashed lines*, boundaries of measured values.

In homes, time-average background magnetic field levels are low. At the homes of 382 Canadian children the arithmetic mean value of magnetic fields (121 nT), determined from 2 consecutive 24 h measurements, was almost three orders of magnitude below the reference level. The span was 10–800 nT. The corresponding mean of electric field strengths, 14 V/m, was 360-fold below the reference value 5,000 V/m with a span of 0.82–65 V/m (Deadman et al., 1999). On the basis of the magnetic field measurements in children's sleeping rooms of 1,835 German residences, the 50 Hz median was 30 nT during daytime and 22 nT during nighttime (Schüz et al., 2000). Background magnetic field levels tend to be about fivefold higher in North America than in Europe, probably because of differences in power supply (more overhead wires, and lower household voltages consequently causing higher electric currents), higher power consumption and different grounding practices (Linet et al., 1997; UKCCS, 1999; Kavet et al., 2000).

Despite the rapid growth of RF-EMF emitting technologies, little is known about every day population exposure to such fields. Radio and TV transmitters are sparse because they expose large areas and, therefore, operate with high power. Mobile telecommunication antennas are forming a dense network of antennas with low output power and directional antenna characteristics. Since propagation is ruled by optical laws shadowing, scattering and multiple reflection considerably influence fields inside and outside buildings. In contrast to power line ELF magnetic fields, distance to transmitters is not an adequate surrogate for exposure levels. Relatively highest exposures are associated with direct visibility of the antenna. Determination of the general public exposure around mobile telecommunication base-stations resulted in maximum intensity values 2 orders of magnitude below limits and a span reaching down to 8 orders of magnitude (Bornkessel et al., 2007). Measurements around radio broadcast transmitters resulted in a maximum frequency-weighted sum of spectral components about 3 orders of magnitude below ICNIRP's reference level (Schubert et al., 2007).

Mobile phone handsets can approach SAR basic restriction levels up to 70% (BfS, 2008). However, this value is measured under worst case operation condition with maximum output power and continuous (pulsed) transmission. In every day use continuous power adjustment and discontinuous transmission mode considerably reduce real exposure. Studies have shown that this reduction effect critically depends on the network provider. Depending on network providers the proportion of calls with highest handset power levels was found to be 57.2% or 6.2%, respectively (Berg et al., 2004).

6. PERCEPTION

In recent centuries, numerous studies have been performed to investigate the hypothesis of self-declared hypersensitivity to EMF exposures and to clarify whether EHS are indeed able to perceive and/or react to environmental EMF exposure at environmental levels well below existing limits.

6.1. Adults

In the ELF range, both electric and magnetic fields interact with the body by inducing electric current densities. If EHS reactions are indeed associated with weak environmental fields, it should be expected as a necessary (but not sufficient) precondition that EHS cases should exhibit considerably lower thresholds than normal to perceive electric currents. Therefore, the ability to perceive electric currents was investigated. The normal range of perception of the general population was determined to compare results of EHS cases. Since EHS is not specifically restricted to RF- ELF, results gained in the ELF range should be helpful although not necessarily sufficient to quantitatively identify EHS.

Until recently, data on the ability to perceive electric currents were available only from groups which were small and did not represent the general population. Thompson (1933) reported on perception thresholds measured in 28 women and 42 men having their left hand immersed in a saline solution and contacting live parts (plates, wires). He found that women were about one-third more sensitive than men. Since that time, the factor 0.66 was used to account for women's increased electric sensitivity without further confirmation of such gender-related differences. In two series of experiments, Dalziel (1950, 1954) measured 60 Hz AC electric current perception thresholds of 115 men touching live copper wires. The integrated probability curve of data, pooled from three differently designed test series exhibited that 0.5% of men perceived currents below 400 μA. Osypka (1963) measured 50 Hz current perception thresholds of 50 healthy men aged between 19 and 39 years using two cylindrical handheld electrodes. His results were similar to Dalziel's. Irnich and Batz (1989) investigated 50 Hz electric current perception of 320 male and 166 female students, aged between 19 and 24 years while grasping cylindrical electrodes. In a second series, Batz and Irnich (1996) investigated 68 male and 133 female students putting their hands on flat live plates. The data of both studies were pooled and confirmed the existence of a gender difference; however, this time it was only 0.8-fold. Tan and Johnson (1990) investigated perception of 60 Hz electric currents flowing between two ECG electrodes placed 10 cm apart at one lower arm. They pooled data of an experiment on 38 men and 18 women and another one on 27 men and 14 women and reported considerably lower mean perception thresholds than published before, but no significant gender-related difference. Levin (1991) investigated 18 men and only 2 women with one hand resting on a 5 cm^2 metal plate and touching a live plate with the forefinger of the other hand. Reported perception thresholds were lower than in most other studies.

The inconsistent results reported by these studies could be explained by a study in a representative sample of the general population of 1,071 individuals, among them 349 men and 359 women aged between 16 and 60 years (Fig. 2). Between two paired electrodes 50 Hz electric currents were applied at the lower arm. It could be shown that the span of inter-individual perception thresholds now comprised two orders of magnitude (Leitgeb and Schröttner, 2002; Leitgeb et al., 2005a, b, 2006, 2007). This was considerably higher than the four to tenfold span reported previously (Thompson, 1933; Dalziel, 1959, 1954; Osypka, 1963; Irnich and Batz, 1989; Tan and Johnson, 1990; Levin, 1991; Reilly, 1992). Results confirmed that women (median perception threshold

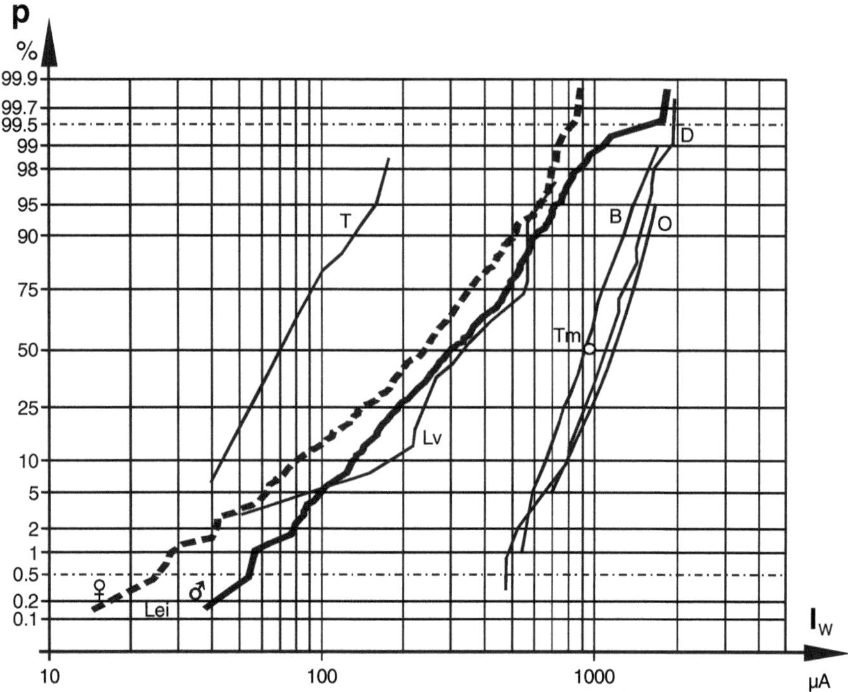

Figure 2. Cumulative probability dist ribution p of 50 Hz electric current perception thresholds of men: $I_{W.}$ D, Dalziel (1954), D_L, Dalziel (1946), B, Batz et al. (1996), Lv, Levin (1991), O, Osypka (1963), T, Tan et al. (1990), Tm, Thompson et al. (1933); *dash-dotted line*, let-go thresholds (Dalziel 1946) and of men ♂ and women ♀, Lei, Leitgeb et al. (2005b).

243 μA) were significantly more sensitive than men (median perception threshold 313 μA). However, quantitative gender-related differences depended on perception probability. At 0.5% perception probability, women's perception thresholds were 0.5-fold lower than men's. At 50% probability this difference was 0.77-fold (Fig. 2).

Cumulative perception probability curves showed that the lowest current level perceived was around 15 μA (Fig. 1). By numerically simulating intracorporal current density distributions, measured perception threshold currents could be associated with subcutaneous electric current densities thresholds. The lowest perceived current was associated with 12.4 μA/cm² (Leitgeb et al., 2006). It is known that apart from vision, excitation of one single cell is hardly sufficient to cause conscious perception. Therefore, stimulation of single cells can already occur below conscious perception levels. Accounting for such subliminal stimulation resulted in an excitation threshold 6.2 μA/cm², which is threefold higher than the 2 μA/cm² basic restriction level of ELF intracorporal current densities (in the CNS). Environmental fields are several orders of magnitude below reference levels and, hence, induce current densities below the lowest stimulation thresholds encountered so far.

In the ELF range both electric and magnetic fields induce intracorporal electric current densities. Since EHS persons exhibit symptoms in the vicinity of field sources emitting ELF *or* RF fields, it could be expected that if hypersensitivity existed, this should be indicated by considerably increased abilities perceiving ELF current densities (reduced perception thresholds). Therefore, specific investigations were conducted on self-declared EHS people. Since EHS is weakly defined, groups were compared which had been recruited by different strategies: The first group (12 men and 25 women) was composed of members of EHS self-aid groups which were most deeply convinced of a causal adverse role of EMFs. The second group (6 men and 23 women) comprised people who responded to advertisements seeking subjects with health symptoms attributed to electrical equipment and EHS patients. The third group (9 men and 15 women) contained worst cases selected from a list of 600 volunteers suffering from sleep disturbances they associated with RF EMF radiation from mobile telecommunication base stations. Electric 50 Hz current perception measurements performed at the lower arms showed that results within and among groups differed widely. All groups exhibited results overlapping the normal range (mean ± standard deviation) with some group members exhibiting lower-than-normal thresholds (Fig. 3). These

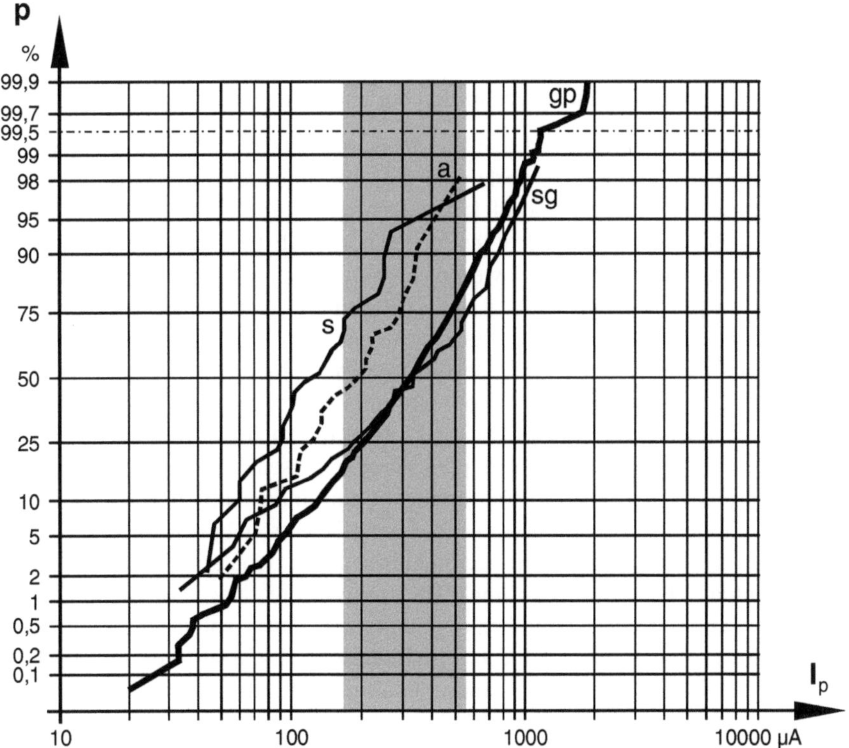

Figure 3. Cumulative frequency p of 50 Hz electric current perception thresholds I_p of pooled data of a 708 person sample of the general population (gp), 37 members of EHS self-aid groups (sg), 29 advertisement-recruited EHS volunteers (a) and 43 individuals suffering from EMF-attributed sleep disturbance (s), *gray*, normal range (Schröttner et al., 2007).

were 21% in the self-aid group 1, 52% in the advertisement responder group 2, and 60% in the RF EMF group 3 (Schröttner et al., 2007). The fact that individuals in the RF EMF group 3 exhibited similar reductions of perception thresholds to those associating their health symptoms to general "electromagnetic pollution" or ELF sources, demonstrated that lower ELF current perception thresholds are also a marker for persons claiming to be hypersensitive to RF EMF. This indicated that EHS exhibit a common signature in terms of increased sensitivity to electric currents.

The fact that EHS individuals did not exhibit perception thresholds orders of magnitude below those of the general population might be explained by different reasons:

> First, this might be due to the fact that the investigated sample of the general population might also have contained several EHS persons which enlarged the span of results. With an estimated prevalence of about 2–5%, the 708-person-sample of the general population could involve 14–35 EHS cases. However, apart from the fact that none of the volunteers had confirmed suffering from EMF-related health symptoms, data of the general population followed a log-normal distribution without any lag separating from EHS-attributable results.

> Second, this demonstrated that EMF-unaffected people might also have an increased ability to perceive electric current densities. Consequently, this ability might be a necessary precondition but not sufficient to develop EHS.

Detailed analysis demonstrated that the measured data of the general population follow a normal distribution overlapped (but not amended) by a second normal distribution at the sensitive end of low perception thresholds attributable to EHS cases (Leitgeb, 1998). The mean of the second normal distribution was only 6.7-fold below the general mean, and the lowest perception threshold found (15 μA) was only 18-fold below the median 270 μA of adults (men and women). These findings do not exhibit the postulated dramatic difference of orders of magnitude which should be expected as a consequence of hypersensitive reactions to environmental EMF several orders of magnitude below exposure limits. Since the span of results observed at EHS individuals did not extend beyond lowest thresholds of the general population the results did not support the hypothesis of hypersensitivity.

In Germany, perception of transcranial stimuli induced by transient magnetic fields was studied in 30 persons with self-reported electromagnetic hypersensitivity (Frick et al., 2005). Controls were recruited based on a population survey involving 758 individuals. From this, two non-EHS groups were selected according to the number of reported nonspecific health complaints. Thirty volunteers were identified with lowest level and 27 subjects with highest level of health complaints (without attribution to EMF). Onset of transcranial magnetic stimulation was identified by magnetically evoked electroencephalographic potentials (MEP). Magnetic stimulation exhibited no significant differences between any group either with regard to magnetic stimulation thresholds or MEG amplitudes. However, the three groups differed significantly with regard to differentiating between sham and true exposure. EHS exhibited the lowest ability while the control subgroup with the highest level of complaints performed best. With regard to complaints levels, EHS exhibited a high complaints level similar to the control group with highest complaints level.

Overall, these investigations demonstrated that people reporting hypersensitivity to electromagnetic fields sources were not able to perceive intracorporal electric current densities sufficiently better to justify the term hypersensitivity. Although there are indications to react more sensitively, observed differences were not large enough to explain EHS reactions to field exposures several orders of magnitude below recommended exposure-limiting reference field levels.

6.2. Children

Children are not just small adults and may respond to EMF exposures differently from adults. They have different susceptibilities during different periods of development they are going through, because of their dynamic growth and developmental processes during pregnancy, after birth, during infanthood and juvenile years. This does not already imply that children are more susceptible to any kind of exposure, but neither does it allow concluding the contrary. It is interesting to note, anyway, that EHS seems to be a phenomenon of adults, although children are supposed to have increased sensitivity to many factors including EMF (Kheifets et al., 2005a, b).

An early study suggesting an association between environmental ELF electric and/or magnetic fields was the epidemiologic study of Wertheimer and Leeper (1979) reporting on a significant increase of risk for childhood leukaemia near power supply wiring. In the meantime, a number of subsequent studies, meta-analyses and pooled analyses have been undertaken (Greenland et al., 2000; Ahlbom et al., 2000). Overall, there are consistent results indicating that the risk of childhood leukaemia might be two times greater for children exposed to 50/60 Hz magnetic fields at levels above 0.3–0.4 μT, which is about 2 orders of magnitude below recommended reference levels (IARC, 2002; ICNIRP, 2003; WHO, 2007) while no consistently elevated risks could be found for adults.

Without an established interaction mechanism or supporting evidence from other studies, in-vitro or in-vivo, and in view of the potential presence of selection bias, misclassification bias, confounding or chance, conclusions from epidemiologic findings remain difficult. In its evaluation the International Agency for Research on Cancer (IARC, 2002) came to the conclusion that, "There is limited evidence in humans for carcinogenicity of extremely low frequency magnetic fields in relation to childhood leukaemia."

If there is indeed a causal relationship, epidemiologic results would indicate that children might have a vulnerability more than 2 orders of magnitude more than that of adults.

Because of ethical reasons, quantitative results on children's sensibility to electricity are sparse. Electric currents perception of children was investigated in 240 pupils (Leitgeb et al., 2006). Overall, 117 girls and 123 boys, aged 9–16 years, were studied as part of demonstrations within their physics lessons. This was done on a voluntary basis with written consent of parents, teachers and heads of schools.

Results showed that girls were more sensitive than adult men. However, their perception ability remained well within the span of women's results. No clear age-dependence could be found for girls. In contrast to this finding, perception thresholds of boys were different. Boys and girls were similarly sensitive to electric currents at ages from 9 to 11 years. However, with age gender differences evolved and boys became more and more insensitive until their perception ability reduced to that of adult men while the sensitivity of the girls remained fairly constant with no significant difference from that of adult women (Fig. 4). These results demonstrated that the widespread precautious assumption that children were much more sensitive than adults could not be confirmed with regard to ELF electric currents.

Since biological interactions are governed by different physical mechanisms in the RF range (where heating replaces stimulation), results and risk factors gained in the ELF range cannot be directly extrapolated to RF electromagnetic fields. This explains why there are no epidemiological studies in the RF range with findings similar to those of ELF magnetic field exposures indicating potentially increased childhood cancer risks. Regarding long-term exposure and limited observation periods of new technologies, concerns about the potential vulnerability of children to RF EMF have been raised, for many reasons. Mobile phones expose their developing nervous system to a higher degree and for a longer lifetime than adults (Kheifets et al., 2005a; Leitgeb, 2008). Increased absorption can be expected because their brain tissue exhibits an increased electric conductivity, RF penetration depths are greater relative to brain structures and their decreased skull thickness, and more flexible pinna are less efficient to keep distance to mobile phone handsets. Figure 5 shows the development of several anatomical parameters with age.

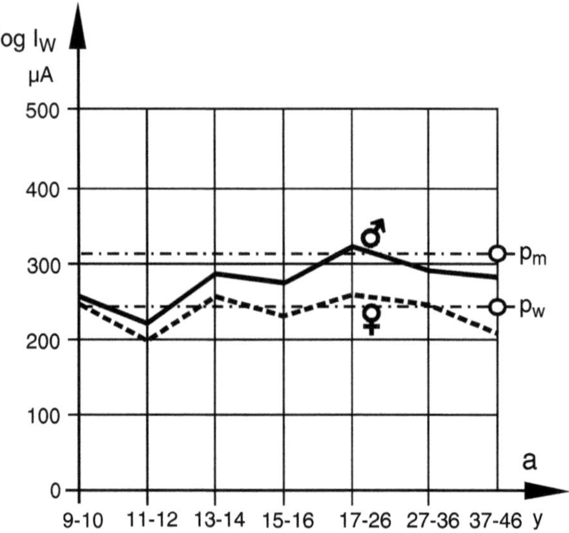

Figure 4. Dependence of electric current perception threshold medians I_w on age classes from children to adults (*full line*, male; *broken line*, female). p_m, median perception threshold of adult men, p_w, median perception threshold of adult women (Leitgeb et al., 2006).

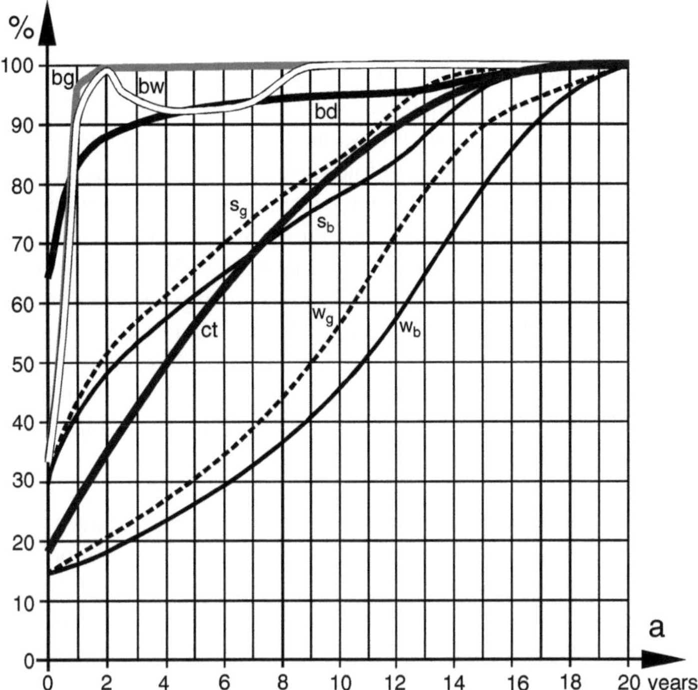

Figure 5. Relative growth curves for anatomical parameters from birth to adult. *bg* gray brain tissue, *bw* white tissue, *bd* brain diameter, *ct* cranial thickness, *sg* stature (girls), *sb* stature (boys), *wg* body weight (girls), *wb* body weight (boys) (Leitgeb, 2008b; Ogden et al., 2002).

During the first 2 weeks after conception the embryo is very sensitive to lethal effects of toxic agents and much less sensitive to induction of malformation ("all-or-none period"). During the following 6–8 week organogenic period, toxic agents with teratogenic potential might cause malformations of the visceral organs. Neuron proliferation, differentiation and migration make the CNS particularly vulnerable during weeks 8–15. During the final foetal period vulnerability to deleterious effects remains high, while it decreases for formerly susceptible organs including the CNS. Although most neurons are already existent at birth, during the first 2 postnatal years the connections grow between neurons, reducing the high water content due to increased nerval myelin in brain tissue (myelination). Because the period from embryonic life to adolescence is characterized by growth and development, deleterious effects could occur at lower levels and be more severe, or lead to effects that would not occur in adults. Therefore, timing of exposure might also be critical. For ionizing radiation, excess risk for leukaemia, brain and thyroid cancer is highest during childhood exposure.

The most relevant effect of RF EMF interaction is heating. Therefore, RF EMF impose heat load to the whole body or locally to sensitive parts. Investigations whether children brains are more susceptible to higher exposure compared with adults resulted in different conclusions. Some groups concluded that energy absorption is not increased (Christ and Kuster, 2005; Wiart et al., 2005; Hadjem et al., 2005), marginal (Andersen, 2003), more likely caused by individual differences in head anatomy and geometry rather than age (Keshvari and Lang, 2005) or larger with decreasing difference from adults towards adolescence (Leitgeb, 2008). Reasons for these different conclusions are manifold such as differences in numerical simulation, anatomic head modelling including the distance-determining pinna, tissue segmentation, electric tissue parameters, modelling the radiating source, simulation parameters (voxel size, meshing, algorithm) and kind of SAR calculation (volume size, geometry), etc. In principle, children's brains are exposed more because of the less efficient spacing of the phone by the more flexible pinna, the smaller skull thickness, the higher absorption coefficient of brain tissue (Gabriel, 2005) and the more unfavourable phone position. Reported differences are not larger than the reduction factor of 5, which had been implemented in guidelines to account for sensitive groups within the general population.

Concerning the use of mobile phones, the main difference between today's children and adults may be the longer lifetime exposure, particularly in view of the increasing prevalence among juveniles and the trend to start using mobile phones at earlier ages, with higher frequency and longer duration per use (Schüz, 2005). Regarding potential long-term health effects and the paucity of data, WHO suggests low-cost precautionary measures are appropriate in particular because some exposures are close to guideline limits.

7. PROVOCATION STUDIES

Apart from perception ability of directly applied electric currents numerous provocation studies, either blind or double-blind, were conducted with EHS to investigate their increased ability to react to field exposures either by detecting them more reliably or developing more symptoms than others (Rubin et al., 2005; Röösli, 2008). Two types of provocation studies were conducted: Laboratory studies with simulated exposures which were most frequent, and field studies with real exposure or where exposure to real environmental fields was varied by shielding (Leitgeb et al., 2008c) or randomly activating mobile phone base stations (Heinrich et al., 2007).

Typically, volunteers were subjected to two different situations with and without field exposure, usually in a random order. However, studies used quite different exposure durations ranging from some seconds to several days.

7.1. ELF Studies

Rea et al. (1991)) reported on 16 students preselected from 100 EHS colleagues also as sensitive to other chemical factors which were responding to AC magnetic fields. In a second series, these 16 students selectively exhibited symptoms during magnetic field exposures at individual "resonant" frequencies (some at 0.1–10 Hz, the majority at 50 Hz) while a healthy control group did not develop symptoms during any of these exposures. At that time the study had gained public attention and had strengthened convictions in EHS. However, it has been scientifically criticised on several methodological grounds such as selection of individuals, reproducibility of exposure and even uncertainty about whether or not it was blind (ICNIRP, 2003). A subsequent study by the same group (Wang et al., 1994) could not replicate the initial findings.

A USA study (Omura et al., 1991) reported on synergistic interaction of EMF with incorporated concentrations of mercury and/or lead. Exposures were associated with health problems and changes of hormones and neurotransmitters such as acetylcholine or thromboxane B2. Changes were reported to follow only 5 min exposure to 10 V/m (60 Hz) electric fields or 60 mT static magnetic fields from magnetic disks or credit cards. They lasted for hours after exposure.

Most early provocation studies concentrated on electric and magnetic fields of VDUs. They could not find any evidence that persons suffering from EHS reacted differently than healthy controls or experience more symptoms during periods when devices were activated (Lonne-Rahm et al., 2000; Flodin et al., 2000; Nilsen, 1982; Swanbeck and Blecker, 1989; Hamnerius et al., 1993, 1994; Sandström et al., 1993; Wennberg et al., 1994). Reactions were found to correlate with belief of the presence rather than the real exposure to fields indicating a nocebo effect (the inverse of placebo in terms of occurrence of adverse rather than benign effects due to beliefs). Comparison of individual's self-classification with measured sensitivities to electric 50 Hz currents demonstrated that convictions of individuals did not significantly correlate with reduced perception thresholds (Leitgeb, 1994). A Swedish study (Sjöberg and Hamnerius, 1995) reported significantly worse symptoms compared with sham in only one out of 10 test series; however, no correction for multiple testing was made. A Norwegian group reported on small delayed beneficial effects of electric VDU shields, however, they were not able to replicate their findings (Oftedal et al., 1995, 1999). Some morphological evidence was reported by Johanssen et al. (2001) who compared cutaneous biopsies of 13 healthy subjects before and after 2 or 4 h exposure to conventional TV or PC screens. Five of the volunteers exhibited an increase in the number of mast cells and their changed distribution in the facial skin while in 2 volunteers a decrease of the mast cell number was found but a shift towards the upper dermis was observed. One day after exposure, the number and location of mast cells were normalized in all subjects.

In Sweden (Lyskov et al., 2001a, b), 20 EHS (15 female, 5 male, mean age 45.8 ± 0.7 years) and 20 healthy controls (15 female, 5 male, mean age 45.0 ± 0.7 years) were exposed to 15 s on/off cycles of 60 Hz/10 μT magnetic fields or sham. The total test period was 40 min. It was divided into two 10-min rest periods and two 10-min periods for performing mathematical tasks. Parameters recorded were EEG, VEP, electrodermal activity, ECG, blood pressure and mathematical performance. Statistical analysis resulted in significant differences between the two groups with regard to heart rate ($p < 0.01$), heart rate variability ($p = 0.02$) and electrodermal activity ($p = 0.04$). However, no corrections were made for multiple parameter statistical testing. The authors concluded that the chosen magnetic field level would not affect EHS or controls, speculating that EHS cases exhibited a shift in baseline values of investigated parameters which could indicate a distinctive physiological predisposition to sensitivity to physical and psychosocial environmental stressors.

In a double-blind Swiss laboratory study (Müller, 2000; Müller et al., 2002), the ability to perceive weak 50 Hz electric and magnetic fields (100 V/m + 6 μT) was tested in 63 subjects (49 self-reported EHS and 14 healthy controls). Fields were applied in randomized sequence (field on/field off) in 2-min intervals. Seven out of all 63 subjects exhibited statistically significant results. However, there was no relevant difference between healthy and EHS subjects, either with regard to field perception or to number and type of symptoms developed during tests. Another part of these investigations concentrated on night-time exposure to 50 Hz magnetic fields of 53 self-declared EHS. Physiological parameters were monitored such as heart rate, breathing, movements and body position (indicating potential attempts to escape exposure). Sleep quality and daytime wellbeing, movement, breathing and heart rate did not show significant changes. However, night-time body position monitoring significantly indicated attempts to move away from the magnetic field zone (Müller, 2000).

In a German study (David et al., 2004), 24 EHS volunteers and 24 healthy controls were randomly exposed to 10 μT/50 Hz magnetic fields for 2 min with 3 min for recovery (two sessions per 10 trials). No significant difference could be found between the two groups.

7.2. RF Studies

7.2.1. RF Field Studies

In Switzerland, during 1992 and 1998, studies on 404 persons living at different distances from a short-wave transmitter antenna were performed assessing somatic and psychosomatic symptoms including sleep quality by questionnaires when the transmitter was switched off for 3 days (Abelin et al., 2005), and in another study after final shut-down some years later (Altpeter et al., 2006). In both cases prevalence of difficulty falling asleep and nocturnal arousals increased with exposure. However, the study suffered from the fact that people could become aware of their exposure and that information exchange among those exposed could not be excluded.

In France (Santini et al., 2003) and Spain (Navarro et al., 2002), inquiries were made in the neighbourhood of mobile phone base stations and results analysed independent of distance to the antenna. Both reported a higher prevalence of symptoms at smaller distances. However, shortcomings like bias, unknown response rates and the inadequate approach using distance as a surrogate for exposure make conclusions invalid.

In Austria (Hutter et al., 2006), 365 persons living in the neighbourhood of mobile phone base stations were investigated. The results were analyzed as a function of distance and measured field levels. Some associations of sleep disorders with measured base station field levels were found but they were also highly significantly associated with the people's concerns.

In Egypt (Abdel-Rassoul et al., 2007), a cross-sectional inquiry study involved 37 people living below and 48 opposite from base stations. It was reported that the prevalence of nonspecific health symptoms such as neurobehavioural complaints (headache, memory changes, dizziness, tremors, depressive symptoms and sleep disturbances) were significantly higher ($p<0.05$) among people living close to base stations compared with 80 matched controls.

In a German field study (Heinrich et al., 2007), for 3 months perception and symptoms were investigated by a daily online questionnaire. Ninety-five employees (67 male, 28 female) were randomly exposed to RF EMF emitted from a mobile phone base station on an office building which was switched on and off for 2–3 day intervals. Operation condition was not identified better than chance, and symptoms developed; however, they were significantly correlated only with the belief of phone operation rather than with real exposure.

In Austria, with a new study design of protection (shielding) from rather than provocation to EMF, 43 volunteers reporting sleep problems due to RM-EMF from mobile phone basestations were investigated in their sleeping rooms at home (Leitgeb et al., 2008c). Sleep quality of volunteers was assessed for ten consecutive nights (with the first night for accommodation) under three test conditions (true-shield, sham-shield and control) selected in random order. Shielding conditions were single-blinded for controlling shielding efficiency, while data analysis was performed double-blind. Sleep quality was assessed by subjective parameters derived from standardised questionnaires and objective parameters from polysomnographic recordings. RF-EMF emmission was continuously recorded frequency-selectively. Pooled analysis did not exhibit statistically significant EMF-dependent sleep parameters changes, either on total RF-EMF emmissions or on base station signals. The majority of volunteer-specific analysis did not show significant effects on sleep parameters. Subjective sleep parameters of several volunteers (16%) exhibited significant placebo effects. However, 9% of volunteers showed consistent statistically significant prolongations of sleep latency times in shielded nights.

7.2.2. RF Laboratory Studies

In the RF range, EHS studies concentrated on exposure to mobile telecommunication fields from handsets or base stations. In Finland (Koivisto et al., 2001), 48 healthy subjects (24 males, 24 females, mean age 26 years, span 28–49 years) were studied in two experiments with 60 min and 30 min exposures to 900 MHz GSM fields from mobile phones, respectively. The reported symptoms of headache, dizziness, fatigue, itching, tingling or redness of the skin, and a sensation of warmth did not reveal any significant differences between exposure and sham.

Hietanen et al. (2002) investigated 20 volunteers (13 women and 7 men) reporting being sensitive to cellular phones (some of them also to other EMF sources).

Volunteers were exposed to one analogue NMT phone (900 MHz) and two digital GSM handsets (900 MHz and 1,800 MHz, respectively) operated at maximum power (1 W_{rms}(cw), 0.25 W_{rms}, and 0.125 W_{rms}, respectively). One exposure (sham or true) lasted for 30 min, followed by 1 h break. Each volunteer was tested three or four times in one day. Blood pressure, heart and breathing rate were monitored. Nineteen of the volunteers reported nonspecific symptoms, most of them related to the head. However, more symptoms appeared during sham exposure. None of the persons could distinguish between sham and real exposure. Higher heart rate and blood pressure at the beginning of a session was attributable to stress. No statistically significant difference was found between sham and real exposure to any cellular phone.

In a study performed in the Netherlands (Zwamborn et al., 2003; HCN-EMFC, 2006), a group of EHS (11 men, 25 female, mean age 55.7 ± 12.0 years) and healthy volunteers (22 men, 14 female, mean age 46.6 ± 16.4 years) were exposed to RF-EMF base station signals emitted by GSM 900 MHz, GSM 1,800 MHz and UMTS antennae with effective electric field strengths of 0.7 V/m (GSM) and 1 V/m (UMTS). No effect on well-being was found in either exposure group at either GSM exposure. UMTS-like signals were associated with a small but statistically significant decrease in well-being after 30 min exposure in both exposure groups; however, the control group was more affected. Cognitive functions were significantly changed during GSM and UMTS exposure, however, with inconclusive patterns of cognitive variables with regard to type of signal and exposed group. These significant differences were found for single parameter testing. After correction for multiple parameters testing, only one significant result remained, namely, the difference in performing memory comparing tests during UMTS exposure. Performance was faster in the control group compared with sham exposure. The comparison between EHS and controls suffered from critically different composition of the two groups.

In a double-blind replication study performed in Switzerland (Regel et al. 2006), 33 persons (14 men, 19 female) with self-reported sensitivity to RF-EMF and a control group of 84 subjects (41 men and 43 female) were exposed to sham and UMTS-like base station signals (1 V/m and 10 V/m). Each exposure lasted for 45 min. In that time two series of cognitive tasks had to be performed starting at the beginning and after 20 min, respectively. Sessions were preceded by one training session and were performed three times at 1 week intervals. All subjects were between 20 and 79-years old (37.7 ± 10.9 years). The results did not show any difference between EHS and controls and no impact on wellbeing or ability to perceive exposure. Cognitive performance was not significantly changed at any field strength after correction for multiple testing.

In the United Kingdom (Eltiti et al., 2007), 44 self-reported sensitive and 114 controls were studied during open (informed) and double-blind provocation with combined 10 mW/cm² base station like GSM signals (5 mW/cm² 900 MHz + 5 mW/cm² 1,800 MHz) and with UMTS signals in comparison to sham. Subjective well-being was assessed by visual analogue scales and symptom scales. In addition, physiological parameters were measured such as pulse, heart rate, and skin conductance. Subjects performed mental arithmetics, digit symbol substitution, and digit span tasks. Exposure lasted for 15 min or 20 min for assessing well-being, 8 min for cognitive tests, and 5 min for on/off perception with 2 min washout intervals in between. During the open provocation, EHS individuals reported lower well-being during both GSM and UMTS signals, and controls developed more symptoms during open UMTS exposure compared with sham. However, double-blind exposure to GSM or UMTS signals did not cause effects in either group. No significant differences were found between EHS and controls.

In Finland (Hietanen et al., 2002), in a double-blind study the ability to detect whether mobile phones were on or off was investigated in 20 volunteers with self-declared sensitivity to mobile phone RF-EMF (7 men, mean age 47.1 years and 13 women, mean age 50.6). Apart from sham, they were exposed to an analogue NMT phone (output power 1 W), a 900 MHz pulsed GSM phone (average output power 250 mW) and a 1,800 MHz pulsed GSM phone (average power 125 mW). Tests lasted for 30 min followed by 1 h break. Blood pressure, heart rate, and breathing were monitored. Three or four tests were performed in random order. Various symptoms were reported, most of them related to the head. Women developed more symptoms than men. No significant difference could be found between sham and exposure; none of the subjects were able to distinguish between sham and real exposure. Overall, no association between exposure to mobile phone radiation and symptoms could be found.

In United Kingdom (Rubin et al., 2006a, b), 60 subjects were investigated who reported getting headache within 20 min mobile phone use (starting with 31 men and 40 female, mean age 37.1 ± 13.2 years) and 60 controls without symptoms (27 men, 33 female, mean age 33.5 ± 10.2 years). Volunteers were exposed to EMF fields emitted from a test mobile phone handset mounted slightly above and behind the left ear. Test conditions were 50 min exposure to 900 MHz GSM and 900 MHz cw signals, causing a local SAR of 1.4 W/kg. For sham exposure a similarly heated dummy handset was used. The main target of investigation was headache. Additional symptoms such as burning sensations, skin sensations, eye pain difficulty concentrating, and dizziness were noted. Volunteers were also asked to guess whether fields were on or off. The study showed that EHS cases developed partly severe symptoms, which for five individuals were the reason to withdraw prematurely. However, since severe symptoms were also developed during sham exposure, no significant difference was found between different exposure conditions. Controls developed almost no symptoms with the exception of some feeling of warmth. No evidence was found indicating that EHS could detect mobile phone signals or that they react to them with increased symptom severity. As sham exposure was sufficient to trigger severe symptoms, psychological factors, in particular nocebo, may play an important role.

In Sweden, 20 subjects experiencing symptoms when using mobile phones were compared with 20 healthy controls (Wilén et al., 2006). Each subject participated in two 30 min tests with sham and true exposure of the head to 900 MHz GSM, $SAR_{1g} = 1$ W/kg, emitted by an indoor base station antenna. No significant differences were found in heart rate, respiration, local blood flow, electrodermal activity, flicker fusion frequency, and short-term memory, except a significant prolongation of reaction time (at the first trial only, it disappeared when the test was repeated) and a shift in heart rate variability toward sympathetic dominance in the autonomous nervous system during flicker frequency and memory tests; however, these appeared in either condition.

In Norway, 42 individuals reporting developing headache when using mobile phones responded to a media call (Oftedal et al., 2007). On the basis of the outcome of an open provocation test 38 subjects were eligible, and finally 17 (5 women and 12 men) mean age 39 years (span 20–58) were included in the study. For exposure wall-mounted base station antennae emitting 900 MHz GSM signals exposed subjects to local $SAR_{10g} = 0.8$ W/kg. One session included one pair of exposures (30 min sham/true). Up to 4 sessions were planned with 2 days in between. In addition to reporting symptoms, heart rate and systolic and diastolic blood pressure were monitored. Fifty-six pairs of trials were conducted. Changes of physiological parameters occurred but did not depend on exposure condition. The degree of reported symptoms was low. If reported, the time course of symptoms was the same for headache and other symptoms and was the same for real and sham exposure. The study gave no evidence that RF-EMF from mobile phones could cause pain or discomfort or influence the measured physiological parameters.

In a Swedish double-blind crossover study, 38 EHS associating headache and vertigo with mobile phone use and 33 healthy controls were randomly exposed for 3 h to GSM handset exposure or sham (Hillert et al., 2008). Encountered symptoms were scored before and after 90 min and 165 min exposure on a 7-point Lickert scale. Neither group could detect RF exposure better than by chance. EHS did not experience more or more severe symptoms. Headache was reported even more frequently by the control group.

To test whether healthy subjects could detect mobile telecommunication RF-EMF, 84 volunteers (57 women, mean age 23.5 ± 5.4 years and 27 men, mean age 26.1 ± 6.1 years) were recruited in Turku, Finland, through advertisements announcing €50 award for good performance (Kwon et al., 2008). A 900 MHz GSM mobile phone handset was mounted in cheek position at the preferred side (17 left, 67 right) causing local SAR_{10g} of 0.86 W/kg. Scores were requested after 5 s, and the following trial was started 1 s after the answer. Tests were made in 6 sessions with 100 trials each. There was a response bias toward “handset off”. Two participants in one session exhibited a high correct score of 97% and 94%, respectively. However, they could not replicate their results and, overall, did not perform better than average. Overall, none of the volunteers were able to win the prize. In spite of the many trials and volunteers, the conclusions from this study are limited because of the extremely short exposure duration and washout period.

Figure 6 shows results of provocation tests demonstrating that EHS did not exhibit increased probability to detect and/or perceive electromagnetic field exposure compared with normal volunteers.

7.3. Neurophysiological Studies

So far, attempts to identify EHS by a characteristic symptom cluster failed (Bergqvist et al., 1997). Reported symptoms comprise a variety of nonspecific health problems

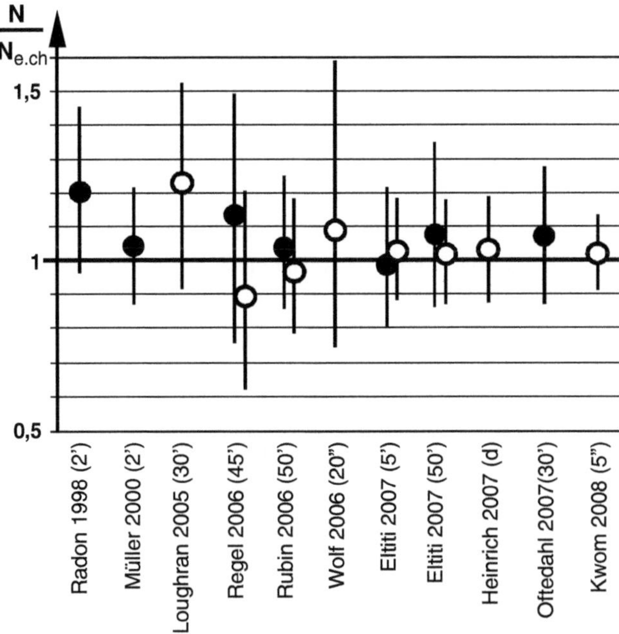

Figure 6. Number of correctly detected field exposures *N* divided by correct answers expected by chance $N_{e.ch}$ of nonsensitive (*open circles*) and EHS volunteers (*full circles*), exposure duration in parentheses (modified from Röösli, 2008).

similar to those known to be associated also with other environmental factors. For this reason, WHO (2005) concluded EHS resembles multiple chemical sensitivities, another disorder associated with low-level environmental exposure to chemicals. The collection of disorders such as dermatological neurasthenic and vegetative symptoms is not part of any recognized symptom. It is shared by other nonspecific medically unexplained symptoms (MUS) associated with external influences summarized as idiopathic environmental incompatibility (IEI).

Quantitative investigations of 94 patients (53 women, 41 men, mean 38 years, span 21–79 years) with health symptoms attributed to dental amalgam or indoor toxins could not substantiate personal convictions, while psychiatric disorders were found in 66% (ICD-10). Somatisation score of 0.9 was considerably higher than the 0.36 found in controls (Kraus et al., 1995).

Lyskov et al. (2001a, b) investigated 20 patients (11 female, 9 male, mean age 47 ± 5 years) with EMF exposure-associated neurasthenic symptoms such as general fatigue, weakness, dizziness, headache, and facial skin (itching, tingling, redness). Their results were compared with those of 20 healthy controls (12 female, 8 male, mean age 44 ± 7 years). Neurophysiological parameters were measured such as blood pressure, heart rate, sympathetic skin response, respiration, flicker fusion frequency, EEG, and visual evoked potentials (VEP). Single-parameter statistical analysis exhibited significant differences of flicker fusion frequency ($p=0.005$), heart rate ($p=0.044$), heart rate variability ($p=0.04$), and sympathetic skin responses such as inset latencies ($p=0.003$), peak latency ($p=0.033$), and amplitude ($p=0.01$). However, no correction for multiple testing was made. If multiple testing of 22 parameters was considered, this would lead to a Bonferroni-corrected p-value of 0.0023 (=0.05/22) with no more significant results remaining. The authors' concluded results indicate that the investigated EHS group exhibited a shift of baseline characteristics of the central and autonomous nervous system indicating a tendency toward hyper-sympathotone hyper-responsiveness to sensory stimulation and probably heightened arousal.

Medical metaanalyses confirmed that medically unexplained functional somatic symptoms are related to but not fully dependent on depression and anxiety (Henningsen et al., 2003). Sometimes medically unexplained symptoms might be associated with objective cognitive abnormalities caused by complex interaction between biological and psychological factors rather than by traditionally defined neurological diseases (Binder and Campbell, 2004).

To clarify whether dysfunctional cortical regulations could play a role in electromagnetic hypersensitivity, cortical excitability was studied in Germany by transcranial magnetostimulation (Landgrebe et al., 2007). Twenty-three individuals with self-reported EMH and two control groups (49 subjects) with low and high levels of unspecific health complaints were investigated. Compared with both control groups, EHS cases showed reduced intracortical facilitation. No differences were seen at motor thresholds and intracortical inhibition. In an extended study (Landgrebe et al., 2008) involving 89 EHS and 107 matched controls, thresholds of perceiving single transcranial magnetic stimulation pulses applied at the dorsolateral prefrontal cortex did not differ. However, discrimination ability was significantly reduced in EHS: 60% of EHS reported sensations during sham compared with 40% of controls. The authors conclude that these results demonstrated cognitive and neurobiological alterations supporting the hypothesis that altered CNS function may account for perceived symptoms in EHS and a higher genuine individual vulnerability.

8. TREATMENT

Although convincing evidence of a causal role of EMF is missing, the fact remains that there are people suffering and exhibiting symptoms. Experience shows that EHS is not suddenly appearing but evolves with time starting with temporary symptoms of unclear origin, seeking causal factors, associating them with EMF, finding reassurance in media, internet, and friend's opinions, and possibly ending with severe symptoms and deep conviction of a causal role of EMF (Hillert, 1998). Case reports demonstrate that afflictions can even be severe enough to make them change their lifestyle, quit their work, and leave urban areas to find relief in housing free from electricity. There is agreement that EHS deserve help.

A systematic review of medical treatments reported that options were limited (Rubin et al., 2006b). The investigation suggested that cognitive behavioral therapy might be effective (Hillert, 2004). Interventions to measure EMFs and taking actions to reduce exposure are assessed controversially. The advantage of responding to the concerns of the patient must be balanced against possible risks of downplaying other potentially relevant factors and inducing fear in yet unaffected persons (Hillert, 1998).

WHO (2005) recommends that rather than focusing on people's perceived need for reducing EMF, treatment of EHS should focus on health symptoms and the clinical picture including

A medical evaluation to identify and treat any specific conditions potentially responsible for the symptoms

A psychological evaluation to identify alternative psychiatric/psychological conditions potentially responsible for the symptoms

An assessment of the site where patients develop their symptoms (workplace and/or home)

Reduction of stress, as appropriate

9. DISCUSSION

Overall, convincing experimental evidence for EHS reactions to environmental EMF exposures is still missing, in the ELF range as well as in the RF range. The EHS hypothesis is challenged by the following arguments:

There is no plausible explanation for the development of similar health symptoms due to exposure to ELF and/or RF EMF. In view of the different underlying physical laws and biological interaction mechanisms of ELF and RF electromagnetic fields, it cannot be explained why EHS should be an overarching phenomenon relevant for the entire frequency range of nonionising technical fields.

Quantitative measurements of sensitivities did not convincingly support the hypothesis that hypersensitive reactions could occur at environmental field levels several orders of magnitude below thresholds for relevant biological responses. Measured differences in sensitivities were not large enough to exceed the reduction margin introduced in exposure limit derivation.

Individuals suffering from EHS did not exhibit perception thresholds of electric and magnetic stimuli below the overall span exhibited by the general population.

Provocation studies demonstrated that subjects with self-attributed EHS were not able to detect exposures better than chance, either in the ELF or in the RF range. When symptoms were developed they were correlated with belief in exposure rather than with real situations. Overall, EHS exhibited a higher false alarm rate than controls. This explains the slightly but insignificantly higher rating of the field-on situation.

Epidemiological studies on childhood leukaemia and environmental magnetic field levels indicated that, if at all, children were more sensitive to EMF. However, EHS remains a phenomenon of adults rather than children.

However, the inhomogeneity of investigated groups prevents a final conclusion whether or not hypersensitivity to electromagnetic fields exists:

Most studies selected volunteers on the weak basis of self-reported sensibility without implementing quantitative or even semi-quantitative identification criteria. Therefore, a negative outcome of provocation studies could still be challenged by assuming inappropriate composition of investigated groups. This applies in particular to volunteers recruited from responders to open calls especially in cases where financial compensation was offered.

Exposure regimes were and still are based on weak grounds. No reliable data exist on response latency. Individual reports vary widely. Therefore, durations of exposures were chosen arbitrarily. In fact, they varied from seconds to hours and days. It is unclear which minimum exposure time would be necessary to develop EMF-related reactions or symptoms.

Likewise it is unclear what minimum recovery time is needed to assure independent results in sequential testing. Therefore, washout intervals between tests were chosen arbitrarily and differed considerably, from seconds to hours. Therefore, crossover artifacts and erroneous scores cannot be excluded from many studies.

It is not even clear whether EHS, if it exists, is a phenomenon of exposure to single subject-specific resonance frequencies, to frequency ranges such as ELF or RF, or specific signal signatures. Therefore, it cannot be finally determined whether or not the chosen exposure conditions were adequate.

Therefore, WHO (2005) concluded that, "EHS is characterized by a variety of nonspecific symptoms that differ from individual to individual. The symptoms are certainly real and can vary widely in their severity. Whatever its cause, EHS can be a disabling problem for the affected individual. EHS has no clear diagnostic criteria and there is no scientific basis to link EHS symptoms to EMF exposure. Further, EHS is not a medical diagnosis, nor is it clear that it represents a single medical problem."

REFERENCES

Abdel-Rassoul, G., El-Fateh, O.A., Salem, M.A., Michael, A., Farahat, F., El-Batanouny, M., and Salem, E., 2007, Neurobehavioral effects among inhabitants around mobile phone base stations. *Neurotoxicology* 28(2): 434–440.

Abelin, T., Altpeter, E.S., and Röösli, M., 2005, Sleep disturbances in the vicinity of the short-wave broadcast transmitter Schwarzenburg. *Somnologie* 9: 203–209.

Ahlbom, A., Day, N., Feychting, M., Roman, E., Skinner, J., Dockerty, J., Linet, M., McBride, M., Michaelis, J., Olsen, J.H., Tynes, T., and Verkasalo, P.K., 2000, A pooled analysis of magnetic fields and childhood leukaemia. *Br J Cancer* 83: 692–698.

Altpeter, E.S., Röösli, M., Battaglia, M., Pfluger, D., Minder, C.E., and Abelin, T., 2006, Effect of short-wave (6–22 MHz) magnetic fields on sleep quality and melatonin cycle in humans: The Schwarzenburg shut-down study. *Bioelectromagnetics* 27: 142–150.

Andersen, V., 2003, Comparison of peak SAR levels in concentric sphere head models of children and adults for irradiation by a dipole at 900 MHz. *Phys Med Biol* 48: 3263–3275.

Batz, L., and Irnich, W., 1996, Untersuchungen zur elektrischen Wahrnehmungs-schwelle und Interpretation der Ergebnisse (investigation on electric perception thresholds and interpretation of results). *Biomed Technik Ergänzungsband1* 41: 556–557.

Berg, G., Schüz, J., Samkange-Zeeb, F., and Blettner, M., 2004, Assessment of radiofrequency exposure from cellular telephone daily use in an epidemiologic study: German validity study of the international case-control study of cancers of the brain – Interphone Study. *J Expos Anal Environ Epidemiol* 14: 1–8.

Bergdahl, J., 1995, Psychological aspects of patients with symptoms presumed to be caused by electricity of visual display units. *Acta Ondoto Scand* 53: 304–310.

Bergqvist, U., Vogel, E., Aringer, L., Cunningham, J., Gobba, F., Leitgeb, N., Miro, L., Neubauer, G., Ruppe, I., Vecchia, P., and Wadman, C., 1997, *Possible health implications of subjective symptoms and electromagnetic fields. CEC DG V report*, Arbetslivinstitutet, Stockholm.

BfS, 2008, Mobile phone handsets SAR values. http://www.bfs.de

Binder, L.M., and Campbell, K.A., 2004, Medically unexplained symptoms and neuropsychological assessment. *J Clin Experim Neuropsych* 26: 369–392.

Bornkessel, C., Schubert, M., Wuschek, M., and Schmidt, P., 2007, Determination of the general public exposure around GSM and UMTS base stations. *Radiat Prot Dosim* 124: 40–47.

Christ, A., and Kuster, N., 2005, Differences in RF energy absorption in the heads of adults and children. *Bioelectromagnetics* 26(Suppl 7): S31–S44.

Dalziel, C.F., 1946, Dangerous electric currents. *Trans AIEE* 65: 579–584

Dalziel, C.F., 1950, Effect of frequency on perception currents. *Trans AIEE* 69: 1162–1168.

Dalziel, C.F., 1954, The threshold of perception currents. *Trans AIEE* ptIIIB: 990–996.

David, E., Reißenweber, J., and Wojtysiak, A., 2004, Provocation studies in electromagnetic hypersensitive persons. *Proc WHO Workshop on EHS, Prague*, 123–133.

Deadman, J.E., Armstrong, B.G., McBride, M.L., Gallagher, R., and Theriault, G., 1999, Exposures of children in Canada to 60 Hz magnetic and electric fields. *Scand J Work Environ Health* 25: 368–375.

Eltiti, S., Wallace, D., Ridgewell, A., Zougkou, K., Russo, R., Sepulveda, F., Mirshekar-Syahkal, D., Rasor, P., Deeple, R., and Fox, E., 2007, Does short-term exposure to mobile phone base station signals increase symptoms in individuals who report sensitivity to electromagnetic fields? A double-blind randomized provocation study. *Environ Health Persp* 115: 1603–1608.

Flodin, U., Seneby, A., and Tegenfeldt, C., 2000, Provocation of electric hypersensitivity under everyday conditions. *Scand J Work Environ Health* 26: 93–98.

Frick, U., Rehm, J., and Eichhammer, P., 2002, Risk perception, somatisation, and self-report of complaints related to electromagnetic fields – A randomized survey study. *Int J Hyg Environ Health* 205: 353–360.

Frick, U., Kharraz, A., Hauser, S., Wiegand, R., Rehm, J., von Kovatsits, U., and Eichhammer, P., 2005, Comparison perception of singular transcranial magnetic stimuli by subjectively electrosensitive subjects and general population controls. *Bioelectromagnetics* 26: 287–298.

Gabriel, C., 2005, Dielectric properties of Biological tissue: Variation with age. *Bioelectromagnetics* 26(Suppl 7): S12–S18.

Greenland, S., Sheppard, A.R., Kaune, W.T., Poole, C., and Kelsh, M.A., 2000, A pooled analysis of magnetic fields, wire codes and childhood leukaemia. *Epidemiology* 11: 624–634.

Hadjem, A., Lautru, D., Dale, C., Wong, M.F., Hanna, V.F., Wiart, J., 2005, Study of specific absorption rate (SAR) induced in two child head models and in adult heads using mobile phones. *IEEE Trans Microwave Theor Techn* 53: 4–11.

Hamnerius, Y., Agrup, G., Galt, S., Nilsson, R., Sandblom, J., and Lindgren, R., 1993, Double-blind provocation study of hypersensitivity reactions associated with exposure from VDUs. *R Swed Acad Sci Rep* 2: 67–72.

Hamnerius, Y., Agrup, G., Galt, S., Nilsson, R., Sandblom, J., and Lindgren, R., 1994, Double-blind provocation study of reactions associated with exposure to electromagnetic fields from VDUs. *Proc Cost 244 Workshop* EHS, Graz, 41–43.

HCN-EMFC, 2006, TNO-study on the effects of GSM and UMTS signals on well-being and cognition. Health Council of the Netherlands publication 2006/11E.

Heinrich, S., Ossig, A., Schlittmeier, S., and Hellbrück, J., 2007, Elektromagnetische Felder einer UMTS – Mobilfunkbasisstation und mögliche Auswirkungen auf die Befindlichkeit – eine experimentelle Felduntersuchung (electromagnetic fields of UMTS base stations and their potential impact on well-being – An experinmental field study). *Umweltmed Forsch Prax* 13: 171–180.

Henningsen, P., Zimmermann, T., and Sattel, H., 2003, Medically unexplained physical symptoms, anxiety and depression: A meta-analytic review. *Psychosom Med* 65: 528–533.

Hietanen, M., Hämäläinen, A.M., and Husman, T., 2002, Hypersensitivity symptoms associated with exposure to cellular telephones: No causal link. *Bioelectromagnetics* 23: 264–270.

Hillert, L., 1998, Hypersensitivity to electricity: Management and intervention programs. *Proc Int Workshop EMF and Non-Specific Health Symptoms*, Graz, 17–30.

Hillert, L., 2004, Cognitive behaviour therapy for patients who report electrical hypertherapy. *Proc WHO Workshop EHS, Prague*, 163–174.

Hillert, L., Berglind, N., Arnetz, B.B., and Bellander, T., 2002, Prevalence of self-reported hypersensitivity to electric and magnetic fields in a population-based questionnaire survey. *Scand J Work Environ Health* 28: 33–41.

Hillert, L., Akerstedt, T., Lowden, A., Wiholm, C., Kuster, N., Ebert, S., Boutry, C., Moffat, S.D., Berg, M., and Arnetz, B.B., 2008, The effects of 884 MHz GMS wireless communication signals on headache and other symptoms: An experimental provocation study. *Bioelectromagnetics* 29: 185–196.

Huss, A., and Röösli, M., 2006, Consultations in primary care for symptoms attributed to electromagnetic fields – A survey among general practitioners. *BMC Public Health* 6: 267–276.

Hutter, H.P., Moshammer, H., Wallner, P., and Kundi, M., 2006, Subjective symptoms, sleeping problems and cognitive performance in subjects living near mobile phone base stations. *Occup Environ Med* 63: 307–313.

IARC, 2002, Non-ionizing radiation, Part 1: Static and extremely low frequency (ELF) electric and magnetic fields. IARC Monographs on the evaluation of carcinogenic risks to humans, Volume 80, IARC, Lyon.

ICNIRP, 1998, Guidelines for limiting the exposure to time-varying electric, magnetic and electromagnetic fields (up to 300 GHz). *Health Phys 74*: 494–522.

ICNIRP, 2003, Exposure to Static and Low Frequency Electromagnetic Fields and Health Consequences (0 Hz–100 kHz). Märkl-Druck, Munich.

IEEE PC95.1-2005, Standard for Safety Levels with Respect to Human Exposure to Radio Frequency *Electromagnetic Fields*, 3 kHz – 300 GHz.

IEEE PC95.6-2002, Standard for Safety Levels with Respect To Human Exposure to Electromagnetic *Fields*, 0–3 kHz.

Irnich, W., and Batz, L., 1989, Die Wahrnehmungsempfindlichkeit gegenüber 50 Hz Wechselspannung bzw. Wechselstrom (perception of 50 Hz AC voltages and currents). *Biomed Techn* 34: 207–209.

Johanssen, O., Gangi, S., Liang, Y., Yoshimura, K., Jing, C., Liu, P.Y., 2001, Cutaneous mast cells are altered in normal healthy volunteers sitting in front of ordinary TVs/PCs – Results from open field provocation experiments. *J Cutan Pathol* 10: 513–519.

Kaune, W.T., Miller, M.C., Linet, M.S., Hatch, E.E., Kleinerman, R.A., Wacholder, S., Mohr, A.H., Tarone, R.E., and Haines, C., 2002, Magnetic fields produced by handheld hair dryers, stereo headsets, home sewing machines, and electric clocks. *Bioelectromagnetics* 23: 14–25.

Kavet, R., Zaffanella, L.E., Daigle, J.P., and Ebi, K.L., 2000, The possible role of contact current in cancer risk associated with residential magnetic fields. *Bioelectromagnetics* 21: 538–553.

Keshvari, J., and Lang, S., 2005, Comparison of radio frequency energy absorption in ear and eye region of children and adults at 900, 1800 and 2450 MHz. *Phys Med Biol* 50: 4355–4369.

Kheifets, L., Repacholi, M., Sounders, R., and van Deventer, E., 2005a, The sensitivity of children to electromagnetic fields. *Pediatrics* 116: 303–313.

Kheifets, L., Sahk, J.D., Shimkhada, R., and Repacholi, M.H., 2005b, Developing policy in the face of scientific uncertainty: Interpreting 0.3 μT or 0.4 μT cutpoints from EMF epidemiologic studies. *Risk Anal* 25: 927–935.

Knave, B.G., Wibom, R.I., Voss, M., Destrom, L.D., and Bergqvist, U., 1985, Work with video display terminals among office employees. I. Subjective symptoms and discomfort. *Scand J Environ Health* 11: 457–466.

Koivisto, M., Haarala, C., Krause, C.M., Revonsuo, A., Laine, M., and Hämäläinen, H., 2001, GSM phone signal does not produce subjective symptoms. *Bioelectromagnetics* 22: 212–215.

Kraus, T., Anders, M., Weber, A., Hermer, P., and Zschiesche, W., 1995, Zur Häufigkeit umweltbezogener Somatisierungsstörungen (on the frequency of environment – Related somatisation disorders). *Arbeitsmed Sozialmed Umweltmed* 30: 147–150.

Kwon, M.S., Koivisto, M., Laine, M., and Hämäläinen, H., 2008, Perception of the electromagnetic field emitted by a mobile phone. *Bioelectromagnetics* 29: 154–159.

Landgrebe, M., Mauser, S., Langguth, B., Frick, U., Hajak, G., and Eichhammer, P., 2007, Altered cortical excitability in subjectively electrosensitive patients. Results of a pilot study. *J Psychosom Res* 62: 283–288.

Landgrebe, M., Frick, U., Hauser, S., Langguth, B., Rosner, R., Hajak, G., and Eichhammer, P., 2008, Cognitive and neurobiological alterations in electromagnetic hypersensitive patients: Results of a case-control study. *Psychol Med* 26: 1–11.

Leitgeb, N., 1994, Electromagnetic hypersensitivity: Quantitative assessment of an ill-defined problem. *Proc Cost 244 Workshop EHS*, Graz, 68–74.

Leitgeb, N., 1995, *Elektrosensibilität. J VEÖ* 98: 51–55.

Leitgeb, N., 1998, Electromagnetic hypersensitivity. Proc Int Workshop EMF and Non-Specific Health Symptoms, Graz, 8–16.

Leitgeb, N., 2008, Mobile phones: Are children at higher risk? *Wien Med Wochenschr* 158: 36–41.

Leitgeb, N., and Schröttner, J., 2002, Electric current perception study challenges electric safety limits. *J Med Eng Technol* 26: 168–172.

Leitgeb, N., and Schröttner, J., 2003, Electrosensibility and electromagnetic hypersensitivity. *Bioelectromagnetics* 24: 387–394.

Leitgeb, N., Schröttner, J., and Böhm, M., 2005a, Does "electromagnetic pollution" cause illness? *Wien Med Wochenschr* 155: 237–241.

Leitgeb, N., Schröttner, J., and Chech, R., 2005b, Electric current perception of the general population including children and elderly. *J Med Eng Technol* 29: 215–218.

Leitgeb, N., Schröttner, J., and Cech, R., 2006, Electric current perception of children: The role of age and gender. *J Med Eng Technol* 30: 306–309.

Leitgeb, N., Schröttner, J., and Cech, R., 2007, Perception of ELF electromagnetic fields: Excitation thresholds and inter-individual variability. *Health Phys* 92: 591–595.

Leitgeb, N., Cech, R., Schröttner, J., Lehofer, P., Schmidpeter, U., and Rampetsreiter, M., 2008a, Magnetic emission ranking of electric appliances. A comprehensive market survey. *Radiat Prot Dosimetry* 129: 439–445.

Leitgeb, N., Cech, R., and Schröttner, J., 2008b, Electric emissions from electric appliances. *Radiat Prot Dosimetry* 129: 446–455.

Leitgeb, N., Schröttner, J., Cech, R., and Kerbl, R., 2008c, EMF-protection sleep study near mobile phone base stations. *Somnology* 12: 234–243.

Levallois, P., Neutra, R., Lee, G., and Hristova, L., 2002, Study of self-reported hypersensitivity to electromagnetic fields in California. *Environ Health Persp* 110 (s4): 619–623.

Levin, M., 1991, Perception of chassis leakage currents. *Biomed Instrum Technol* 25: 135–140.

Linet, M.S., Hatch, E.E., Kleinerman, R.A., Robinson, L.L., Kaune, W.T., Friedman, D.R., Severson, R.K., Haines, C.M., Hartsock, C.T., Niwa, S., Wacholder, S., and Tarone, R.E., 1997, Residential exposure to magnetic fields and acute lymphoblastic leukemia in children. *N Eng J Med* 337: 1–7.

Lonne-Rahm, S., Andersson, B., Melin, L., Schultzberg, M., Arnetz, B., and Berg, M., 2000, Provocations with stress and electricity of patients with sensitivity to electricity. *J Occup Environ Med* 24: 387–394.

Lyskov, E., Sandström, M., and Hansson Mild, K., 2001a, Neurophysiological study of patients with perceived electrical hypersensitivity. *Int J Psychophysiol* 31: 233–241.

Lyskov, E., Sandström, M., and Hansson Mild, K., 2001b, Provocation study of persons with perceived electrical hypersensitivity and controls using magnetic field exposure and recording of electrophysiological characteristics. *Bioelectromagnetics* 22: 457–462.

Mild, K.H., Repacholi, M., vanDeventer E., and Ravazzani, P., Eds., 2006, Electromagnetic hypersensitivity. *Proc WHO Workshop, Prague*, 2004.

Müller, C.H., Krueger, H., and Schierz, C., 2002, Project NEMESIS: Perception of a 50 Hz electric and magnetic field at low intensities (laboratory experiment). *Bioelectromagnetics* 23(1): 26–36.

Müller, C.H., 2000, Projekt NEMESIS: ELF electric and magnetic fields and electrosensibility in Switzerland (in German). Thesis, ETH Zurich No 13903.

Navarro, E.A., Segura, J., Gómez-Perretta, C., Portolés, M., Maestu, C., and Bardasano, J.L., 2002, Exposure from cellular phone base sations. A first approach. *Proc 2nd Int Workshop Biol Eff EMFs, Rhode*, 353–358.

Nilsen, A., 1982, Facial rush in visual display unit operators. *Contact Dermatitis* 8: 25–28.

Oftedal, G., Vistnes, AI., and Rygge, K., 1995, Skin symptoms after the reduction of electric fields from video display units. *Scand J Work Environ Health* 21: 335–344.

Oftedal, G., Nyvang, A., and Moen, B.E., 1999, Long-term effects on symptoms by reducing electric fields from video display units. *Scand J Work Environ Health* 25: 415–421.

Oftedal, G., Straume, A., Hohnsson, A., Stovner, L.J., 2007, Mobile phone headache: A double-blind sham-controlled provocation study. *Cephalalgia* 27: 447–455.

Ogden, C.L., Kuczmarski, R.J., Flegal, K.M., Mei, Z., Guo, S., Wei, R., Grummer-Strawn, L.M., Curtin, L.R., Roche, A.F., and Johnsson, C.L., 2002, Centers for Disease Control and Prevention 2000 Growth Charts for the United States: Improvements to the 1977 National Center of Health Statistics Version. *Pediatrics* 109: 45–60.

Omura, Y., Losco, B.M., Omura, A.K., Yamamoto, S.Y., Ishikawa, H., Takeshige, C., Shimotsuura, Y., and Muteki, T., 1991, Chronic or intractable medical problems associated with prolonged exposure to unsuspected harmful environmental electric, magnetic or electromagnetic fields radiating in the bedroom or workplace and their exacerbation by intake of harmful light and heavy metal from common sources. *Int J Acupunct Electrotherapeut Res* 16: 143–177.

Osypka, P., 1963, Messtechnische Untersuchungen über Stromstärke, Einwirkungsdauer und Stromweg bei elektrischen Wechselstromunfällen an Mensch und Tier (metrological investigations in humans and animals on amplitude, exposure time and pathway in AC electric current accidents). Bedeutung und Auswertung für Starkstromanlagen. *Elektromedizin* 8: 153–214.

Preece, A.W., Kaune, W., Grainger, P., Preece, S., and Golding, J., 1997, Magnetic fields from domestic appliances in the UK. *Phys Med Biol* 42: 67–76.

Rea, W.J., Pan, Y., Fenyves, E.J., Sujisawa, I., Samadi, N., and Ross, G.H., 1991, Electromagnetic field sensitivity. *Bioelectricity* 10: 241–246.

Regel, S.J., Negovetic, S., and Röösli, M., 2006, UMTS base station-like exposure, well being and cognitive performance. *Environ Health Perspect* 114(8): 1270–1275.

Reilly, J.P., 1992, *Electrical Stimulation and Electropathology*. Cambridge University Press, Cambridge.

Röösli, M., 2008, Radiofrequency electromagnetic field exposure and non-specific symptoms of ill health: A systematic review. *Environ Res* 107: 277–287.

Rubin, G.J., Javati, D.M., and Wessely, S., 2005, Electromagnetic Hypersensitivity: A systematic review of provocational studies. *Psychosom Med* 67: 224–232.

Rubin, G.J., Hahn, G., Everitt, B.S., Cleare, A.J., and Wessely, S., 2006a, Are some people sensitive to mobile phone signals? Within participants double blind randomised provocation study. *BMJ* 332(7546): 886–891.

Rubin, G.J., Das Munshi, J., and Wessely, S., 2006b, A systematic review of tretments for electromagnetic hypersensitivity. *Psychother Psychosom* 75: 12–18.

Sandström, M., Stenberg, B., and Hansson-Mind, K., 1993, Experiences of provocations with electric and magnetic fields. *R Swed Acad Sci Rep* 2: 62–66.

Santini, R., Santini, P., LeRuz, P., Danze, J.M., and Seigne, M., 2003, Survey study of people living in the vicinity of cellular phone base stations. *Electromagn Biol Med* 22: 41–49.

Schreier, N., Huss, A., and Röösli, M., 2006, The prevalence of symptoms attributed to electromagnetic field exposure: A cross-sectional representative survey in Switzerland. *Soz Präventiv Med* 51: 202–209.

Schröttner, J., Leitgeb, N., and Hillert, L., 2007, Investigation of electric current perception thresholds of different EHS groups. *Bioelectromagnetics* 28: 208–213.

Schubert, M., Bornkessel, C., Wuschek, M., and Schmidt, P., 2007, Exposure of the general public to digital broadcast transmitters compared to analogue ones. *Radiat Prot Dosimetry* 124: 53–57.

Schüz, J., 2005, Mobile phone use and exposures in children. *Bioelectromagnetics* 26(Suppl 7): S45–S50.

Schüz, J., Grigat, J.P., Störmer, B., Rippin, G., Brinkmann, K., and Michaelis, J., 2000, Extremely low frequency magnetic fields in Germany. Distribution of measurements, comparison of two methods for assessing exposure, and predictors for the occurrence of magnetic fields above background level. *Radiat Environ Biophys* 39: 233–240.

Schüz, J., Petters, C., Egle, U.T., Jansen, B., Kimbel, R., Letzel, S., Nix, W., Schmidt, L.G., and Vollrath, L., 2006, The "Mainzer EMF-Wachhund": Results from a watchdog project on self-reported health complaints attributed to exposure to electromagnetic fields. *Bioelectromagnetics* 27: 280–287.

Sjöberg, P., and Hamnerius, Y., 1995, Study of provoked hypersensitivity reactions from a VDU. Proc *Electromagnetic Hypersensitivity*, Copenhagen.

Swanbeck, G., and Blecker, T., 1989, Skin problems from visual display units. Provocation of skin symptoms under experimental conditions. *Acta Derm Venerol* 69: 46–51.

Tan, K.-S., and Johnson, D.L., 1990, Threshold of sensation for 60 hz leakage current. Result of a survey. *Biomed Instrum Technol* 24: 207–211.

Thompson, G., 1933, Shock threshold fixes appliance insulation resistance. *Electrical World* 101: 793–795.

UKCCS Investigators, 1999, Exposure to power-frequency magnetic fields and the risk of childhood cancer. *Lancet* 354: 1925–1931.

Ulmer, S., and Bruse, M., 2006, Supplementary Information on Electromagnetic Hyopersensitive. Final report German Mobile Telecommunication Research Program. http://www.emf-forschungsprogramm.de

Wang, T., Hawkins, L.H., and Rea, W.J., 1994, Effects of ELF magnetic fields on patients with chemical sensitivities. *Proc Cost 244 Workshop EHS, Graz*, 123–132.

Wennberg, A., Franzen, O., and Paulsson, L.E., 1994, Electromagnetic field provocations of subjects with electromagnetic hypersensitivity. *Proc Cost 244 Workshop EHS, Graz*, 133–139.

Wertheimer, N., and Leeper, N., 1979, Electric wiring configurations and childhood cancer. *Am J Epidemiol* 109: 273–284.

WHO, 2005, Electromagnetic fields and public health. *Electromagnetic Hypersensitivity. Fact Sheet No*. 296.

WHO, 2007, Environmental Health Criteria 238. Extremely Low Frequency Fields, Geneva.

Wiart, J., Hadjem, A., Gadi, N., Bloch, I., Wong, M.F., Pradier, A., Lautru, D., Hanna, V.F., and Dale, C., 2005, Modeling of RF head exposure in children. *Bioelectromagnetics* 26(Suppl 7): S19–S30.

Wilén, J., Johansson, A., Kalezic, N., Lyskov, E., and Sandström, M., 2006, Psychophysiological tests and provocation of subjects with mobile phone related symptoms. *Bioelectromagnetics* 27: 204–214.

Zwamborn, A.P.M., Vossen, S.H.J., van Leersum, B.J.A., Ouwens, M.A., and Makel, W.N., 2003, Effects of global communication system radio-frequency fields on well-being and cognitive functions in human subjects with and without subjective symptoms. TNO Report 2003(FEL03-C148): 1–89.

Chapter 6

Occupational Exposure in Wireless Communication

Kjell Hansson Mild and Jonna Wilén

ABSTRACT

Today we are exposed to electromagnetic fields from the use of wireless communication devices almost everywhere. However, occupational exposure where there is a possibility to exceed the international guidelines occurs only in work very near mobile phone base stations, and this exposure can easily be dealt with in practice in the form of instructions and administrative measures. All other devices produce exposure well below present guidelines. This low-level exposure has been discussed from a health perspective, and in this paper the exposure from sources such as mobile phones, cordless phones, WiMax, WLAN and base station antennas is discussed. The problem of exposure assessment for epidemiological studies is also dealt with in a general manner.

1. INTRODUCTION

During recent decades there has been a rapid development in the use of wireless communication. Today, we use wireless not only for our mobile or cordless phones but also for computers, and with Blue-tooth technology communication between devices is increasing rapidly. This leads to exposure of the users to the electromagnetic fields emitted from these devices.

K.H. Mild and J. Wilén Department of Radiation Physics, Umeå University, SE, 901 87, UMEÅ, Sweden, e-mail: kjell.hansson.mild@radfys.umu.se

J.C. Lin (ed.), *Advances in Electromagnetic Fields in Living Systems, Health Effects of Cell Phone Radiation, Volume 5*,
DOI 10.1007/978-0-387-92736-7_6,

Several workplaces use only cellular or cordless phones (DECT as well as IP phones) instead of landline phones, and this leads to both active and passive exposure to microwaves during the working day for employees. Very few workplaces offer hands-free devices to employees, although the WHO, the Nordic radiation protection authorities and the Swedish work environmental board all recommend them.

In this paper we will concentrate on the occupational exposure in connection with wireless communication, which means that we will concentrate on handheld phones and work very close to base station antennas.

Before we enter a discussion on occupational exposure we need to clarify a few points, namely, what we mean by the terms 'worker' and 'working environment'. This has been dealt with in IEC and CENELEC documents whose definitions are as follows:

Worker: *Any person employed by an employer, including trainees and apprentices but excluding domestic servants (from 89/391/EEC).*

Employer: *Any natural or legal person who has an employment relationship with the worker and has responsibility for the undertaking and/or establishment (from 89/391/EEC).*

The exposure of the employee that takes place at the location where that employee carries out his or her work is thus called 'occupational exposure'. The equipment at the workplace combined with its use is called the working environment. To make a comprehensive assessment, the following information about this working environment must be available: the type of equipment that generates the electromagnetic fields, the type of work carried out by the worker and the circumstances under which the equipment is used. In the exposure assessment there is a need to recognize the complexity of the exposure assessment situation by carefully inspecting the working environment and consulting the local safety engineer about the exposure problem and conditions. There is also a need to consult the relevant CENELEC standards and NIOHS document.

However, many of the standards from IEC and CENELEC are product and emission standards, not exposure standards. They are aimed at compliance of the product to the standard, and if the equipment is used as given in the emission standard then they can also serve as exposure measurement standards. But in many situations the exposure is different from the setup in the emission standard, so a special exposure assessment has to be made.

2. MOBILE PHONES

Different countries have different phone operating systems, and slightly different frequencies are also used. The Nordic countries were among the first in the world to introduce cellular phones. The analogue [Nordic Mobile Telephone (NMT) System] phones operating at 450 MHz were introduced in Sweden in 1981. First they were used in cars with fixed external antenna, but from 1984 portable NMT 450 phones became available on the market. The next generation of analogue phones using 900 MHz (NMT 900) was used in Sweden between 1986 and 2000. The digital system [global system for mobile communication (GSM)] started in 1991 and has during recent years dramatically increased to be the most common phone type. This system uses dual band, 900 and 1,800 MHz, for communication. From 2003, the third generation of mobile phones, 3G or universal mobile telecommunication system (UMTS) mobile phone using wideband code-division multiple-access (WCDMA), has started operating in Sweden at 1,900 MHz. Other countries are using slightly different systems.

2.1. Technical Data on Mobile Phones

The NMT analogue telephones operated with a maximum power of 1 W and very seldom down-regulated this. The NMT system in the Nordic countries was closed down in 2000. The GSM 900 phones are operating with a maximum of 0.25 W but can down-regulate the power to a few mW depending on the distance to the base station. The new 1,800-MHz system has a maximum power output of 0.125 W, and this can also be down-regulated to some mW. The phone adjusts the output power to provide sufficient signal strength at the base station for acceptable quality of connection, while at the same time keeping the output power as low as possible to minimize interference and increase battery life.

Persson et al. (2002) measured the output power from GSM 900 phones and found that mixed use typically corresponded to an average output power of approximately 20% of the maximum available output power in the sample measurements using test phones primarily in a large city area (Stockholm). In a predominantly indoor GSM office environment the corresponding typical value was noted to be about 4% of the maximum available output power.

The GSM system operates with eight time slots. Each of the slots has burst duration of 0.577 ms; intermittency between bursts: 20 ms of the basic frame (4.61 ms) were on and one off (slot 7). The GSM handset phone uses one of the eight slots whereas the base station uses seven of the eight slots of the basic frames (one is used for system info). The two pulse-modulated signals include the same ELF modulation components (2, 8, 217, 1,736 Hz and the corresponding harmonics), but the spectral power of these components is considerably higher in the handset signal. This signal structure results in the spectral components of 2, 8 and 217 Hz, plus the corresponding harmonics. The burst and the intermittency between the bursts led to additional components at 1,733 Hz and 50 kHz. There is also a so-called DTX function, meaning that when you are talking on the phone the transmission rate is 217 Hz whereas when listening the rates goes down to 2 Hz. See also Kuster et al. (2004).

The 3G phone technology adjusts the output power rapidly to keep the received power at the base station at approximately the same level as for other connected phones. In a recent study by Persson et al. (2005) the output power was recorded and found to be below 0.01 mW in indoor and urban areas. The output power was slightly higher in suburban and rural areas but in all environments the output power was less than 1 mW for more than 90% of the measurement points. The maximum output power was almost never used.

Table 1. Mobile phone net frequencies and power

	Mobile phone		Base station	
Mobile net	Frequency (MHz)	Max power mean (W)	Frequency (MHz)	Typical antenna power (W)
GSM 900	880–915	0.25	925–960	20
GSM 1800	1,710–1,785	0.125	1,805–1,880	15
UMTS/WCDMA	1,920–1,980	0.125	2,110–2,170	10

Of interest in this context are also desktop cordless phones. First the analogue system in the 800–900-MHz RF range was used, but since 1991 digital cordless telephones (DECT) that operate at 1,900 MHz are on the market. The output power of a cordless phone is lower than the maximum output power of a GSM phone (0.01 W compared with 0.25 W) but since cordless phones are not able to down-regulate the output power as GSM phones do, and since the total time spent on a cordless phone is probably still longer than on mobile phones (Hardell et al., 2006) the exposure from cordless phones can not be neglected. However, on the market now there are DECT phones that have a built-in function for down-regulation.

For a more comprehensive technical report on mobile communication systems the reader is referred to a recent publication from the UK National Programme (MTHR, 2007).

2.2. RF Exposure from GSM Phones

Use of mobile and desktop cellular telephones results in exposure to microwaves. The exposure is characterized through the 'specific absorption rate' (SAR), expressed as W/kg. The basic restriction for radio frequency exposure is given as SAR, which denotes how much energy per time and mass unit is absorbed by the tissue, and SAR is proportional to the square of the electric field (E^2).

Most present Western standards and guidelines are based on SAR. In the ICNIRP guidelines from 1998 the limits are set in SAR values, and for the frequency range 10 MHz to 10 GHz the allowed SAR value is 0.4 W/kg whole-body exposure. For parts of the body the maximum allowed value is 10 W/kg with an averaging mass of any 10 g of contiguous tissue. For the general public these values are reduced by a factor of 5 giving the maximum whole-body SAR at 0.08 W/kg and local maximum at 2 W/kg. These values have been adopted by the European Union as the limits for mobile phones that are sold within the EU. In other countries other limits are given and are slightly different. Table 2 shows a summary of the values for handheld phones:

In the new IEEE standard C95.1 from 2005 the values for localized exposure are relaxed to 2 W/kg, and for persons in controlled environments 10 W/kg is allowed. SAR is averaged over a 10-g tissue defined as a tissue volume in the shape of a 10-g cube. For extremities and pinnae these values are increased to 4 and 20 W/kg, respectively.

Table 2. SAR limits for handheld phones in different countries

Region/ country	Reference to SAR measurement protocol	Reference to SAR limit	Limit
Europe	European Specification ES 59005 (1998)	ICNIRP guidelines (ICNIRP, 1998)	2.0 W/kg in 10-g of tissue
Australia	Australian Communications Authority (ACA) Standard (ACA RS, 1999)	Australian Standard AS/ NZS 2772.1	1.6 W/kg in 1 g of tissue
US	Federal Communications Commission (FCC) Guidelines (FCC, 1997)	American Standard ANSI C95.1 (ANSI, 1992)	1.6 W/kg in 1 g of tissue
EU	EU council recommendation for general public (1999)	ICNIRP Guidelines (ICNIRP, 1988)	2.0 W/kg in 10 g of tissue

When a person is exposed in the far field from the source it is possible to get information about the exposure from a measurement of the power density. For the action levels for occupational exposure the aforementioned values can be given as power density, and the ICNIRP values are $f/40$ W/m^2 for frequencies between 400 and 2,000 MHz (f in MHz), and from 2 to 300 GHz the value is 50 W/m^2. The corresponding values from IEEE are $f/30$. Both are to be time-averaged over 6 min. For the general public the maximum allowed power density is, according to ICNIRP, $f/200$ W/m^2 and the IEEE limit is the same except that the time is averaged over 30 min instead of the usual 6-min period. However, note that it is not meaningful to try to measure power density from a handheld mobile phone. The exposure assessment in this case has to be done in a standardized way in laboratory measurements; see Kuhn and Kuster (2007) for more details.

The anatomical area with the highest exposure is the ipsilateral (same) side of the brain as used during the call. If a hands-free device is used and the cellular telephone is placed at another part of the body, that anatomical area receives the highest radio frequency (RF) exposure.

Because of different antennas and device sizes, different phones deposit the radio frequency (RF) energy at different anatomical localizations in the head. Hence, the SAR distribution as well as the maximum SAR_{1g} and SAR_{10g} (averaged over 1 or 10 g, respectively) differ between different phone models [Fig. 2 in Wilén et al. (2003)], and also between NMT, GSM and cordless phones, where NMT phones have higher maximum SAR values than GSM phones and the latter in general have higher maximum SAR values than cordless phones due to the difference in output power.

Kuster (1997) measured 16 different European digital phones and found a very wide variation in the SAR values. The phone with the lowest value, when averaged over a 10-g tissue, had a SAR of 0.28 W/kg and the one with the highest value had 1.33 W/kg; all normalized to an antenna input power of the maximum 0.25 W. If the averaging was done over a 1-g tissue the span was from a low of 0.42 W/kg to a high of 2.0 W/kg. The SAR measurements were made under normal user conditions. However, when the phone is slightly tilted toward the head of the user the value can go from 0.2 to 3.5 W/kg. Thus, for different phones under maximal output, we have a factor of about 5 between the extremes, and the personal handling of the phone adds a factor of tenfold or more.

There is a large spread in SAR values for different phones. Lists of values are published on various web pages, and you can find values ranging from 0.1 to 1.8 W/kg. However, it is not enough to know the SAR value of a mobile phone to say how it will perform in practice. One of the important parameters in this respect is the so-called TCP value – telephone communication power – that measures how much of the phone output power is available for communication. It is usually measured in a revertebration chamber with a phantom with the phone positioned at the ear. Then, the power that is not absorbed by the phantom or lost as mismatch in the phone is measured in Watts. The Swedish TCO union has long advocated that phones should have SAR values less than 0.8 W/kg and TCP values above 0.3 W. You can view the TCP values also as percentage of the total output power; for the GSM 900 phones the specified value is 2 W peak power (average 0.25 W) and for GSM 1800 it is 1 W peak power (average 0.13 W). The TCO demand thus states that for GSM 900 phones at least 15% of the output power should be used for communication, and for GSM 1800 the value is 30%. TCO has been testing phones for SAR and TCP for many years. Figure 1 shows the latest results of these tests, and as can be seen not all phones meet the TCO demand. If you are using a phone with a low TCP value the phone has to fire up the output power to establish a connection, whereas a phone with a high TCP value can make the connection with the base station with a much lower power output, and thus a lower exposure to the user. Thus, a phone with a low SAR value might not be the best choice since if the TCP value is low the phone must operate at a high-output level. The phone that would give the lowest exposure is the one with a low SAR value and a high TCP value.

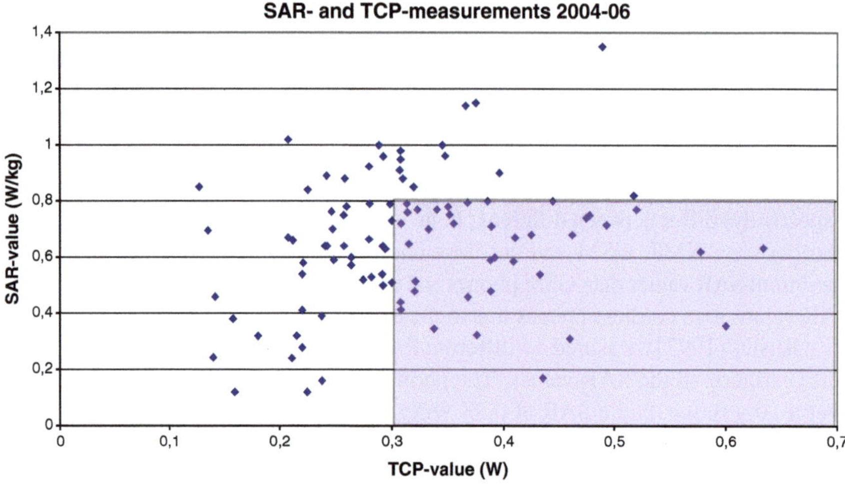

Figure 1. The results of measurements on GSM mobile phones 2004–2006 showing the TCP and SAR values (courtesy of TCO development).

Anger (2003) reported on measurement of SAR on 21 different phones and his result is similar. The SAR ranged from 0.3 to 1.7 W/kg over 10 g tissues. He also reported the TCP value, that is, how much of the output power that can be used for communication, and he found values ranging from only about 5% in the lowest phone to just less than 50% in the 'best' phone. Thus, more than half of the output power from the phone is lost: some due to mismatch in the phone and the antenna and some deposited as SAR in the user.

The handling of the phone will affect the SAR significantly (Kuster, 1997) and it is easy to see that the maximum SAR_{1g} value is not valid for each user in the study. It is very difficult, probably impossible, to make any relevant adjustments for this in epidemiological studies.

2.3. ELF Magnetic Fields from GSM Phones

The first measurements of low frequency magnetic fields from GSM phones were done by Linde and Hansson Mild (1997) and Andersen and Pedersen (1997). Since then several others have published data on the magnetic field: Jokela et al. (2004), Ilvonen et al. (2005) and Straume et al. (2007).

In principle, the phone draws current from the battery pack in a pulsed mode, which gives rise to magnetic fields near the phone. In full-peak power transmitting mode there is a current with a main frequency of 217 Hz and an amplitude in the order of 1 A, and this gives rise to low-frequency pulsed DC magnetic fields with peak magnetic flux density of the order of tens of μT. Jokela et al. (2004) recently measured seven different GSM phones and looked at the frequency content of the magnetic pulse and found that a considerable amount is found in the low kHz range. It is even found that some phones exceed the ICNIRP guideline reference values when the multiple frequency formula is applied, but calculations show that the basic restrictions are not exceeded. For the NMT phones the magnetic field from the battery current is to be regarded as pure DC fields.

2.4. Exposure to Mobile Phones in Epidemiological Studies

2.4.1. Combined RF Exposure from Several Phone Models

Most users of mobile phones have not been using just one single telephone, and it is even more likely that if they have been using a mobile phone for more than a few years they have changed the phone a few times, and also possibly the system they use: NMT, GSM or 3G. Many users will also have been using different phone systems such as analogue and digital, and probably many of them have also been using a cordless phone at home or at work. In the epidemiological studies on mobile phone use and brain tumours this has not been taken into account, and the main reason for this is that at the moment it is not clear how to combine the use of different phones with different power output, systems, frequencies and anatomical SAR distribution into one exposure and dose measure. The difficulties lie in the fact that we do not know the interaction mechanism(s) between the electromagnetic fields emitted from the phone and the biological organism. However, in spite of this we need to start a discussion on how to take into account the use of several phones.

The most obvious way to combine different phones is to add the total time in hours for use of each phone without assigning each a different weight. Another way of adding different phone types could be to give each phone system a score, based on the mean output power of each system. The Nordic Model Telephones (NMT) are operating with a maximum power of 1 W and very seldom down-regulate this. The Group Special Mobile (GSM) 900 phones are operating with a maximum of 0.25 W average but can down-regulate the power to a few mW depending on the distance to the base station, with a typical value of 0.1 W. The cordless phones operate at 10 mW. One selection of weighting factors according to mean output power of the phones could then be NMT = 1, GSM = 0.1 and cordless = 0.01. In Hansson Mild et al. (2004, 2005) both of these methods were applied, but the effect of the model based on scores was small, probably due to the fact that most of the subjects had used an analogue phone, which would dominate the total exposure. In the recent French part of the Interphone (Hours et al., 2007) they used number of phones used as a two-sided exposure measure; that is, only one phone or two or more.

For people who have used more than one device during their lifetime, another problem also arises: how to integrate the, probably different, SAR distributions from different devices.

One method could be to use the specific absorption (SA), expressed in J/kg instead of SAR. By integrating the estimated 'use time' with the SAR value for each device the mean SA for all devices can then be used as a dosimetric quantity (approaches other than mean values can also be interesting).

In an epidemiological study of subjective symptoms among mobile phone (MP) users, Hansson Mild et al. (1998) used several factors to assess the exposure and estimate dose. The study was set up to see if there were differences between users of the analogue (NMT) and the digital (GSM) systems. GSM users reported warmth sensation on the ear and behind or around the ear less frequently than NMT-users. There was also a statistical association between both *calling time* and *number of calls* per day and the occurrence of warmth sensation as well as headache and fatigue both among NMT users and GSM users. When calling time per day was used as exposure parameter it was found that people using phones for 15–60 min per day were 1.6 times more likely to complain of fatigue and 2.7 times more likely to complain of headache than people who used their phones for 2 min or less per day. Users of phones who talked more than 60 min per day were 4.1 times more likely to complain of fatigue and 6.3 times more likely to have a headache than those who talked for less than 2 min per day.

In a follow-up to this Wilén et al. (2003) looked at approximately 2,500 of the users in the study, distributed over four different phone models. Based on the distribution of the SAR values over the area (Fig. 2), exposure was assessed in three different areas/volumes: above the ear, on the ear and below the ear. On each site the dose was expressed as specific absorption (SA), expressed in J/kg instead of SAR. By integrating the estimated *use time* with the SAR value for each device the mean SA for all devices was obtained. Two other different dosimetric quantities were also

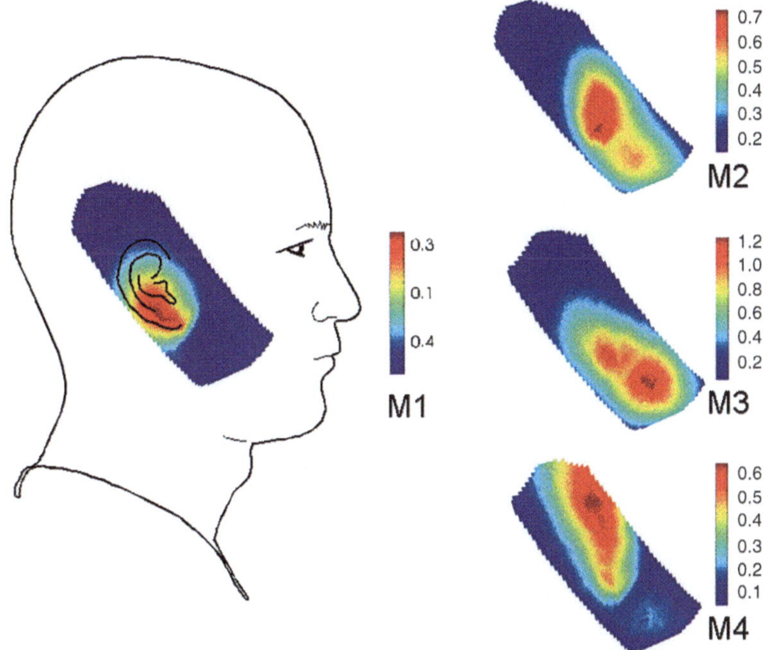

Figure 2. Schematic of the distribution of specific absorption rate (SAR) (W/kg) for the devices used in the study (M1–M4). The figures show the measured SAR on the right hand side of the phantom [from Wilén et al. (2003)].

used: specific absorption per day (SAD) and specific absorption per call (SAC). Both SAD and SAC are time-integrated quantities expressed in J/kg, but with different time scales; SAD was calculated using the total calling time per day while SAC used the average calling time per call. The results indicated that SAR values > 0.5 W/kg may be an important factor in the prevalence of some of the symptoms, especially in combination with long calling times per day.

Morrissey (2007) discussed the variable output from the phones and how to assess this in epidemiological studies. In the majority of epidemiological studies investigating correlations between long-term low-level radiofrequency (RF) exposure from mobile phones and health endpoints they have followed a case-control design, requiring reconstruction of individual RF exposure. For instance, 'time of use' has been used as an exposure surrogate obtained from questionnaire information or billing records (note that billing records often only give outgoing calls and not incoming). His point is that the variability in mobile phone transmit power is not taken into account here. There is a variability in output power (a) during a single call, (b) between separate calls, (c) between averaged values from individuals within a local study group and (d) between average values from groups in different geographical locations. He further points out that to identify dose response and statistical correlations between mobile phone use and subtle health endpoints is a significant challenge to the researcher.

See also Erdreich et al. (2007) for a discussion on what factors (in US) most affect the energy output. It was found that the factors of most importance were study area, followed by user movement and location (inside or outside), use of a hands-free device and urbanicity, although the two latter factors accounted for trivial parts of overall variance. The average energy output rates were usually less than 50% and were generally comparable to the standard deviation. These results provide information applicable to improving the precision of exposure metrics for epidemiological studies of GSM mobile phones and may have broader application for other mobile phone systems and geographic locations.

2.4.2. Time Scale

The time scale for the integration is also important. In Wilén et al. (2003), SAD and SAC were used as time-integrated quantities expressed in J/kg, but with different time scales; SAD was calculated using the total calling time per day and SAC used the average calling time per call. If possible long-term effects are studied, perhaps a lifetime dose is most relevant. How to achieve a dosimetric measure of this with any good precision is an open question.

In epidemiological studies the important consideration, which time scale to use, should be based on hypotheses; for instance, in studies on brain tumours and mobile phone use the concept of induction and progression of the studied endpoint variable should be clearly considered.

Since the calling time in epidemiological studies is probably based on subjective judgment, the possible recall bias for the reported calling time, as has been discussed by Hillert et al. (2003), is also valid for the calculated SA values.

3. DECT

DECT stands for digital enhanced cordless telecommunications, which is a standard for cordless phones. These phones and their base stations operate in the frequency range 1,880–1,900 MHz and are spread over ten frequencies. DECT utilizes 24 time frames of which the first 12 are used for the uplink from a portable terminal to the access point, and the last 12 to the downlink. The use is such that the portable terminal transmits in one time frame and gets answered from the access point exactly 12 time frames later. The output power is peak 250 mW, giving an average of about 10 mW. The system has no power regulation capabilities, and thus it always operates at full power. The reach is about 50 m indoor and 300 m outdoor.

The access point (base station) is always transmitting when turned on, thus, even when no phone call is made there is a standby signal.

3.1. Discussion of Mobile Phones

The weighting methods described earlier (NMT = 1, GSM = 0.1 and cordless phones = 0.01) or the use of measured SAR values as weighting parameters are based on the assumption of linear dose-response where possible threshold effects are not taken into account. For example, does one need to exceed a certain SAR value/power level before the potential risk increase? Also the fractionation of the absorption is neglected, which means that, for example, one 10 min call is equal to ten 1 min calls.

However, one needs a clear hypothesis about how the absorption of RF from mobile phones could affect the endpoint variable in terms of anatomical localization of the absorption, the duration of the exposure and the induction and progression of the endpoint variable before choosing an appropriate dosimetric quantity. Obviously on theoretical grounds using sum of use of the different phone types is not an appropriate method to combine exposure to these RF fields. Using a weighting factor, as before, may be appropriate until proper dosimetry is available.

The time spent on mobile or cordless phones varies between countries and not much information on usage is available. In the epidemiological studies on mobile phone use and risk for brain tumours we can find some information. Hardell et al. (2006) have shown data on the total lifetime use of mobile phones among their cases and controls. They found that among the cases 20 % have been using the phones for more than 2,000 h, and the corresponding percentage for controls is only 7.3 %. In the recently published study on the French part of Interphone (Hours et al., 2007) the upper quartile for usage is 260 h. This difference is to be expected since the Nordic countries have been using mobile phones for much longer than other countries.

4. WIRELESS IP

Many companies and also individuals are now using wireless Internet Protocol telephony. This works either in a company with access points spread out in the office or at home with an antenna attached to the PC. The phone then communicates with an IEEE 802.11 protocol. However, the phones operate on a fixed output power with no regulatory possibilities during a call. Typically the operating frequency is 2.4 GHz and the output power is 100 mW. The latter can be adjustable in fixed steps, if set before the call by the system operator, but typically the highest is used. It is difficult to find reports on the SAR values of these phones from the manufacturers. In one case we saw a value of 0.7 W/kg_{10g} for 0.1 W output. If this is to be compared with a GSM phone with a maximum output power of 0.25 W the SAR value has to be upgraded 2.5 times, and this will take it to one of the highest SARs noted. Note that the devices are operating at different frequencies and a comparison is not straightforward.

5. TETRA

Terrestrial trunked radio (TETRA) is a modern digital private mobile radio system designed to meet the requirements of professional users, such as the police and fire brigade. The current frequency allocations in the UK are 380–385 and 390–395 MHz for the public sector network. Tetra is a TDMA (time multiplex) system where four calls share the same 25-kHz radio channel. Since this is a digital technique data can also be transferred. The transmission is done with a pulse rate of 17.6 Hz.

Dimbylow et al. (2003) studied the exposure from a Tetra handset with finite-difference time-domain (FDTD) calculations of the specific energy absorption rate (SAR) from a representative TETRA handset in an anatomically realistic model of the head. The calculations of SAR in the head were performed for positions of the handset in front of the face and at both sides of the head. (The TETRA can be used both as a telephone and as a walkie-talkie.) The representative TETRA handset considered operating at 1 W in normal use showed compliance with both the ICNIRP occupational and public exposure restrictions. The handset with a monopole antenna operating at 3 W in normal use showed compliance with both the ICNIRP occupational and public exposure restrictions. The handset with a helical antenna operating at 3 W in normal use will show compliance with the ICNIRP occupational exposure restriction but will be over the public exposure restriction by up to approximately 50% if kept in the position of maximum SAR for 6 min continuously.

6. WIMAX

Worldwide interoperability for microwave access (WIMAX) is used for wireless data transmission. The technique is used to give broadband access to a fixed receiver placed on buildings (clients) with an antenna facing the WIMAX base station. The distance between the base and the client can be up to 30 km. The frequency used is around 3.5 GHz, and the base station antenna is transmitting with a power of about 0.6 W and the client uses 0.1 W.

Recent measurements, performed by the Swedish National Radiation Protection Authority, from a city in Northern Sweden, showed that the environmental levels are very low. All measured values were below 0.01 $\mu W/m^2$, and thus only a fraction of the total environmental level including radio/TV broadcasting and mobile phone communication where levels of the order of tenths of mW/m^2 was found.

There are no values available for exposure directly in the vicinity of the base station antenna for occasional occupational exposure directly in front of the antenna. However, in view of the low total power it is not likely that the levels could reach levels of significance for violations of the exposure guidelines.

7. WLAN

Wireless local area networks (WLANs) make it possible for wireless communication between a user's computer and an access point (AP) (wireless router, gateway). In a recent paper Foster (2007) presented 356 measurements from 55 locations worldwide from both APs and laptops. The frequencies covered 75 MHz–3 GHz. The frequencies used for WLAN are around 2.4 and 5.1–5.8 GHz. Measurements were conducted under conditions that would result in the higher end of exposures from such systems. Where possible, measurements were conducted in public spaces as close as practical to the Wi-Fi access points. The output power is limited to a maximum isotropic transmitter power output of 0.1–1 W, depending on country of operation. In practice, the laptop computer usually has an output power of the order of tens of mW.

The levels found by Foster from the APs were of the order of up to $\mu W/m^2$ and at 1 m from the laptops when uploading large files; the median power density was a factor of 10 higher. In all cases, the measured Wi-Fi signal levels were very far below international exposure limits (IEEE C95.1-2005 and ICNIRP) and in nearly all cases far below other RF signals in the same environments. Important limiting factors are the low operating power of client cards and access points, and the low duty cycle of transmission that normally characterizes their operation.

8. BASE STATIONS

The cellular telephone communicates with a base station usually located at some distance, the antenna of which typically is on the top of a building or on a mast. The proliferation of radio base stations in urban environments, particularly on building rooftops and on facades, has resulted in an increasing number of RF EMF sources at different frequencies, output power levels and types. Radiofrequency emissions vary depending on the design and power of the base station. The transmission powers are relatively low, usually less than 40 W. Generally, most powerful antennas are sited on highest places, such as on broadcasting towers. On the other hand, base station antennas situated inside the buildings have lowest transmission powers, usually less than 1 W. These low-power antennas can be considered to be safe from the point of occupational exposure at any distance.

Occupational exposure is possible during maintenance of base stations, as well as during construction and similar tasks on the roof in close proximity to the antenna. Antennas should be mounted so that the general public can not access the area where corresponding exposure limits may be exceeded.

People working with antenna installation and maintenance for the network operators receive information concerning their exposure to RF fields, but it may also be the case that other workers can temporarily have their workplace in the near vicinity of an antenna system, without having received any, or limited information about the RF exposure situation. Some examples are maintenance and operator personnel, janitors, painters, chimney sweepers, window cleaners, technical staff dealing with repair or maintenance of roof, roof-racks and gutter or with heating, ventilation and air conditioners, etc., and generally workers who have to work on roof tops and building facades, as well as others who could be called on to perform work on buildings.

Modern base station antennas are sector antennas that transmit only in a forward direction. If you are working behind the antenna it is likely that the radiofrequency fields are at the background level due to the very good directivity of the transmission. Antennas are often sited on the edges of roof tops, so they can be approached only from behind. Another general installation place for the antenna is on small mast, chimney or higher on the building wall. In this case the antenna will send the transmission over the heads of people working below antennas. It is therefore safe to work below the antennas. The same principle applies if you are working higher than the antenna. In that case the transmission goes below you and the power densities at the work location are well below allowed levels.

A compliance boundary can be used to describe the RF exposure and outside this boundary the exposure is below the relevant limits. The size and shape of the compliance boundaries vary with frequency, output power level and antenna type. Figure 3 shows a simplified compliance boundary determined with respect to the ICNIRP general public reference levels for a typical 3G/WCDMA base station antenna (14.5–18 dBi antenna gain, 60–130 cm height) with emitted output power levels up to 25 W. For this configuration, the compliance boundary has the shape of a cylinder with a diameter of 3 m and a height corresponding to the antenna height plus 20 cm (10 cm above and 10 cm below). The cylinder starts 10 cm behind the back of the antenna. Using the basic restrictions, expressed in terms of SAR values for whole-body and local exposure, it has been shown by numerical calculations with a human whole-body model that the diameter of the compliance boundary is less than 1 m for the same type of base station configuration (Nordström, 2004). When applying the ICNIRP occupational exposure limits the size of the boundary is further decreased to some tenths of a meter. An example of such a calculation is given in Fig. 4 from Nordström (2004).

The electric field due to a 20 W, 2,100 MHz, 3G antenna, located 80 cm in front of the human body model and computed SAR distribution are illustrated in Fig. 5.

Figure 3. Compliance boundary: Roof-mounted 3G (WCDMA, 2,100 MHz) antenna cylindrical compliance boundary (for ICNIRP occupational and general public limits) diameter: 3 m, height: antenna height + 20 cm (courtesy of C Törnevik, Ericsson Systems, Sweden).

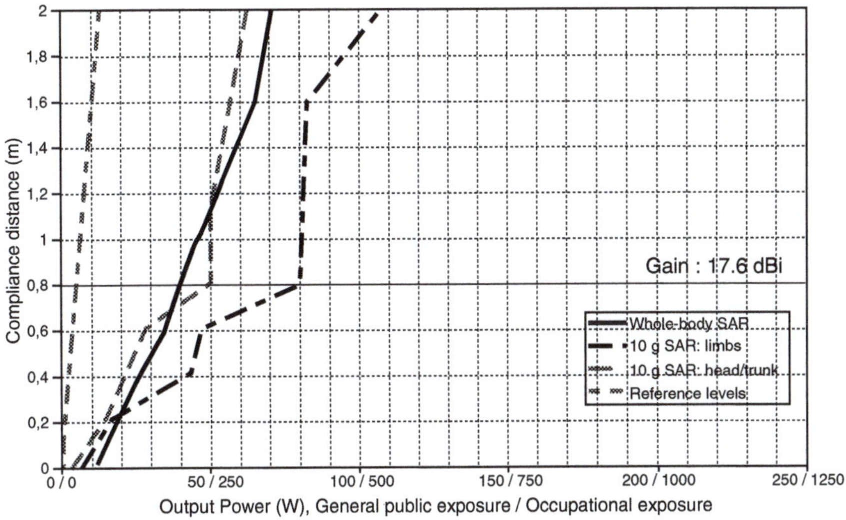

Compliance distance as a function of output power for the 17.6 dBi antenna.

Figure 4. Example of a compliance boundary with regard to the ICNIRP reference levels for the general public [from Nordström (2004)].

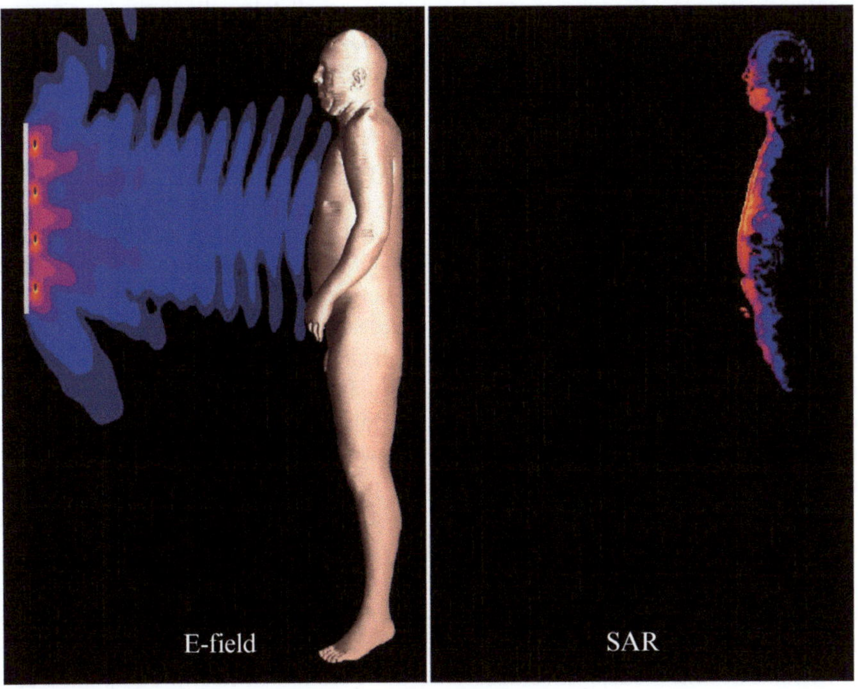

20 W 2100 MHz 3G antenna 80 cm in front of body
10 g SAR = 1.6 W/kg, whole-body SAR = 0.04 W/kg
SAR below ICNIRP occupational basic restrictions at 20 cm distance

Figure 5. Illustration of the electric field pattern of a 3G antenna, 80 cm in front of the body model and induced SAR distribution inside the body. (courtesy of C Törnevik, Ericsson Systems, Sweden).

Note that the 10-g SAR = 1.6 W/kg, whole-body SAR = 0.04 W/kg. These SAR values are below ICNIRP occupational basic restrictions at 20-cm distance. (courtesy of C Törnevik, Erresson systems, Sweden).

Possible co-location of antennas at a site and additional exposure from other RF sources might change the size and dimension of the compliance boundary. Further, the occasionally complex geometry of reflecting areas from roof surfaces can change the compliance boundary determined in free space although it can be shown that reflecting objects outside of the main beam of the antenna (defined by the −3 dB beam widths) only have minor effects. In Europe, standards are under development describing the requirements and procedures to verify that the general public has no access to areas where the exposure, including exposure from other radio sources, might exceed the limits when a base station is put into service in its operational environment.

When considering occupational exposure it is often of interest to investigate the time-averaged exposure when, for example, a chimney sweeper needs to quickly pass the area directly in front of an antenna. The ICNIRP exposure guidelines specify a 6 minute averaging time, and due to the high directivity of most base station antennas, which means that the radiated energy is concentrated to a main beam in front of the antenna, the time-averaged exposure is well below the limits for a person passing with normal walking speed in front of a typical antenna.

However, there could arise situations where workers need access to areas with exposure levels exceeding the general public limits. For such situations, there is a need for a careful investigation and clarification of operational procedures of control, measure-ments and instructions, elaborated and aimed to protect the health and safety of workers.

Brochures or similar materials should be made available for technical staff repairing buildings, which provide relevant information about EMF exposure from base station antennas, precautions to be undertaken before and during the execution of tasks at the roof of buildings, working instructions, access restrictions and possible warnings, etc. Many network operators already provide such information to landlords and owners of the buildings on which the antennas are installed. There may be a need to standardize this type of information to workers. Additionally, there may be a need, within this scope, to review the situation of workers wearing pacemakers, microprocessor-controlled medication dosage devices, metallic prostheses, etc.

A matter that may also need to be clarified in some countries is that of legal responsibilities. Although employers normally have the safety responsibility of their employees, contractual forms between various possible employers (building owners, subcontractors, etc.) and workers may result in an unclear understanding of who is responsible to provide the information to workers.

Some recent studies have addressed the question of health complaints and living near mobile phone base stations. To assess the exposure various measures have been used. Santini et al. (2003) in France obtained information from 530 people responding to a questionnaire about non-specific health symptoms and about distance to the base station. Respondents living at various distances (less than 300 m) from base stations were compared with a reference group living more than 300 m from the base stations. Santini et al. also asked the subjects to estimate the distance to the nearest base station. However, it has been shown by, among others, Schuz and Mann (2000) that distance to a base station cannot be regarded as a relevant surrogate for RF exposure. See also some recent measurements by Neubauer et al. (2007) where the complexity of exposure near base stations is discussed.

Measurements of exposure are a better surrogate than distance to the base station. However, it needs to be pointed out that due to reflection and multipath signals the power density levels of one BCCH of a GSM station varied up to two orders of magnitude within restricted areas (about 1 m^3) (Neubauer et al., 2007). Table 3 below shows some examples from Neubauer et al.

Regarding measurement, not only the spatial averaging needs to be taken into account but also the temporary variation. There will be variation in the output power of the base station on a daily as well as the weekly basis, all depending on the number of ongoing calls on the station. The BCCH will always be broadcasting independent of calls, but the traffic channel fills up on demand, which will be different at different times of the day and on weekends versus weekdays.

Recently, Neitzke et al. (2007) studied base station exposure and concluded again that distance is not a good measure for exposure. However, it could be stated that for residents far away from a base station the exposure could be predicted to be low.

If we look at the absorption of microwaves from a base station it can be found in textbooks on dosimetry that for frequencies of 1–3 GHz the whole-body SAR is approximately 3 mW/kg per incoming W/m^2. From the earlier Table 3 we can see that with the maximum value of power density of 18 mW/m^2, the whole-body SAR is of the order of 60 μW/kg, which is to be compared with the limit of 80 mW/kg; the real exposure even in this case is more than 1,000 times lower than the limit. The average exposure to the general public is in the range of 0.01 mW/m^2 giving a SAR value of 30 nW/kg. The localized exposure can be a few times higher than this, and it is thus clear that the exposure from handsets is by far the most dominant source of exposure for the general public: see also Kuhn and Kuster (2007).

Table 3. Exposure from mobile phone base stations [From Neubauer et al. (2007)]

Scenario	Distance (m)	S_{max} (mW/m^2)	S_{min} (mW/m^2)	S_{avg} (mW/m^2)	Ratio between S_{max} and S_{min}
GSM 900, no direct view	200	0.000048	0.0000003	0.0000051	173
GSM 900, direct view	60	0.3774	0.01194	0.08825	32
DCS 1800, no direct view	2,500	0.0009	0.00003	0.00019	31
GSM 900, direct view	12	17.7618	0.06051	2.42606	294

Maximum, minimum averaged power density and ration between maximum and minimum of the signal of the respective BCCH channel. In the first column 'scenario' the respective frequency band and the conditions of view (line of sight or no line of sight from the measurement location to the base station) are given. In the second column information on the distance between measurement location and base station is given

9. DISCUSSION

Looking at occupational exposure from wireless communication it is clear that the dominating sources are handheld phones – both mobile or cellular phones and cordless phones. It may even be that cordless phones produce a higher total exposure than mobiles, since the output power is not adjusted in cordless phones as it is in mobile phone systems; see Persson et al. (2002). In addition, the time spent on the phone is usually longer on cordless phones.

From a precautionary point of view the use of hands-free devices is to be recommended for both mobiles and cordless phones, something that many authorities now recommend.

In the studies of brain cancer and mobile phone use all of the problems discussed earlier apply, but since for this type of disease it is the exposure 5–10 years ago or more that is of interest, exposure assessment becomes an even greater problem than for acute effects. Most users of mobile phones have not been using just one single telephone, and it is even more likely that if they have been using a mobile phone for more than a few years they have changed phones a few times. Many users will also have used different phone systems such as analogue and digital, and many of them have probably also been using a cordless phone at home or at work. The problem we are facing then is: how to integrate the – probably different – SAR distributions from the different devices.

The epidemiological studies on mobile phone use and brain tumours have not taken this into account, and the main reason is that at the moment it is not clear how to combine the use of different phones with different power output, different systems, different frequencies and different anatomical SAR distribution into one exposure and dose measure. The difficulties lie in the fact that we do not know the interaction mechanism(s) between the electromagnetic fields emitted from the phone and the biological organism.

For mobile phone studies we need to be more detailed with the use of SARs and not just use the highest value found anywhere near the phone, paying no attention to the anatomical localization. We need more information from medical/biological experts as to what sites are of interest for which symptom and/or disease.

It is difficult to see how the actual exposure (if measured as SAR) can be proxied by 'billing record', a procedure that in Europe would only give the total time for outgoing calls, not showing incoming calls or the power settings of the phone. Instead of billing records or estimates of SARs, exposure in the epidemiological studies was assessed by transmitter system and the estimated number of minutes on the phone per day and the number of calls per day.

For future studies of MP users it would be useful to perform a study of the base station regulation for a number of users. This would not be possible to do retrospectively, but it could be done prospectively for a selected number of users. For these it would be possible to obtain detailed records of their phone use (with their permission, of course), both number of calls, length of each call and the actual power settings of the phone as ordered by the base station. These records could then be compared with the subjective estimates of number and time of calls from the users themselves.

As shown with examples from the RF epidemiological studies none of these have related the effect to the SAR and its distribution in the body. However, this issue is of great importance for continued research in bioelectromagnetics. One of the questions we need to address is, for instance, how time factors into the connection between exposure and dose, and here we need to distinguish between different aspects of time: very short times (order of minutes, daily averages) and total time in the actual occupation (number of years with exposure).

Persson et al. (1997) studied the effect of various exposure times and power densities on the blood–brain-barrier changes in rats. They obtained the specific absorption (SA) from the known SAR distribution and the exposure time. SAs are expressed with the unit J/kg, which in ionizing radiation is better known as Gray (We are thereby not implying that the effects are the same for the two different types of 'radiation'). The paper does not give any details on what combinations they used, so this does not help us in answering the question brought up in connection with the use of mobile phones: Is there a difference in the effect of one 10-min call and ten 1-min calls? We think it would be of value to look into how the dose concept has developed regarding exposure to ionizing radiation. We need to better understand things like dose-rate, fractioned dose, etc., for non-ionizing radiation also.

10. CONCLUSIONS

Occupational exposure from wireless communication devices is high from handset mobile and cordless phones, and the health issue regarding the effects of long-term exposure is far from settled. The exposure can be reduced by using hands-free device or blue-tooth solutions. Exposure from base station to mobile phones is low and only when working very close to the antenna is there a need for precautionary measures. The exposure from other sources for wireless communication such as WLAN and WiMax is very low.

REFERENCES

Andersen, J.B., and Pedersen, G.F., 1997, The technology of mobile telephone systems relevant for risk assessment. *Radiat Prot Dosimetry* 72: 249–257.

Anger, G., 2003, SAR and emitted power for 21 mobile phones. Swedish Radiation Protection Authority SSI Rapport: 01 January 2002 (in Swedish).

ANSI, Standard for Safety Levels With Respect to Human Exposure to Radio Frequency Electromagnetic Fields, 3 kHz to 300 GHz, (IEEE Standard C95.1), New York 1992.

Dimbylow, P., Khalid, M., and Mann, S., 2003, Assessment of specific energy absorption rate (SAR) in the head from a TETRA handset. *Phys Med Biol* 48: 3911–3926.

Erdreich, L.S., Van Kerkhove, M.D., Scrafford, C.G., Barraj, L., McNeely, M., Shum, M., Sheppard, A.R., and Kelsh, M., 2007, Factors that influence the radiofrequency power output of GSM mobile phones. *Radiat Res* 168(2): 253–261.

FCC, "Evaluating compliance with FCC guidelines for human exposure to radiofrequency electromagnetic fields," Federal Communications Commission, Washington, D.C, OET Bull. 65, Aug. 1997.

Foster, K.R., 2007, Radiofrequency exposure from wireless LANs utilizing WiFi technology. *Health Phys* 92: 280–289.

Hansson Mild, K., Oftedal, G., Sandström, M., Wilén, J., Tynes, T., Haugsdal, B., and Hauger, E., 1998, Comparison of symptoms experienced by users of analogue and digital mobile phones. A Swedish–

Norwegian epidemiological study. Swedish National Institute for Working Life, Arbetslivsrapport: 23, ISSN 1401-2928.

Hansson Mild, K., Wilén, J., Carlberg, M., and Hardell, L., 2004, "Exposure" and "dose" in mobile phone health studies. In: Mobile Communication and Health: Medical, Biological and Social Problems. International WHO Conference, Moscow, 22–24 Sept., pp. 70–79.

Hansson Mild, K., Carlberg, M., Wilén, J., and Hardell, L., 2005, How to combine the use of different mobile and cordless telephones in epidemiological studies on brain tumours? *Eur J Cancer Prev* 14: 285–288.

Hardell, L., Hansson Mild, K., and Carlberg, M., 2006, Pooled analysis of two case-control studies on use of cellular and cordless telephones and the risk for malignant brain tumours diagnosed in 1997–2003. *Int Arch Occup Environ Health* 79: 630–639.

Hillert, L., Ahlbom, A., Feychting, M., Jarup, L., Larsson, A., Neasham, D., and Elliott, P., 2003, Mobile phone use: Validation of exposure assessment. The Bioelectromagnetics Society Annual Meeting, Hawaii, USA.

Hours, M., Bernard, M., Montestrucq, L., Arslan, M., Bergeret, A., Deltour, I., and Cardis, E., 2007, Cell phones and risk of brain and acoustic nerve tumours: The French interphone case-control study. Revue d'Epidemiologie et de Santé Publique doi: 10.1016/j.respe.2007.06.22.

Ilvonen, S., Sihvonen, A.P., Kärkkäinen, K., and Sarvas, J., 2005, Numerical assessment of induced ELF currents in the human head due to the battery current of a digital mobile phone. *Bioelectromagnetics* 26: 648–656.

International Commission on Non-Ionizing Radiation Protection (ICNIRP), 1998, Guidelines for limiting exposure to time-varying electric, magnetic, and electromagnetic fields (up to 300 GHz). *Health Phys* 74: 494–522.

Jokela, K., Puranen, L., and Sihvonen, A.P., 2004, Assessment of the magnetic field exposure due to the battery current of digital mobile phones. *Health Phys* 86: 56–66.

Kuhn, S., and Kuster, N., 2007, Experimental EMF exposure assessment. In: Polk, C., and Postow, E., Eds., Handbook of Biological Effects of Electromagnetic Fields, 3rd Ed. Bioengineering and Biophysical Aspects of Electromagnetic Fields. CRC Press/Taylor and Francis, Boca Raton, FL, pp. 381–409.

Kuster, N., 1997, Swiss tests show wide variation in radiation exposure from cell phones. Microwave News Nov/Dec, pp. 1, 10–11.

Kuster, N., Schuderer, J., Christ, A., Futter, P., and Ebert, S., 2004, Guidance for exposure design of human studies addressing health risk evaluations of mobile phones. *Bioelectromagnetics* 25: 524–529.

Linde, T., and Hansson Mild, K., 1997, Measurement of low frequency magnetic fields from digital cellular telephones. *Bioelectromagnetics* 18: 184–186.

Morrissey, J.J., 2007, Radio frequency exposure in mobile phone users: Implications for exposure assessment in epidemiological studies. *Radiat Prot Dosimetry* 123(4): 490–497.

MTHR, 2007, Progress report from the UK's Mobile Telecommunications and Health Research (MTHR) *Programme. Progress report* 2007.

Neitzke, H.P., Osterhoff, J., Peklo, K., and Voigt, H., 2007, Determination of exposure due to mobile phone base stations in an epidemiological study. Radiat Prot Dosimetry doi: 10.1093/rpd/ncm371.

Neubauer, G., Feychting, M., Hamnerius, Y., Kheifets, L., Kuster, N., Ruiz, I., Schuz, J., Uberbacher, R., Wiart, J., and Roosli, M., 2007, Feasibility of future epidemiological studies on possible health effects of mobile phone base stations. *Bioelectromagnetics* 28(3): 224–230.

Nordström, E., 2004. Calculation of specific absorption rate (SAR) for typical WCDMA base station antennas at 2140 MHz. LiTH-IFM-EX-04/1355-SE, MSc Thesis report, Linköping University.

Persson, B., Salford, L., and Brun, A., 1997, Blood–brain barrier permeability in rats exposed to electromagnetic fields used in wireless communications. *Wireless Netw* 3: 455–461.

Persson, T., Törnevik, C., Larsson, L.E., and Lovén, J., 2002, GSM mobile phone output power distribution by network analysis of all calls in some urban, rural and in office networks, complemented by test phone measurements. Bioelectromagnetics Society 24th Annual Meeting, 23–27 June, Quebec City, QC, Canada, pp. 182–183.

Persson, T., Hamberg, L., Törnevik, C., and Larsson, L.E., 2005, Measured output power distribution for 3G WCDMA mobile phones. Bioelectromagnetics 2005. A joint meeting of the Bioelectromagnetics Society and the European Bioelectromagnetics Association, Dublin, pp. 473–474.

Santini, R., Santini, P., Le Ruz, P., Danze, J.M., and Seigne, M., 2003, Survey of people living in the vicinity of cellular phone base stations. *Electromagn Biol Med* 22: 41–49.

Schuz, J., and Mann, S., 2000, A discussion of potential exposure metrics for use in epidemiological studies on human exposure to radiowaves from mobile phone base stations. *J Expo Anal Environ* Epidemiol 10(6, Part 1): 600–605.

Straume, A., Johnsson, A., Oftedal, G., and Wilén, J., 2007, Frequency spectra from current- vs. magnetic flux density measurements for mobile phones and other electrical appliances. *Health Phys* 93(4): 279–287.

Wilén, J., Sandström, M., and Hansson Mild, K., 2003, Subjective symptoms among mobile phone users – A consequence of absorbtion of radiofrequency fields? *Bioelectromagnetics* 24: 152–159.

Chapter 7

Dosimetry and Temperature Aspects of Mobile-Phone Exposures

Paolo Bernardi, Stefano Pisa, Marta Cavagnaro, Emanuel Piuzzi, and James C. Lin

ABSTRACT

Dosimetric assessment is an import subject in studying the effects from exposure to wireless communication devices. It provides a quantitative measure for epidemiological studies and in the development of exposure guidelines. Dosimetry may be accomplished either numerically or experimentally, or by a combination of both since each technique has its own advantages and drawbacks. While numerical dosimetry forms the focus of this chapter, some descriptions of experimental dosimetry are included to illustrate the complementary nature of numerical investigations and experimental studies, especially in testing against compliance of mobile phones with exposure guidelines. Numerical dosimetry requires the use of detailed anatomical models of the human head for mobile phones used in a conventional manner or the use of human torsos in the case of mobile terminals of various communication systems that involve body-worn devices. Moreover, detailed models of the mobile phone are often required to account for the antenna structure, phone case, and internal components of the device. The dosimetric quantity, specific absorption rate (SAR), and

---

P. Bernardi, S. Pisa, M. Cavagnaro, and E. Piuzzi Department of Electronic Engineering, Universita' di Roma "La Sapienza", Rome, Italy
J. C. Lin Department of Electrical and Computer Engineering and Department of Bioengineering, University of Illinois – Chicago, Suite 1020 SEO (MC 154), 851 South Morgan Street, Chicago, IL, 60607-7053, USA, e-mail: lin@uic.edu

J.C. Lin (ed.), *Advances in Electromagnetic Fields in Living Systems, Health Effects of Cell Phone Radiation, Volume 5*,
DOI 10.1007/978-0-387-92736-7_7,

associated induced temperature increments in tissues may be computed using different analytical and numerical techniques. One of the most widely applied numerical techniques is the FDTD method; it will be briefly discussed in this chapter. Recent dosimetric research will be summarized, including the influence of different metallic implants worn by mobile phone users and the environment in which exposure occurs, such as inside a vehicle. Some topics may be of interest to the general public, research scientists, or cell phone manufacturers and operators because of their importance in mobile phone compliance testing. Other topics discussed will address the specific concerns of mobile phone use by children. Among the topics of technical interest are the influence of the pinna on computed SAR, effect of averaging procedures on SAR values, and the variation of results due to the uncertainties associated with the dielectric parameters used to characterize human tissues.

1. INTRODUCTION

In the study of the interaction between electromagnetic fields (EM) and biological systems, dosimetric aspects are of particular interest and importance. Dosimetry considers the coupling of electromagnetic energy between the external source and the exposed subject. It deals with the determination of the electromagnetic field induced inside the subject and the distribution of absorbed energy in tissues. Temperature increases are also important because they are produced inside tissues by electromagnetic power dissipation.

The development of theoretical and numerical dosimetry began in the 1960s, by considering human exposure to the radiofrequency electromagnetic fields present, at that time, in the environment; namely, fields produced by radio and TV broadcasting antennas and radar systems. In these studies, human models constituted by simple geometrical shapes like spheres, prolate spheroids, and ellipsoids and models constituted by homogeneous or stratified tissue exposed to uniform plane waves impinging on them with different possible polarizations have been adopted (Lin, 1986). These models were suitable to evaluate human exposure in the far field of the sources, which is the case of the radar and the broadcasting type electromagnetic systems. Even though very simplified, they allowed elucidation of the basic concepts and key parameters shedding light on the physics of the problem. They have provided useful information concerning human exposure, which were generally valid within a good level of approximation. The models provided results concerning the dependence of power absorbed by the human body with the frequency of the impinging EM field, the presence of a resonance frequency at which the human body shows an absorption peak, the influence of the polarization of the impinging field and of the dimensions of the exposed body, and, finally, the distribution of absorbed energy within the body, with the presence of hot spots inside the body itself, sometimes located deep inside because of diffraction effects resulting from tissue composition and geometry.

The development of electromagnetic technologies used in modern communication and wideband data transmission systems that began around the 1990s and still continues today at an increasing pace has led to the use of systems in which exposed subjects are located in the radiating near-field of the sources or are even placed in direct contact with the body, as in the case of the mobile phone for cellular telecommunications. In these cases two fundamental aspects, which are not present for far-field exposure, must be taken into account. The presence of the exposed body modifies the radiating properties of the antenna such as input impedance, efficiency, bandwidth, antenna current, and radiated power. Also, a large portion of the radiated power is absorbed by the exposed body, with a highly nonuniform distribution inside its volume.

In this chapter, the topic of numerical dosimetry is presented in some detail. While some descriptions of experimental dosimetry are included, they are intended to illustrate the need for numerical investigations in experimental studies designed for testing compliance of mobile phones with exposure guidelines. Numerical dosimetry requires the use of detailed anatomical models of the human head in the case of mobile phones used in a conventional manner, or the use of human torso for the case of mobile terminals of different communication systems that are becoming more widespread and for recent developments involving body-worn devices. Moreover, detailed models of the mobile phone are often required to take into account the antenna structure, phone case, and internal components of the device.

Specific absorption rate (SAR) and induced temperature increments in tissues produced by exposure to cellular mobile phones may be computed using different analytical and numerical techniques. One of the most widely applied numerical techniques is the finite difference time domain (FDTD) method; it will be briefly summarized in this chapter. The results from recent dosimetric research will be discussed, including the influence of different metallic implants worn by mobile phone users and the environment in which exposure occurs, such as inside a car. Many of the topics may be of special interest to the general public, research scientists, or cell phone manufacturers and operators because of their importance in mobile phone compliance testing. Other topics discussed will address the specific concerns of mobile phone use by children. Among the topics of technical interest are the influence of the pinna on computed SAR values, the effect of averaging procedures on numerical SAR, and the variation of results due to the uncertainties associated with the dielectric parameters used to characterize human tissues.

Accordingly, we will begin with the topic of dosimetric quantities and computational methods, which will include computational models for mobile phones and anatomical models of human bodies, i.e., the head and the torso, and results of dosimetry studies on SAR computations under normal usage positions of the mobile phones, i.e., next to the head, and other usage positions of mobile phones for body-worn devices. We will review the related issues of standardization of dosimetry procedures and the effect of metallic implants or wearable metallic elements on exposure and SAR as well as that due to partially enclosed environments on RF energy absorption. The topic of temperature increments will be covered in a separate section for normal and other usage positions of the mobile phones.

This will be followed by a discussion on the influences of variations of tissue thermal parameters on computed temperature elevations. In the last section, the regulatory and public health-related issues will be discussed, which would include topics such as the effect of the pinna on computed SAR values, dielectric property of head-equivalent homogeneous tissue phantoms, exposure of child- and adult-sized head models, and effect of averaging procedure on computed SAR values.

2. DOSIMETRIC QUANTITIES AND COMPUTATIONAL METHODS

2.1. Dosimetric Quantities

To characterize electromagnetic absorption in biological bodies several quantities are used such as the induced electric field, power deposition, or specific energy absorption. The type of quantity chosen to characterize the interaction between electromagnetic fields and biological tissues usually depends on the target effect to be studied. When considering electromagnetic fields at radio frequencies (RF) or microwave frequencies (MW), the specific absorption rate (SAR) or the specific absorption (SA) is normally used. SAR (W/kg) is defined as the time derivative of the incremental energy absorbed by (or dissipated in) an incremental mass contained in a volume of a given density, while SA (J/kg) is the total amount of energy deposited or absorbed and is given by the integral of SAR over a finite interval of time (Lin, 2007). Once the electromagnetic field induced into a biological body is known, the SAR can be derived according to the following formula:

$$\mathrm{SAR}(x,y,z) = \frac{\sigma(x,y,z)E^2_{\mathrm{rms}}(x,y,z)}{\sigma_{\mathrm{m}}(x,y,z)} \quad \text{(W/kg)}, \tag{1}$$

where σ represents the tissue bulk electrical conductivity (S/m) and ρ_{m} the tissue mass density (kg/m^3). SAR has a direct link to the temperature increase in the biological body. When a very short period of exposure is considered, to prevent significant convective or conductive heat contribution to tissue temperature rises, SAR and temperature increase (ΔT) are linked by a direct proportionality law, according to:

$$\mathrm{SAR} = C\frac{\Delta T}{\Delta t}, \tag{2}$$

where C is the specific heat capacity of tissue (J/kg °C), and Δt is the duration (s) over which ΔT is measured.

In safety guidelines, at RF and MW, SAR has been chosen as the basic parameter to be limited to prevent detrimental effects on human health (ICNIRP, 1998; IEEE, 2005), since at these frequencies the most significant effects of electromagnetic fields on human systems have been found to be associated with temperature increase. Moreover, SAR has been considered as a convenient quantity for comparing effects observed under various exposure conditions (ICNIRP, 1998), as well as for extrapolating experimental results from tissue to tissue, animal to animal, animal to human, and human to human exposures (Lin, 2000a,b).

When quantifying the energy absorption or power deposition in the human body, the electromagnetic problem governed by Maxwell's equations must be solved. Once the SAR distribution has been evaluated, the corresponding temperature increase and distribution can be evaluated by solving the bioheat equation (BHE).

2.2. Computational Methods

The numerical methods used to evaluate the power absorption into biological bodies include the quasistatic impedance method, the method of moments, the finite element method (FEM), and the FDTD method (Lin and Bernardi, 2007). Hybrid methods derived from the combination of these methods and other methods used for electromagnetic propagation characterization (e.g., the ray tracing technique) are also used.

Among the cited methods, the FDTD is the most often used. For more details, please see more specific books on this subject (Kunz and Luebbers, 1993; Taflove and Hagness, 2000).

This method, first proposed by Yee (1966), is based on the substitution of each partial derivative in the Maxwell's equations in time domain with its finite difference representation. This substitution leads to a set of six equations where each field component is evaluated at a point in time as a function of the adjacent components evaluated in preceding points of time. To this end, space and time are divided into discrete intervals in which the electromagnetic field is supposed constant. With reference to space, this leads to the definition of a unit cell (referred to as Yee's cell) in which the electromagnetic field is supposed constant. To improve the method's precision without adding complexity to the obtained formulas, the E-field and H-field components are placed in different positions within the Yee's cell, as illustrated in Fig. 1; similarly, they are evaluated at half-time steps.

Being a method in time domain, a stability condition is needed to make sure the equations converge to a solution. This condition is usually referred to as the Courant condition and it limits the time step (Δt) as a function of the space steps (Δx, Δy, and Δz), according to:

$$\Delta t \le \frac{1}{v\sqrt{\frac{1}{\Delta x^2}+\frac{1}{\Delta y^2}+\frac{1}{\Delta z^2}}}, \tag{3}$$

where υ represents the field-velocity in vacuum. Similarly, to ensure the obtained solution is close to the actual one, an accuracy condition must be satisfied, i.e.,

$$\max\,(\Delta x, \Delta y, \Delta z) \ll \delta \min \tag{4}$$

$\lambda_{\min}$ is the minimum field wavelength in the considered problem, found in the tissue with maximum dielectric permittivity.

Finally, to limit the domain under study, absorbing boundary conditions (ABCs) are needed. Several ABCs have been proposed through the years (e.g., Mur's ABC, Higdon, Retarded time, and so on). Nowadays, the most often used ABCs are those

Figure 1. Yee's cell.

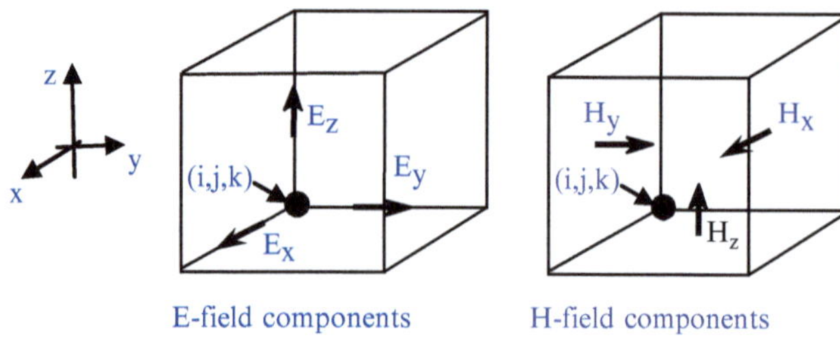

proposed by Berenger (1994) and known as perfectly matched layer (PML), or its evolutions (e.g., UPML; Gedney, 1996).

To evaluate the temperature distribution obtained in a biological body when exposed to an electromagnetic field, the BHE, with its initial and boundary conditions, must be solved (Pennes, 1948). Within the BHE, the electromagnetic field becomes an exogenous heat source, responsible for the alteration of temperature inside the exposed subject from the resting temperature distribution. As a consequence, the implicit assumptions are that the electromagnetic problems and thermal problems are independent, and that the electromagnetic transients are irrelevant with respect to the thermal ones, so that the two problems can be investigated sequentially.

The BHE can be solved by developing an explicit finite difference scheme to directly link the electromagnetic and the thermal problems (Bernardi et al., 2003). The finite difference scheme, in fact, can be associated with the FDTD method and allows one to use the same numerical model for the biological body in the electromagnetic and thermal problem solution. To obtain the finite-difference explicit formulation of the BHE, a thermal balance approach is used in each cell of the biological model, according to the formula:

$$Q_{\mathrm{tot}}^{n,n+1}(i,j,k) = \Gamma(T^{n+1}(i,j,k) - T^{n}(i,j,k))\ (\mathrm{J}), \tag{5}$$

where $Q_{\mathrm{tot}}^{n,n+1}(i,j,k)$ represents the heat accumulated (or lost) in the cell in the unit time interval, $(T^{n+1} - T^{n})$ is the cell temperature variation in the same time interval, and Γ is the thermal capacitance of the cell (Lin and Bernardi, 2007).

Since an explicit scheme is developed for the time evolution of the temperature, a stability condition is needed. This condition, obtained through Fourier's analysis, requires that the time step Δt must satisfy the following condition (Wang and Fujiwara, 1999):

$$\Delta t \leq \frac{1}{6\dfrac{K}{C\rho\delta^2} + \dfrac{B}{2C\rho}}, \tag{6}$$

where K (W/(m °C)) is the tissue thermal conductivity, B (W/(°C m^3)) represents capillary blood perfusion proportional to blood flow, C (J/(kg °C)) is the tissue specific heat, ρ (kg/m^3) the tissue density, and δ is the linear dimension of the cell (m).

The explicit finite-difference formulation of the BHE becomes computationally unaffordable when very small cell sizes are used or when high thermal conductivity materials are present in the domain under study. In fact, in these cases an extremely small time step Δt is obtained from (6) and an alternate-direction implicit (ADI) formulation can be used (Pisa et al., 2003). The ADI solution of the BHE is obtained by extending to the three-dimensional case the one-dimensional Crank–Nicholson's scheme, which averages the outcome from the explicit and implicit formulation to obtain second-order accuracy both in space and in time variables (Ozisik, 1985). In particular, Crank–Nicholson's scheme can be extended to the three-dimensional case by using a sequence of approximate Crank–Nicholson's solutions along the three axes, the last one being used as the final estimate for the temperature distribution at the successive time step (Lin and Bernardi, 2007). This scheme is rigorous and valid for parabolic equations; however, due to the presence of the term related to blood flow, the BHE is not exactly parabolic. In such cases the ADI scheme can still be applied, but it loses its unconditional stability. Nevertheless, for time steps of the order of a few seconds, the scheme has proved to be stable in typical applications (Pisa et al., 2003).

3. MOBILE PHONE AND ANATOMICAL MODELS

Mobile phone and anatomic models must be accurately chosen to correctly compute the SAR distribution inside the head of a cellular phone user. Both these models have been improved tremendously in recent years, starting from very simple structures (spheres and half-wavelength dipoles) and evolving into very accurate geometries (anatomical head of 1-mm resolution and computer-aided-design (CAD) files reconstructed cellular phone models). This evolution will be described in the following paragraphs together with a description of the torso models to be used for body-worn devices.

3.1. Mobile Phone Models

The evaluation of SAR distribution inside the head of a cellular phone user is an essential task to assess possible health hazards, perform compliance testing, and correctly interpret epidemiological studies.

In numerical dosimetry for mobile communications, some initial studies have been performed using the multipole method (Kuster and Balzano, 1992; Kuster, 1992) and the method of moment (Chuang, 1994); but in recent years, the use of the FDTD has dominated over the other numerical methods due to its simplicity and ability to treat highly nonhomogeneous structures.

In an FDTD code, cellular phones can be modeled simply as half-wavelength dipoles or as quarter-wavelength monopoles over a box, but more complex and realistic phone geometries can also be considered, as will be described in the following.

Three different numerical models of the dipole have been adopted. The first model is the "infinitely thin wire" approximation, obtained by setting to zero the tangential electric field component along the dipole's axis, with the exception of the feeding gap (Martens et al., 1995; Bernardi et al., 1996). The second is the "thin" model in which a static approximation is applied and the electric and magnetic field components are assumed to vary as $1/r$ near the wire, where r is the distance from the wire center (Dimbylow, 1993). This approach allows one to take into account the radius of the wire. The third is the "thick wire" approximation, obtained by assigning the copper conductivity value to each cell belonging to the dipole, with the exception of the antenna feed-point (Chen and Wang, 1994).

To model a half-wavelength dipole, an odd number of cells are considered to produce symmetrical arms around a central gap. As an example, for a phone operating at 1,710 MHz the wavelength in air is 17.5 cm, hence $\lambda/2 = 8.75$ cm. For a cell side of 2.5 cm the dipole length can be modeled by 35 cells, with the central one corresponding to the feeding gap (Nikita et al., 2000).

A better phone model is constituted by a quarter-wavelength monopole over a box (Toftgard et al., 1993; Jensen and Rahmat-Samii, 1994). These antenna models can be used as a rough model of the retractable antenna, which was in nearly all of the early cell-phone handsets. As an example, in the framework of the "Cellular phone standard" (CEPHOS) Project the phone case was modeled as a conducting box of 120 mm (length) × 55 mm (width) × 20 mm (depth) with the monopole antenna centered on the upper side of the box (Nikita et al., 2000). The front face of the metal box was covered with a Plexiglas dielectric insulator of 0.5-cm thickness, and the size of the feeding gap was set to 0.25 cm.

More recently, the need for more compact terminals and dual-band operation has given rise to new antenna types. In particular, two types of antennas have been developed: planar-integrated antennas and helical antennas. While half-wavelength dipoles and monopoles can be easily implemented inside an FDTD code, modeling of helix or planar antennas can become a rather difficult task.

The difficulties in modeling helical structures with the FDTD method were revealed in some studies. For example, only rather large structures have been studied employing a pure FDTD scheme in which the electric field components along the helix wires are set to zero (Bernardi et al., 1996; Cavagnaro and Pisa, 1996; Colburn and Rahmat-Samii, 1998; Caputa et al., 2000; Rowley et al., 2002; Dimbylow et al., 2003; Koulouridis and Nikita, 2004). For smaller structures, published reports had either employed a stack of electric dipoles and magnetic loop sources with relative weights obtained from analytical expressions for the helix far-field (Lazzi and Gandhi, 1998) or a hybrid MoM/FDTD technique (Mangoud et al., 2000; Cerri et al., 1998). While these reports show some problems and drawbacks, investigations using FDTD, properly modified through the use of a graded mesh, have obtained good agreement with MoM and experimental results (Bernardi et al., 2001a, b; Pisa et al., 2005b). In these studies, both near-field and radiation patterns of dual band cell phones equipped with a helical antenna have been reproduced.

Simple planar antennas can be modeled by zeroing the tangential E-field component on the patch surface (Jensen and Rahmat-Samii, 1995). However, the geometries employed in commercial phones are not planar and conformal graded mesh is necessary for their modeling (Pisa et al., 2006). Planar antennas can be mounted on top, lateral, or back sides of the phone (Jensen and Rahmat-Samii, 1995; Rowley and Waterhouse, 1999; Bernardi et al., 2000; Li et al., 2000). Shorted patch antennas typically have a 10-dB bandwidth between 5 and 10% that can be increased to about 12% by parasitically coupling another printed radiator in the vertical direction (stacked patch). For comparison, the bandwidth is about 30% for the monopole antenna (Rowley and Waterhouse, 1999). For a cell phone equipped with a planar antenna, an important consideration is the influence of the hand wrapped around the handset. In this case, the hand has a detuning effect on the antenna resonant frequency and causes a reduction of the bandwidth, which is evident when the hand masks the antenna.

Another important task for an accurate evaluation of the power absorption in the head is the model adopted for the phone case. The typical approach followed in the literature, consists of representing the case as a box, i.e., a plastic-coated metal parallelepiped. To model the correct shape of cell phones, both CAD files (Tinniswood et al., 1998a) and topometric sensors (Schiavoni et al., 2000) have been used. However, in most studies, the internal structures of the phone have been modeled simply as a homogeneous perfect conductor. Recently, CAD files have also been used to model the internal structures (printed circuit board, battery, keypad, and buttons) of the phone (Chavannes et al., 2003). An alternate approach to a suitable numerical model of the mobile phone was proposed by Pisa et al. (2005a). It starts with a simplified model which includes only the main phone parts (antenna, keyboard, internal box, and plastic coating) having "realistic" dimensions and electric properties. The "realistic" parameters are then tuned using an optimization procedure, which minimizes a function that depends on the differences between the measured and simulated electric and magnetic fields in front of the phone and on the SAR inside a cubic phantom. For the applicability of the proposed optimization method, a numerical model of a commercial phone, operating at 900 MHz, was implemented and the power deposition in the VH model of the human head was computed (Pisa et al., 2005a). The results show that using the wrong phone model could lead to overestimating the SAR averaged over 1 g or 10 g up to three times.

3.2. Head Models

A cellular phone usually operates in close contact with the human head whose presence influences the performance of the phone antenna. Moreover, a significant fraction of the phone radiated power is absorbed inside the head tissues.

Spherical head models have largely been used for testing the input impedance and the radiation pattern of the phone antenna (Toftgard et al., 1993) and for defining canonical problems (Nikita et al., 2000; Koulouridis and Nikita, 2004).

For more accurate cellular phone dosimetric evaluations, anatomical and heterogeneous head models have been developed. These models consist of large datasets obtained from magnetic resonance imaging (MRI), computer tomography (CT), and anatomical images. To be used for computational electromagnetic dosimetry, these digital data sets must be converted to a so-called "segmented" version. A segmented model is a model where every pixel, usually called in such models as "voxel" (volume element), does not contain information about the color (like in digital images) but rather contain a label which is uniquely associated with a given tissue. In such a way, it is possible to know which tissue fills each of the model voxels and hence assign the correct complex permittivity values to be used in numerical simulations.

Segmentation of the original image sets is a complex and time-consuming activity, which is difficult to do with only automatic procedures, such as contour recognition algorithms, and inevitably requires intervention by experts in human anatomy.

A millimeter-resolution model of the human head has been developed from the MRI scans of a male volunteer of height 176.4 cm and weight 64 kg (Gandhi and Chen, 1995). The MRI scans were taken with a resolution of 3 mm along the height of the body and 1.875 mm for the orthogonal axes in the cross-sectional planes. The MRI sections were converted into images involving 15 head tissue types. This model has been used to calculate the electromagnetic absorption in the human head, neck, and shoulders for cellular telephones operating at frequencies of 835 and 1,900 MHz (Gandhi et al., 1996; Lazzi and Gandhi, 1997) and for EM energy absorption of various antennas for the handheld transceiver of a 6-GHz personal communication network (Gandhi and Chen, 1995).

Another fine resolution head model has been constructed from a set of serial MRI slices from a single male subject (Dimbylow and Mann, 1994). The MRI data have been segmented into 4 million 1-mm voxels in the shape of cubes and with a tag which identifies one of the 14 different tissue types. To obtain a model more easily tractable for FDTD calculations, the original model was rescaled to 2 mm resulting in about half a million voxels in the head. This model has been used to study SAR distributions in the head for mobile communication transceivers operating at 900 MHz and 1.8 GHz (Dimbylow and Mann, 1994), and for the assessment of SAR in the head of users of terrestrial trunked radio apparatus operating between 380 and 390 MHz (Dimbylow et al., 2003).

More recently, a high-resolution human model has been obtained using the "Visible Human Project," developed by the National Library of Medicine (Ackerman, 1998). The Visible human is a three-dimensional digital image library representing an adult human male and female. The data set for both the male and the female include photographic images obtained through cryosectioning of human cadavers and digital images obtained through CT and MRI of the same cadavers. In particular, the photographic images represent a highly accurate and realistic counterpart of the anatomical cross sections contained in human anatomy atlases. The male data set, the first to be constructed, consists of 1,871 digital axial images obtained at 1.0-mm intervals, with a pixel resolution of 1 mm, while the female one contains 5,189 digital axial images, obtained with a finer spatial step of 0.33 mm. The segmentation procedure has been carried out for the male model by researchers at the Air Force Research Laboratory, Brooks Air Force Base, TX, USA (Mason et al., 2000). The final segmented model, made freely available to the scientific community (ftp://starview.brooks.af.mil), comprises $586 \times 340 \times 1{,}878$ voxels with a resolution of $1 \times 1 \times 1$ mm^3, and is segmented in about 40 different tissue types. The model has been widely used to study head exposure to electromagnetic fields radiated by cellular phones (Bernardi et al., 2001a, b; Gjonaj et al., 2002; Wang et al., 2004; Pisa et al., 2005a), and is now being included in many commercially available electromagnetic simulation tools with capabilities for dosimetric evaluation.

Other head models have been developed at the Zurich University Hospital (Burkhardt and Kuster, 2000; Hombach et al., 1996), at the Osaka University

(Hirata et al., 2000), and at the Yale University (Zubal et al., 1994; Okoniewski and Stuchly, 1996).

A simplified physical model of the human head of the user of handheld radio transceivers (the specific anthropomorphic mannequin (SAM)) has been proposed by IEEE (IEEE Standard 1528, 2003) and IEC (IEC 62209-1, 2005) standards for compliance testing. SAM has been also adopted by the European Committee for Electrotechnical Standardization (CENELEC) (EN 50361, 2001), the Association of Radio Industries and Businesses in Japan (ARIB STD-T56, 2002), and the Federal Communications Commission in the USA (FCC, 2001).

SAM was developed by the IEEE Standards Coordinating Committee 34, Subcommittee 2, Working Group 1 (SCC34/SC2/WG1); it has a lossless plastic shell and an ear spacer. Because current technology does not allow reliable measurement of the SAR in small complex structures such as a simulated pinna, SCC34/SC2/WG1 chose to use a thin lossless ear spacer on SAM to maximize the energy reaching the brain and minimize the measurement uncertainty. The SAM shell is filled with a homogeneous fluid having the electrical properties of average head tissue at the test frequency. The dielectric properties of the fluid were based on calculations to give conservative spatial-average SAR values averaged over 1 g and 10 g for the test frequencies. A primary design goal for SAM was that "SAM shall produce a conservative SAR for a significant majority of persons during normal use of wireless handsets." To determine the extent to which SAM is truly conservative, various investigators have used computational radio frequency (RF) dosimetry to compare the SAR in SAM and in anatomically correct models of the human head (Gandhi and Kang, 2004; Christ et al., 2005; Kainz et al., 2005; Beard et al., 2006).

3.3. Torso Models (Body-Worn Devices)

Wireless devices are commonly used in contact with the ear. However, when they are used with a headset, they can be positioned at different body locations. The SAR evaluation for these body-worn devices can be numerically accomplished using a heterogeneous phantom that correctly reproduces the human anatomy. To assess experimentally the compliance of body-worn devices, flat phantoms have been suggested.

Among whole-body anatomical models there are: the Visible Human data set (Ackerman, 1998), the Utah University model (Tinniswood et al., 1998b), the Norman model (Dimbylow, 1998), the Trento model (Mazzurana et al., 2003), and the Japanese models (Nagaoka et al., 2004). Some of these have already been used to investigate the SAR distribution inside the body produced by half-wavelength dipoles placed at various body positions (Kang and Gandhi, 2002; Troulis et al., 2003; Onishi et al., 2005; Bernardi et al., 2005; Christ et al., 2006).

With reference to experimental compliance, the United States Federal Communication Commission (FCC) recommends the use of flat phantom models to test handset and push-to-talk devices that transmit in body-worn operating configurations (FCC, 2001). These phantoms should have internal dimensions at least twice the corresponding dimensions of the test device, including its antenna. The document recommends a shell with thickness of 2.0 ± 0.2 mm, dielectric constant less than 5.0, and loss tangent not exceeding 0.05. The tissue-simulating liquid must fill the phantom for at least 15 cm ± 0.5 cm and should have dielectric properties that simulate body tissues at various frequencies. In the FCC document, the flat phantoms are also suggested for testing compliance of devices used in front of the face. In fact, the use of anatomical models in these cases could give rise to E-field probe boundary errors, due to the curved region of these phantoms such as the nose, lips, eyes, etc.

Recently, the International Electrotechnical Commission (IEC) initiated studies on a SAR measurement procedure for mobile wireless communication devices used in close proximity to the human body that employs a flat phantom (IEC-622092, 2007).

Since the number of existing anatomical computer models is small, and their quality is often not sufficient to study the effects of large variations of tissue thicknesses, a planar three-layer body model has been proposed (Christ et al., 2006). This model is constituted by a low water content tissue layer (fat) embedded between two high water content tissues (skin and muscle). The spatial distribution chosen can be considered as representative for all body regions, since the electrical properties of fat are similar to those of bone, whereas muscle tissue has properties similar to most of the inner organs of the body. A thickness range of the skin from 0.4 to 2.6 mm is considered and a thickness range from 0 to $\lambda/2$ is assumed for the fat layers to cover all possible absorption phenomena. The muscle tissue layer is used to terminate the model assuming that possible effects of reflections from a fourth layer can be neglected because of the low reflection coefficient between adjacent tissues with high water content and because of the high attenuation in the muscle layer.

4. RESULTS OF DOSIMETRY STUDIES

Numerical studies on the power absorption in head models due to the field emitted from mobile communication equipment began to appear in the scientific literature in the early 1990s. However, in the late 1970s and the 1980s, several studies were conducted on near-field exposure from an experimental point of view (e.g., Balzano et al., 1978; Chatterjee et al., 1985; Stuchly et al., 1987). In fact, in those years, the first portable equipment began to appear and the potential hazard of an electromagnetic source positioned in the vicinity of the human body started to emerge.

With reference to numerical studies, until the 1990s, dosimetry works were devoted to the study of the whole human body exposed to a field far from the electromagnetic source, and as such represented by a plane wave (Dimbylow and Gandhi, 1991; Gandhi et al., 1992). Nevertheless, some numerical studies were conducted on localized exposures such as those linked to RF sealers (Chen and Gandhi, 1989; Chen et al., 1991).

The improvement of the FDTD method, particularly with reference to the development of suitable ABCs (Mur, 1981; Berenger, 1994) and the great expansion in speed and memory of portable computers, led to an increase in the number of numerical dosimetry studies, so that at the beginning of the 1990s many works finally appeared on electromagnetic power absorption in the head.

Numerical studies were used to compare the computed SAR values with the basic restrictions on local SAR set by international safety guidelines (ICNIRP, 1998; IEEE, 2005). Many studies were also devoted to the standardization of numerical procedures, as well as to the definition of a worst-case condition, and to the evaluation of experimental procedures for compliance studies. Finally, in the last decade the attention has been focused on specific points as, e.g., the influence of the pinna on SAR distribution.

Nowadays, after a great deal of work on cellular phones, the attention has turned mainly to radio base stations as the electromagnetic field source to be considered in dosimetry studies. At the same time, many works have appeared in the literature on the temperature increments caused in the human head by the electromagnetic field emitted from a cellular phone and on the link between SAR distribution and temperature increments.

In the following sections, the fundamental process in dosimetry studies devoted to cellular phones will be reported, with an attempt to give a rationale for the observations made in the last 20 years. Separate subsections are provided for special exposure conditions such as cellular phones used in a partially closed environment, as well as for particular projects devoted to dosimetry studies such as those carried out by the European Community (the COST 244 and COST 281 actions and the CEPHOS Project). Finally, a separate section is devoted to studies which reported, together with the SAR data, the corresponding temperature increments obtained in the head of a cellular phone user.

4.1. SAR Results

Most dosimetry studies have been conducted by the FDTD method. Since the FDTD cell dimension is fundamental to understanding the reported data, specific indications are given in the following. Of course, the FDTD implementations used are different from group to group since they include both self-developed and commercial codes, and different ABCs are used. In particular, in early works of Mur, second-order boundary conditions (Mur, 1981) or retarded-time ABC (Bernsten and Hornsleth, 1994) were usually used, while after the publication of Berenger's work (Berenger, 1994), PML ABC or its evolutions (e.g., Uniaxial PML – (Gedney, 1996)) began to be used.

As already cited, numerical dosimetry studies on mobile phones were usually conducted to evaluate RF absorption in the head for compliance reasons or to evaluate possible risk of the electromagnetic exposure. Such studies are used to compare different mobile phone antenna or corresponding SAR data with the basic restrictions set in safety guidelines (ICNIRP, 1998; IEEE, 2005). However, some studies were also devoted to the head's influence on phone performance, realizing that the electromagnetic power absorbed into the head represented an unwanted power loss in the communication link.

4.1.1. Normal Usage Positions of the Mobile Phones: Next to the Head

Research on power deposition into the human head due to the field emitted from mobile communication equipment originated from earlier findings in studies on plane wave radiation. In fact, having found that even when the incident field is a uniform plane wave the resulting power deposition can be highly nonuniform, and that resonance phenomena can occur in the head. The question became how does power deposition behaves when the electromagnetic field was due to a localized source placed next to the human body (Dimbylow and Gandhi, 1991). Moreover, a phone placed next to the head is close to the eye, which is a very sensitive organ from the thermal point of view, and is without tissue layers to protect it from external agents.

At the beginning, phones were represented by simple dipoles or monopoles on a conducting box. This was due to both simulation reasons and because the first mobile phones were equipped with antenna which behaved mostly as dipoles or monopoles. Such models for the phone radiating element are very often used nowadays also, since they represent a sort of canonical exposure source.

To study possible safety issues on the eye, a vertical dipole, operated at 892 MHz and 1.89 GHz, located in front of the left eye at different distances from the eye surface was studied (Dimbylow, 1993). Phantoms representing an adult and a 1-year-old child were developed from anatomical cross sections and were made by seven different tissues comprising brain, bone/fat, muscle, skin, blood, air, and eyes. The eye was modeled as a sphere, with four types of tissue (lens, humor, a composite lens/humor, and a composite sclera/humor). The adult head model had a cell size of 3.17 mm, and it was downscaled by a factor of 0.7 to obtain the 1-year-old child head model (cell size 2.22 mm). The tissue electrical properties were obtained from a previous work (Dimbylow and Gandhi, 1991), those for the brain being the average of the two sets previously investigated.

The SAR as averaged over 1 g ($SAR_{1\,g}$), over the whole eye (9 g for the adult and 3 g for the infant), over 100 g, and the percentage of irradiated power deposited in the head were evaluated for 1.0 W of radiated power. Some of the data are shown in Table 1, where the monotonic decrease of SAR with distance can be noted. It was also observed that the power deposition became more superficial as the frequency was raised, and that the percentage of deposited power was about 75 % of the irradiated. However, the hot spot obtained with the plane wave irradiation was not found with the dipole (Dimbylow, 1993).

A comparison between the SAR values obtained when a dipole or a monopole over a box was considered next to the head was also performed (Dimbylow and Mann, 1994). A model of the head developed from MRI images of an actual head, with 2-mm resolution and ten different tissues whose dielectric properties were the same as those used by Dimbylow and Gandhi (1991). The monopole was quarter-wavelength long and was placed at the center or at the back corner of a metallic box (15-cm tall, 6-cm wide, and 2.4-cm deep). The antenna was placed near the ear in vertical or horizontal alignment and in front of the eye with a vertical alignment, at varying distances from the head. Results were reported for two frequencies (900 and

Table 1. SAR as averaged over 1 g and averaged over the whole eye for a dipole near an adult and 1-year-old child head model for several dipole–head distances

Head–dipole separation (cells)	Adult				1-year-old child			
	892 MHz		1.89 GHz		892 MHz		1.89 GHz	
	$SAR_{1\,g}$ (W/kg)	Eye (W/kg)	$SAR_{1\,g}$ (W/kg)	Eye (W/kg)	$SAR_{1\,g}$ (W/kg)	Eye (W/kg)	$SAR_{1\,g}$ (W/kg)	Eye (W/kg)
1	30.4	12.1	62.3	29.4	28.6	16.7	89.8	62.1
2	22.9	10.1	47.6	26.2	19.6	12.6	74.5	55.2
4	15.0	7.47	22.9	15.3	13.4	9.04	55.6	41.7
6	10.6	5.70	12.3	8.87	10.1	7.05	39.4	29.5
8	7.82	4.44	7.33	5.46	7.72	5.54	24.2	18.3
10	5.71	3.38	4.55	3.48	6.25	4.57	15.5	11.8
12	4.22	2.59	3.63	2.42	4.99	3.71	9.45	7.30
14	3.20	2.01	3.08	1.78	4.03	3.04	6.66	5.20
16	2.47	1.59	2.69	1.39	3.30	2.53	4.93	3.86

1.0-W radiated power. Note that the cell side is 3.17 mm in the adult model and 2.22 mm in the child model (data from Dimbylow, 1993)

1,800 MHz) and for 1.0 W of radiated power. In general, with the exception of the shorter distances, the dipole calculations produced SAR values higher than those from the monopole (Dimbylow and Mann, 1994).

Other works reported the SAR in the human head when exposed to the field emitted from a mobile phone modeled as an equivalent dipole antenna (Chen and Wang, 1994; Gandhi et al., 1996). Chen and Wang (1994) studied a 10.5-cm long dipole, operating at 835 MHz ($\lambda_{835} \approx 36$ cm), with an equivalent resistor of 120Ω located at the center gap between the two arms, and radiating 600 mW. The head model was developed with a resolution of 5 mm but no geometrical details are given in the paper; the dielectric properties were obtained from the literature. Considering a dipole–head distance varying between 1.0 and 2.5 cm, maximum SAR values (i.e., averaged over a single FDTD cell, equivalent to 0.125 g) between 1.23 and 2.63 W/kg were found. No averaging was made.

The effect of antenna length and its positioning on the power deposited and SAR distribution for the various regions of the head and the neck was studied through several simulations with an antenna on the box (Gandhi et al., 1996; Watanabe et al., 1996).

In Gandhi et al. (1996), a monopole antenna $\lambda/4$ or $3\lambda/8$ long, placed on a metal box of dimensions 2.76×5.53×15.3 cm, covered by a 1-cell layer of plastic coating of effective dielectric constant lower than for rubber to account for the real plastic coatings thinner than 1 cm, was considered at 835 and 1,900 MHz. The head model was derived from MRI images with a resolution of 1.974×1.974×3 mm and 15 different tissues. Two orientations were considered for the handset, one vertical relative to the head (tilt angle of 0°) and the other with a tilt angle of 30° relative to the head.

Several simulations, concerning the head of child, different models for the ear, and the possible influence of using the dielectric constant values measured by

Gabriel et al. (1996) instead of the values previously used in the literature, were performed. With reference to the influence of the antenna length on the obtained SAR values, the absorption in the head and the neck and the peak 1 g SARs were lower for the $3\lambda/8$ antennas than for the $\lambda/4$ antennas both at 835 and 1,900 MHz. This can be readily explained since the $3\lambda/8$ antenna is 50% longer than the $\lambda/4$, so the center of the antenna, where the highest value of the current occurs, is farther away from the head than the shorter antenna. Similarly, the $\lambda/4$ antenna gives a lower SAR value when held in a tilted position at 835 MHz, again being held farther away from the head, while this does not happen at 1,900 MHz probably because the antenna becomes too small at this frequency.

The same conclusions were obtained studying the absorption in a heterogeneous realistic head model (2.5-mm resolution, seven tissues), based on an anatomical chart of a Japanese adult head (Watanabe et al., 1996), and in a very recent paper (Ali et al., 2007). In Watanabe et al. (1996) the handheld portable radio was assumed to be a $13.5 \times 4.25 \times 2.5$ cm conducting box with either a half-wavelength dipole antenna or a $\lambda/4$-wavelength monopole antenna on its upper plane and operated at 900 MHz or at 1.5 GHz. Ali et al. (2007) also studied the influence of the wire radius on power deposition, finding that differences up to an order of magnitude in the radius of the dipole antenna wire did not significantly affect the SAR, although they did significantly affect the antenna bandwidth.

Detailed models of the mobile phone were developed by several authors (Jensen and Rahmat-Samii, 1995; Bernardi et al., 1996, 2000, 2001a,b; Li et al., 2000).

The monopole, side-mounted PIFA, and back-mounted PIFA resulted in very similar values of the peak SAR both when vertically oriented and when rotated by 60° to place the phone between the ear and the mouth (Jensen and Rahmat-Samii, 1995). An exception was the side-mounted PIFA geometry where the rotated handset placed the antenna nearly in contact with the ear tissue, resulting in a higher peak SAR compared to the result for the upright handset. Simulations on a top-mounted bent inverted F antenna (BIFA), in which the antenna element is mounted on the back of the handset, showed a substantial reduction in the peak SAR. Thus, pointing toward designing the mobile phone antenna and its placement within the handset to minimize SAR (Jensen and Rahmat-Samii, 1995). These results, with particular reference to the high power deposition linked to the PIFA antenna, were confirmed by others (Bernardi et al., 2000; Li et al., 2000).

Realistic models developed for mobile phones also include a mobile phone equipped with a sleeve dipole antenna and with a whip antenna (Bernardi et al., 1996), and helical antennas (Bernardi et al., 2001; Lazzi and Gandhi, 1998). The whip antenna consists of a monopole with a very thin helix to shorten its length and a coil at its base. When the monopole is collapsed, the coil forms together with the metal in the radio case a RF radiator, so that the phone could be used both with the monopole extracted or collapsed into the radio case (Bernardi et al., 1996).

The helical antenna, operating at 835 MHz, induces about 50% absorption of the total radiated power into the head. When placed in contact with the ear, a peak SAR of 0.98 W/kg averaged over 1 g for a radiated power of 250 mW was found; while the peak SAR as averaged over 10 g ($SAR_{10\,g}$) was 0.63 W/kg. Pressing the

Table 2. Comparison between CAD-derived phone models and their equivalent plastic covered metal box model (from Gandhi et al., 1999)

		$SAR_{1\,g}$	Power absorbed in the head (%)	Power absorbed in the hand (%)
Phone no. 1 (frequency: 835 MHz, radiated power: 600 mW)	CAD model	2.17	23.8	17.2
	Plastic covered metal-box model	2.39	26.7	17.8
Phone no. 2 (frequency: 1,900 MHz, radiated power: 125 mW)	CAD model	0.78	32.1	17.9
	Plastic covered metal-box model	0.64	31.4	18.1

phone model against the ear increased the SAR values to 1.62 W/kg and 0.97 W/kg for $SAR_{1\,g}$ and $SAR_{10\,g}$, respectively. Finally, tilting the phone, to put it into a realistic usage position, led to the values of 0.86 W/kg for $SAR_{1\,g}$ and 0.49 W/kg for $SAR_{10\,g}$. It is worth noting that in Lazzi and Gandhi (1998), a value of 1.6 W/kg for $SAR_{1\,g}$ was obtained at 835 MHz (for a radiated power of 250 mW), by using a completely different model for simulating the helical antenna.

Other developments of realistic phone models for numerical simulations used CAD data sometimes provided by mobile phone manufacturers (Gandhi et al., 1999). In Table 2 comparisons between the SAR values obtained from two phone models derived from CAD data, and with an idealized box model counterpart, are provided (Gandhi et al., 1999). Note that CAD data have been used to describe the external shape of several phone models (Schiavoni et al., 2000).

Since CAD data of mobile phones are not generally available, a realistic numerical model has been proposed, starting from the external shape and size of the mobile phones and using measured SAR and near-zone electric and magnetic fields (Pisa et al., 2005a, b). Numerical simulations comparing the SAR values obtained in a human head model from the optimized phone model with those obtained by a nonoptimized phone model showed that the nonoptimized model can lead to SAR values as much as three times the values obtained from the optimized model (Pisa et al., 2005a).

Studies devoted to the analysis of the influence of the head model on power deposition were often carried out by developing a head phantom for use in experimental procedures. This phantom, in fact, should be of simple shape, possibly homogeneous, and able to not underestimate the power deposition. To find a simple shape for the head model, two models of the human head were used, one obtained by CT with 26 different tissues and a resolution up to 3.4 mm, and the other obtained from MRI data with 7 tissues and a resolution of 5 mm, and compared with a box or a sphere model (Okoniewski and Stuchly, 1996). The numerical analysis considered a monopole of length 8.5 cm centered over a metal box of 15 (length) × 6 (width) × 3 (depth) cm, which was in turn covered with a dielectric 2-mm thick and with dielectric constant equal to 2.0. It was found that a box model of the human head provides grossly distorted and unreliable results for the antenna radiation pattern, while a spherical model provides results that are relatively close to those obtained with a relatively simple, but more realistic, head model. Moreover, the SAR values obtained with spherical or simplified head models that do not include the ear, are greater than those obtained from head models that include the ear, and a hand holding the handset absorbs a significant amount of antenna output power. Other works on the possibility of defining simple head shapes to be used in experimental compliance studies can be found in later sections of this chapter.

To study the effect that the head has on mobile phone performance, a $\lambda/4$ monopole over a metallic box ($2.5 \times 6 \times 15$ cm) was considered next to a head represented by a homogeneous muscle sphere with a radius of 9 cm (Toftgard et al., 1993). Also the hand, modeled as a homogeneous muscle brick ($\varepsilon_r = 50.5$, $\sigma = 1.2$ S/m at 914 MHz; $\varepsilon_r = 49.0$, $\sigma = 1.6$ S/m at 1,890 MHz), 10-cm wide and 2-cm thick and wrapped around the lower part of the telephone on the side furthest from the head was taken into account. The results showed that when the antenna is placed near the head model, its resonance frequency drops; the radiation pattern changes significantly, with a shadow effect in the direction of the person; a considerable cross-polarization takes place; and about 45% of the power is lost in the hand and the head at both frequencies. On average, the authors obtained a system loss of 3–4 dB in the link budget, due to the presence of the person talking on the mobile phone. They concluded that when antennas are designed for handheld portable telephones, it is necessary to take into account the presence of the person next to the antenna, since this presence affects input impedance, far-field radiation pattern, radiation efficiency, and the near field of the antenna. It is worth noting that, as a matter of fact, the evolution in mobile communication antenna design closely followed this indication.

The results of the monopole antenna performances, with a noticeable change in the antenna radiation pattern when the antenna is used next to a head, were confirmed in the study of Jensen and Rahmat-Samii (1995)). In this study also, the side-mounted PIFA, introduced in the mobile phone design to replace the monopole with more conformal, less obtrusive element, and top-mounted BIFA were studied with respect to their performances. The PIFA antenna showed a detuning effect on the resonant frequency, and particularly significant was the high impedance mismatch occurring when the hand begins to mask the antenna; the input impedance detuning due to the presence of the hand was found also for the BIFA antenna, although less severe than for the PIFA (Jensen and Rahmat-Samii, 1995).

Finally, the influence of the chosen model for the head (simple sphere or the inhomogeneous realistic model) on the antenna performance was also investigated, finding that the head model exercises little influence on the antenna input impedance, while the radiation characteristics show an increased sensitivity to the presence of the head as well as the physical model used. In particular, the inhomogeneous head model results in somewhat higher absorption levels compared to the simpler spherical head model. This difference occurs because the sphere is physically smaller and has a permittivity/conductivity distribution that reduces the electromagnetic penetration depth compared to the inhomogeneous model.

With reference to the influence of the hand on power deposition, the detuning effects found on very compact antennas such as the PIFA and the BIFA were not obtained for antennas such as the dipole, the monopole, the sleeve dipoles, and the whip antenna (Dimbylow and Mann, 1994; Bernardi et al., 1996). However, at 900 MHz the effect of the hand (8-cm high, 2-cm thick, covering the back and two sides of the box; the electrical properties taken to be 2/3 muscle and 1/3 bone) was small and in most cases the SAR in the head decreased; while at 1.8 GHz the effect was more pronounced, and an increase of the local SAR in the head was also found, counterintuitively to what was expected (Dimbylow and Mann, 1994).

One of the major problems in comparing dosimetry results is the lack of reproducibility of the different studies due to shortcomings in the description of the exposure situation considered. In fact, in the beginning, the influence on the obtained results of the different modeling elements, such as the model of the phone, the model of the head (geometry and electrical properties), the numerical code, and the phone positioning with respect to the head, was not so clear. The consequence is a great difficulty in comparing the different numerical studies. As an example, taking into account the works from Dimbylow and Mann (1994) and Chen and Wang (1994), although in both cases a dipole is used, the dipole FDTD model is different ($\lambda/2$ in Dimbylow and Mann (1994) and a shorter dipole in Chen and Wang (1994)), the head model is different (from MRI data in Dimbylow and Mann (1994), self-developed with no details given in Chen and Wang (1994)) and, last but not least, different dielectric properties for the human tissues were used. These difficulties spur the development of specifically devoted projects such as the COST 244 and the CEPHOS, which will be discussed in a separate subsection.

With reference to the dielectric properties, Gabriel and colleagues measured the dielectric properties of freshly excised animal tissues and interpolated their results with a Cole–Cole analysis (Gabriel et al., 1996). Besides the breadth of the obtained data, they made their work available through an FCC Web site (Gabriel, 1996); so that since 1996 most of the work done in numerical dosimetry uses the dielectric properties available on that site (see http://www.fcc.gov/fcc-bin/dielec.sh).

An indication of the importance of a clear definition of the positioning of the phone next to the ear can be derived by considering the influence of the rotation of the axis of the phone with respect to the head on the obtained SAR values. In Table 3, the $SAR_{1\,g}$ and $SAR_{10\,g}$ obtained for a $\lambda/4$ monopole over a plastic covered box ($2.96 \times 5.73 \times 15.5$ cm) placed next to the ear with different tilting angles and for two

Table 3. SAR values obtained for various tilt angles of the phone with respect to the head for a monopole over a plastic covered metal box (from Gandhi et al., 1999)

	835 MHz – 600 mW		1,900 MHz – 125 mW	
Tilt angle	$SAR_{1\,g}$	$SAR_{10\,g}$	$SAR_{1\,g}$	$SAR_{10\,g}$
0°	2.93	1.41	1.11	0.59
20°	2.7	1.33	1.08	0.56
30°	2.44	1.21	1.08	0.57
45°	2.14	1.1	0.85	0.42
30° + 9°	2.31	1.08	1.20	0.44

Table 4. SAR data for a monopole (covered by a rubber slab) over a metallic PCB, covered by a plastic box (from Kainz et al., 2005)

	Cheek position		Tilted position	
Frequency	$SAR_{1\,g}$ (W/kg)	$SAR_{10\,g}$ (W/kg)	$SAR_{1\,g}$ (W/kg)	$SAR_{10\,g}$ (W/kg)
835 MHz	7.5	5.2	5.0	3.4
1,900 MHz	9.2	5.3	13.1	7.3

frequencies of operation (835 and 1,900 MHz) are reported (Gandhi et al., 1999). From the table, it can be noted that the SAR values reduce as the phone position is tilted starting from the vertical alignment (0°). The last row in the table gives SAR data when the monopole is placed in the so-called cheek position which corresponds to a double rotation of the phone to place its microphone next to the mouth.

A formal definition was proposed for the positioning of mobile phones next to the head models (Kainz et al., 2005). This definition was based on anatomical characteristics of the head and tested by evaluating SAR data in 14 different head models and comparing them with the SAM model used in EN 50361-2001 (CENELEC, 2001) and IEEE 1528-2003 (IEEE, 2003) standard procedures (Kainz et al., 2005). The phone model used was a generic monopole on a flat metallic plate ($40 \times 1 \times 100$ mm) representing the PCB, inside a plastic box ($42 \times 21 \times 102$ mm). The monopole, covered by a rubber slab, was 71-mm long for 835 MHz and 36-mm long for 1,900 MHz. The comparison among the different head models focused on the importance of the modeling of the pinna for the evaluation of the maximum SAR. This point will be discussed later. In Table 4, the SAR as averaged over 1 g and 10 g of tissue, obtained at the two frequencies of 835 and 1,900 MHz, for 1.0 W of radiated power is presented.

Results from different authors on SAR values in the human head when a mobile phone is placed next to it are summarized in Tables 5 and 6. The antennas considered are the simple dipole and the monopole over a box, operating at 900 MHz (Table 5) and at 1,800 MHz (Table 6), and radiating 1.0 W of power.

4.1.2. Other Usage Positions of Mobile Phones: Body-Worn Devices

With the development of new types of mobile phones able to perform several tasks, the phone usage position changed from the ear-position to different ones; very often the phone is held in the hand and kept in front of the face, or positioned within a pocket and the conversation carried on using hands-free kits. Correspondingly, some researchers addressed the problem of power deposition into the human body when the phone is placed in a position other than next to the ear.

Kang and Gandhi (2002) compared SAR data obtained in the chest when the phone is placed in a shirt pocket, with SAR data obtained in the head when the same phone is held next to the ear. In this case they considered a body model from the chest to the waist and considered different phone–body distances to

Table 5. SAR values in the human head when a mobile phone is placed next to it (Frequency 900 MHz, Radiated power 1 W)

Reference	Phone model	Head model, resolution	Distance phone-ear (cm)	$SAR_{1\,g}$	$SAR_{10\,g}$	SAR eye	Percentage of absorbed power
Bernardi et al. (2000)	Dipole	MRI, 2 mm (16 tissues)	Contact	4.56	3.2		38.8
Dimbylow and Mann (1994)	Dipole	NMR 2 mm (10 tissues)	1.4	3.36	2.55	0.030	41.8
Dimbylow and Mann (1994)	Dipole	NMR 2 mm (10 tissues)	2	2.56	1.92	0.034	35.0
Dimbylow and Mann (1994)	Dipole	NMR 2 mm (10 tissues)	3	1.71	1.25	0.038	26.7
Dimbylow and Mann (1994)	Dipole	NMR 2 mm(10 tissues)	4	1.20	0.866	0.041	21.0
Dimbylow and Mann (1994)	Dipole	NMR 2 mm(10 tissues)	5	0.876	0.631	0.043	17.0
Gandhi et al. (1996)	Dipole	NMR 1.974 × 1.974 × 3 mm (15 tissues)	1.38	35–45[a]		0.033 (lens av.)	52.0
Bernardi et al. (2000)	Monopole + box	MRI, 2 mm (16 tissues)	Contact	3.6	2.15		36.5
Dimbylow and Mann (1994)	Monopole + box[b]	NMR 2 mm (10 tissues)	1.4	3.62	3.04	0.061	38.5
Dimbylow and Mann (1994)	Monopole + box[b]	NMR 2 mm (10 tissues)	2	2.14	1.67	0.051	31.5
Dimbylow and Mann (1994)	Monopole + box[b]	NMR 2 mm (10 tissues)	3	1.20	0.935	0.041	23.4
Dimbylow and Mann (1994)	Monopole + box[b]	NMR 2 mm (10 tissues)	4	0.891	0.612	0.035	18.4
Dimbylow and Mann (1994)	Monopole + box[b]	NMR 2 mm (10 tissues)	5	0.668	0.444	0.030	14.7
Gandhi et al. (1996)	Monopole + box[b]	NMR 1.974 × 1.974 × 3 mm (15 tissues)	1.38	3.45–41.5		0.044 (lens av.)	51.7
Okoniewski and Stuchly (1996)	Monopole + box + dielectric cover 2-mm thick	MRI, 5 mm (7 tissues)	3	2.42	1.4		28.0
Okoniewski and Stuchly (1996)	Monopole + box + dielectric cover 2-mm thick	CT, 5 mm (26 tissues)	3	1.1	0.8		23.0

[a]Different averaging volume, see text
[b]Monopole at the center

Table 6. SAR values in the human head when a mobile phone is placed next to it (Frequency 1,800 MHz, Radiated power 1 W)

Reference	Phone model	Head model, resolution	Distance phone–ear (cm)	SAR 1 g	SAR 10 g	SAR eye	Percentage of absorbed power
Dimbylow and Mann (1994)	Dipole	NMR 2 mm(10 tissues)	1.4	6.20	4.26	0.014	30.9
Dimbylow and Mann (1994)	Dipole	NMR 2 mm (10 tissues)	2	4.40	3.00	0.015	23.4
Dimbylow and Mann (1994)	Dipole	NMR 2 mm (10 tissues)	3	2.84	1.95	0.017	17.3
Dimbylow and Mann (1994)	Dipole	NMR 2 mm(10 tissues)	4	2.03	1.39	0.019	14.1
Dimbylow and Mann (1994)	Dipole	NMR 2 mm(10 tissues)	5	1.64	1.12	0.022	12.8
Gandhi et al. (1996)	Dipole	NMR 1.974 × 1.974 × 3 mm (15 tissues)	1.38	4.4–6.5[a]		0.016 (lens av.)	46.4
Dimbylow and Mann (1994)	Monopde ho\ t)	NMR 2 mm (10 tissues)	1.4	5.63	3.84	0.015	28.1
Dimbylow and Mann (1994)	\lonopolei-box ()	NMR 2 mm (10 tissues)	2	3.56	2.45	0.013	21.2
Dimbylow and Mann (1994)	Monopole+box[b]	NMR 2 mm(10 tissues)	3	1.93	1.32	0.012	14.2
Dimbylow and Mann (1994)	Monopolc-box()	NMR 2 mm (10 tissues)	4	1.23	0.830	0.012	10.6
Dimbylow and Mann (1994)	Monopole-t-box()	NMR 2 mm (10 tissues)	5	0.902	0.606	0.013	8.9
Gandhi et al. (1996)	Monopole + box	NMR 1.974 × 1.974 × 3 mm (15 tissues)	1.38	4.24–6.96[a]		0.020 (lens as.)	45.4

[a]Different averaging volume, see text
[b]Monopole at the center

take into account possible different clothing thicknesses. It was found that the SARs in the chest when the telephones are placed with the antennas closer to the body ("back" position) could be up to 2.1 times higher at 835 MHz and up to 5.8 times higher at 1,900 MHz with respect to the SARs found when the antennas are placed far from the body ("front" position). This can be expected since turning the phone with the antenna against the body can cause the antenna to be up to 12–16 mm closer to the body than in the opposite placement. Moreover, both the peak 1 g and 10 g SARs reduced monotonically with increasing separation from the planar phantom, and when the telephones were placed against a model of the human head with a 6-mm plastic spacer in the shape of the pinna (as recommended by US FCC (2001) and CENELEC (2001)) the SARs were close to SARs obtained using a flat phantom with a 6-mm base thickness.

Similar results were obtained with the phone placed at the waist of a human body model with or without a hands-free connection (Troulis et al., 2003). In particular, an increase of the SAR values and deposited power was found when the phone is placed at the waist of the body with respect to the case in which the phone is placed close to the ear.

4.1.3. Standardization of Dosimetry Procedures: COST 244 and CEPHOS Projects

Since 1971, the European Community has funded projects under the umbrella of the "European Cooperation in the Field of Scientific and Technical Research," to encourage the exchange of information and collaboration among European researchers (http://www.cost.esf.org/). Among the several projects (called "actions") funded, one was devoted to the interaction of electromagnetic fields with living systems. This action started in 1992 with the name of "Biomedical Effects of Electromagnetic Fields," also referred to as COST 244, and was renewed under the name of COST 244bis in 1996. The COST 281 actions, "Potential Health Implications from Mobile Communication Systems" can be seen as its natural progression.

Within the COST 244 project, the research on mobile communication was supported by a working group (WG3 – Systems Application & Engineering). In particular, having ascertained the researchers' difficulties in comparing results from different dosimetry studies, starting from 1994, the WG3 activity focused on comparing all numerical and experimental results obtained by cellular phone dosimetry studies to this end, *a physical canonical problem* and *a numerical canonical problem* were proposed to all interested researchers (D'Inzeo, 1994). The numerical problem consisted of a simple, homogeneous or two layered phantom (cube or sphere) placed in close proximity to a $\lambda/2$ dipole or $\lambda/4$ monopole on a conducting box. The frequencies considered were 900 and 1,800 MHz, and SAR values in some particular points of the phantom and radiation pattern were specified (Bach-Andersen et al., 1994).

The results of the canonical exercise showed a good agreement in the normalized data, while considerable differences were obtained in the absolute data, e.g., in the maximum SAR. These rather confusing results initiated a thorough analysis of all the possible error sources which led to the conclusion that the differences could be explained by the different choices and approximations made by the research groups in developing the numerical solution; e.g., in modeling the radiating antenna.

In 1997, the research on comparisons between dosimetry studies became a part of the European Commission Fifth Framework Programme. The project, funded under the acronym of CEPHOS – Cellular Phones Standards, was carried out between 1997 and 1999 by 15 organizations from six European Countries: Denmark, Finland, France, Greece, Italy, and UK. Among the several tasks of the CEPHOS Project some were devoted to the comparison of numerical and experimental studies, considering both the canonical problems and more realistic situations. The canonical problems were derived from the COST 244 *physical* and *numerical canonical problem.*

This research has shown the fundamental roles played not only by the numerical code but also by the way the numerical head models are developed. A simple example is the sphere: both the COST 244 and the CEPHOS projects defined a sphere of 20 cm in diameter; however, when the sphere is numerically realized, it can be made of an odd number of rows or an even number of rows, in which case the two central rows have the same length. This apparent minor difference leads to detectable differences in the local SARs.

Another aspect of the CEPHOS project was the comparison of results for the numerical canonical problems (Nikita et al., 2000) where the data obtained from the different research groups were reported on several parameters such as antenna input impedance, radiation pattern, radiated power, and local and averaged SAR. The comparison was made among different FDTD implementations, as well as among different models for representing the antenna and the head. Any differences may be regarded as variability of numerical studies. In particular, it was found that the results were not sensitive to the ABCs used, while the variability in the peak local SAR and in the 10 g averaged SAR were mainly related to the phantom and the source modeling. As an example, the uncertainty related to the source modeling (e.g., choosing a thin wire model or a thick model for the metal wire) in predicting the averaged SAR values was of the order of 12–15%. The position of maximum local SAR value was found to be independent of both source modeling and ABCs while, in evaluating the position of maximum averaged SAR values, an uncertainty of the order of 1–3 cells (0.25–0.75 cm) was observed according to the source model used. The total uncertainty in computing the power absorbed by the head was mainly related to phantom modeling. Concerning the prediction of the input impedance, source modeling was found to result in large differences while it was observed that the simulated feed-point impedance, in the presence of the head, is almost independent of the ABCs used. With regard to radiation patterns, a difference of the order of 1–2 dBi was observed in the main field polarization in the direction where the head is located, associated with the use of different source models. However, it was observed that the radiation patterns were not dependent on the ABCs used, when a sufficient distance between the boundary and the near-to-far field transformation surface was used. The conclusion is large differences in numerical results could be produced depending on the numerical phantom used and on the source modeling.

The details of the procedure used for SAR averaging and the subvolumes (1 g or 10 g) to consider for evaluating the average SARs, particularly at points close to the head surface, were not clearly defined in the safety guidelines, at first. This led to different averaging procedures being used by different research groups, and the resulting variability on the obtained data was difficult to resolve. Later the exact averaging procedure used and whether averaging over a tissue mass lower than the reference mass had been performed for points lying close to the head boundary was deemed a fundamental requirement when judging numerical dosimetry studies (Nikita et al., 2000).

The importance of the "canonical problems" developed within the framework of the COST 244 and CEPHOS projects can be seen considering the many works which were devoted to the open points indicated by the projects, and also considering that the canonical problems are still used as a reference to validate numerical results (Anderson, 2003). Finally, although beyond the scope of this chapter, it is worth citing here that great effort has also been devoted to the intercomparison of experimental compliance procedures (Davis et al., 2006).

4.1.4. Effect of Metallic Implants or Wearable Metallic Elements on RF Absorption

The effect of metallic implants inside the head of a mobile phone user, or wearable metallic elements such as spectacles and earrings, on the SAR distribution has also been studied. However, most studies relating to metallic implants used a far-field exposure condition, being concerned about the exposure of workers to high level electromagnetic fields, and only few dealt with a near-field exposure.

When considering the effect of the presence of a metallic implant on RF absorption, the size and the position of the implant with respect to the incident RF field, frequency, and polarization play a crucial role. Usually, an implant dimension close to the field half-wavelength can induce higher current on the metallic object, and as such can have a higher influence of RF absorption in its surroundings (Cooper and Hombach, 1996; Virtanen et al., 2006). Moreover, it has been found that the thinner the element, the greater the increase in local SAR values. A $SAR_{1\,g}$ increase of a factor of 2 and $SAR_{10\,g}$ increase of a factor of about 1.3 have been reported as a worst case, studying metallic disk, pin, or filament in a lossy homogeneous sphere placed next to a radiating dipole antenna (Cooper and Hombach, 1996). However, the enhancement was found only when the structures had sizes close to the field half-wavelength, were parallel to the source, and close to the sphere surface facing the antenna

A numerical study of common metallic implants (a skull plate, a bone plate, fixtures, brace, an earring, and ear tubes), with dimensions chosen according to assumed resonance conditions, evaluated the SAR distribution for a dipole antenna with $0.47\lambda_{air}$ length and operated at 900, 1,800, or 2,450 MHz (Virtanen et al., 2007). In most of the cases, the implant affected the SAR distribution in the head, with the exception of the brace and the ear tubes that had no relevant SAR enhancement; the SAR was higher around the implant than in the corresponding location without the implant. Generally, the effect of implants was strongest at 900 MHz and, with particular reference to mass averaged SAR (10 g, 1 g), the effect was strongest with the earring, skull plate, and fixtures (Virtanen et al., 2007). The increase in SAR values, when metallic objects attached to the human ear, has been confirmed by others both numerically and experimentally (Fayos-Fernandez et al., 2006).

Metallic implants usually strongly modify the electromagnetic field distribution close to the implant itself, so that when the SAR is averaged over 1 g or 10 g, the value is greatly reduced. A maximum increase of 3 in the averaged SAR was obtained for some worst case situations such as changing size, shape, and orientation of a pin or a ring implanted in a cylindrical two-layer phantom (Virtanen et al., 2005).

An early study on the influence of such an object on the RF energy absorption was performed in two dimensions considering a plane wave impinging on a human eye with a glass layer in front to simulate the lens of a spectacle. An increase in the SAR averaged over the whole eye of approximately 31% has been reported for 18 GHz (Bernardi et al., 1998). More recently, an investigation was performed on the shape of metallic spectacles to maximize power deposition when an electromagnetic plane wave impinges from the front of a human head. No general rule was found on the link between spectacles shape and power deposition. As an example, at 1.8 GHz, square spectacles gave higher SAR values as averaged over the whole eye than circular spectacles for most of the considered sizes (from 34 to 42 mm), while at 2.4 GHz the opposite was found, i.e., circular spectacles gave higher SAR than square ones for most sizes (Whittow and Edwards, 2004). Comparing power deposition due to the plane wave with that obtained from a vertical dipole placed 8 cm from the nose gave similar results showing that the effects of spectacles are more dependent on the size and shape of the frames, the frequency, and the polarization of the impinging electromagnetic field than on the specific electromagnetic source (Whittow and Edwards, 2004; Edwards and Whittow, 2005).

4.1.5. Effect of Partially Closed Environments on RF Absorption

The issue of exposure of the head to the field emitted from a mobile phone and the possible changes the absorption could have when the phone is used in a closed environment were also studied. For example, when the mobile phone is used inside a car, metallic walls are present near the electromagnetic source which could change its radiating properties and perhaps cause a higher absorption in the human head (Bernardi et al., 1996; Dominguez et al., 2002). In numerical studies, the environmental effects have been simulated through partially or totally reflecting walls located in various positions with respect to the phone. It was found that the presence of a horizontal reflecting wall placed 6 cm above the head, representing the roof of a car, decreases the SAR values in the part of the head directly exposed to the phone antenna, while it increases the SAR values in the part not directly exposed. The presence of a vertical glass wall parallel to the antenna, 3.5 cm back from the feeding point of the antenna, does not significantly influence the SAR distribution,. However, when the wall is a perfect conductor, it raises the SAR values everywhere in the head (Bernardi et al., 1996). Similar conclusions were reached by Dominguez et al. (2002), who also considered the presence of three reflecting walls (one parallel to the phone close to it, another parallel to the phone but placed on the opposite side of the head, and the last one on top of the head) at the same time. It was found that the SAR distributions do not differ much from the case in which the three walls are considered separately.

4.2. Temperature Increments

An important outcome of RF power deposition in biological systems is the dissipation of the absorbed energy which causes the temperature to increase. Indeed, RF energy absorption becomes a source of heat. In the human body, the time constant for bulk tissue heating is of the order of 6–7 min (Bernardi et al., 1998; Riu and Foster, 1999). This time constant is much longer than the time constant associated with electromagnetic phenomenon. Moreover, even for local heating by focal RF power deposition, the thermal time constant would be long compared to the near instantaneous RF energy absorption. Thus in numerical simulations, it is possible to first solve the electromagnetic problem, and then the thermal one, taking into account the instantaneous SAR distribution (Lin and Bernardi, 2007). Note that the temperature increments associated with RF absorption induced by mobile phones are sufficiently small that they would not be able to change the dielectric properties of tissues, which is known to be temperature dependent.

4.2.1. Normal Usage Positions of the Mobile Phones

The FDTD method was applied by Wang and Fujiwara (1999) to solve the electromagnetic and thermal problems for a human head obtained from an anatomical chart of a Japanese adult head, with a 2.5-mm resolution and six different tissues. The phone was modeled as a quarter-wavelength monopole antenna placed on the center of the top face of a metal box (12-cm high, 4-cm wide, and 2.5-cm deep) covered with a dielectric insulator of 2.5-mm thickness and relative permittivity of 2. The phone was positioned with a vertical alignment at the side of the head, both next to the ear and pressed against the ear. The antenna output power was set to 0.6 W at 900 MHz and 0.27 W at 1.5 GHz. Temperature increments were evaluated using the BHE in which the metabolic heat was not considered.

Computed results showed that a phone operating for 3 min yields a temperature rise over 60% of the steady-state value, and 6–7 min operation yields a temperature rise of approximately 90% of the steady-state value. At 900 MHz, maximum temperature rises of 0.16 and 0.18°C were obtained for the phone touching or pressed against the ear, respectively. Correspondingly, at 1.5 GHz, the maximum temperature rises were 0.13 and 0.15°C, respectively. These peak temperature rises all occurred in the ear region. In the brain, the maximum temperature rises obtained were 0.09°C at 900 MHz, and 0.07°C at 1.5 GHz, for the phone pressed against the ear (Wang and Fujiwara, 1999).

Temperature increments similar to those reported by Wang and Fujiwara (1999) were obtained by Van Leeuwen et al. (1999) for a half-wavelength dipole antenna operating at 915 MHz, vertically oriented and positioned 2 cm from the head. The effects of individual vessels were included in thermal modeling of the head. The maximum temperature rise in the brain was about 0.12°C, and was associated with a maximum averaged SAR over an arbitrarily shaped 10 g volume of approximately 1.6 W/kg. Comparing the normal head with a head with a thicker skin, it was seen that the maximum temperature rise in the skin was significantly higher in the latter. The maximum temperature rise in the brain, however, was almost equal to that in the normal head. The similarity of the temperature increments obtained to those of Wang and Fujiwara (1999) shows that the use of the classical BHE instead of a more refined discrete vasculature is sufficient for this kind of problem.

Maximum temperature increases in the brain of about 0.1°C were also obtained for a cellular phone in a normal operating condition, using a quarter-wavelength antenna over a metallic box, and the standard BHE with metabolic processes included (Wainwright, 2000). The numerical solution for temperature rises was developed with the FEM, while the electromagnetic problem was solved by the FDTD method. The head model (eight tissues) used in the FDTD method was fitted into the FEM grid. After considering several phone–head positions, and both 900 and 1,800 MHz, a maximum temperature increment of 0.8°C was obtained in the skin and 0.4°C in the brain for 1.0 W of radiated power.

A comparison of the maximum temperature increase obtained in the head, brain, and lens has been performed for different phones: half-wavelength dipole, quarter-wavelength monopole, whips, and planar inverted F antenna (PIFA) – the last three mounted on a plastic covered metal box (Bernardi et al., 2000). For all phones considered, with the case kept in direct contact with the ear, half or more of the available power (600 mW) was deposited inside the head. The maximum temperature increase in the head was obtained for the PIFA antenna, and was equal to 0.39°C when the phone was held in a vertical position and 0.43°C when the phone was tilted. The corresponding maximum temperature increases in the brain were 0.16°C and 0.14°C, respectively. The whip antenna and the monopole gave similar results (about 0.25°C in the head and 0.13°C in the brain); while the dipole gave 0.33°C in the head and 0.19°C in the brain when vertically oriented (Bernardi et al., 2000).

Simulations performed while halving the blood perfusion into the brain led to a slight increase of the brain temperatures, showing that, due to the superficial nature of the power deposition, local blood perfusion plays a relatively minor role. Moreover, simulations performed while halving the surface heat transfer coefficient led to a significant increase of the temperature in the outer layers, and to an increase of about 1.5°C on the skin surface. It was then postulated that this could explain the perceived "heating" effect sometimes reported by mobile phone users, due to the reduction of convective cooling when the phone and hand are placed over the ear (Wainwright, 2000). This was subsequently confirmed by others (Bernardi et al., 2001b; Gandhi et al., 2001).

In particular, in Bernardi et al. (2001b), a dual band phone equipped with a monopole and an helix antenna was considered in a vertical position, with the ear piece aligned with the auditory canal and the phone case in direct contact with the ear, and in a double-tilted position ("cheek" position) with a more realistic alignment of the phone with the human ear and mouth. Temperature increments were evaluated at 15 min after power-on for SAR alone, contact between the phone and the ear, i.e., blocking heat convection due to the contact, and for the dissipation of the power amplifier, assuming a 50% efficiency, i.e., adding a power deposition of 250 mW at

Table 7. Temperature increments in the head for different cell-phone heat sources (from Bernardi et al., 2001b)

Frequency	Cell-phone heat source	*T* maximum (°C)	*T* maximum in the brain (°C)
900 MHz	SAR (P_{rad} 250 mW)	0.136	0.023
	Contact phone-head	1.543	0.012
	Contact phone-head + power dissipation	1.544	0.012
	SAR + contact phone-head + power dissipation	1.581	0.023
1,800 MHz	SAR (P_{rad} 125 mW)	0.085	0.011
	Contact phone-head	1.543	0.012
	Contact phone-head + power dissipation	1.543	0.012
	SAR + contact phone-head + power dissipation	1.549	0.012

900 MHz and 125 mW at 1,800 MHz uniformly distributed inside the upper part of the phone. Table 7 shows the results obtained.

From the reported data it can be noted that the maximum temperature increase in the head is obtained when the blocking of heat convection due to the presence of the phone in direct contact with the skin is considered. Adding to this contact the power dissipation in the phone amplifier does not change the obtained data, while the maximum temperature increase in the head due to the electromagnetic energy absorption is at least one order of magnitude lower. For the maximum temperature increase in the brain, the different heat sources give comparable results. It is worth noting that the maximum values reported in the last column in Table 7 are obtained in different parts of the brain (with the contact effect heating the lower external brain region, and the SAR heating the upper external brain region) so that the two data sets do not add when the heating causes are considered acting at the same time (Bernardi et al., 2001b).

One interesting aspect of temperature evaluation which has attracted some attention in recent years is the correlation between maximum temperature increase and local SAR. The proportionality between temperature and SAR can be derived from the BHE under steady-state conditions. The BHE is given by

$$C\rho\frac{\partial T}{\partial t} = \nabla\cdot(K\,\nabla T) + \rho\,\mathrm{SAR} + A_0 - B_0(T - T_\mathrm{B}), \tag{7}$$

which, under steady-state conditions, reduces to

$$0 = \nabla\cdot(K\,\nabla T) + \rho\,\mathrm{SAR} + A_0 - B_0(T - T_\mathrm{B}). \tag{8}$$

Subtracting from (2) the common terms for the exposed situation to the unexposed situations, i.e., metabolic heat production (A_0) and fixed blood temperature (T_B), the following equation is obtained

$$B_0 T - \nabla\cdot(k\,\nabla T) = \rho\,\mathrm{SAR}. \tag{9}$$

Equation (9) shows a direct proportionality between temperature increase and specific energy absorption. However, as can be seen from (7), tissue heating is influenced not only by the power dissipated in the local tissue volume, but also by how the RF absorption is distributed in the surrounding area, the thermal characteristics of the tissue and its neighbors, and finally, by the heat exchange with the external environment (Bernardi et al., 1998). Moreover, the influence of shape and size of the averaging volume on the relationship between local SAR and temperature increments has also been suggested (Wainwright, 2000).

The correlation between local SAR and temperature increase has been extensively investigated by considering different configurations of a dipole next to the human head with varying frequency, polarization, distance from the head, feeding point position, head models, phone models, and dielectric properties (Hirata and Shiozawa, 2003; Hirata et al., 2006; Fujimoto et al., 2006). It was concluded that the maximum temperature increase in the head and brain are reasonably correlated with the peak SARs averaged over 1 g or 10 g of tissue.

Note that the studies evaluating temperature increases in the head due to the electromagnetic field emitted from cellular phones (Wang and Fujiwara, 1999; Van Leeuwen et al., 1999; Bernardi et al., 2000; Wainwright, 2000, Gandhi et al., 2001) agree in finding a temperature rise within the brain of about 0.1–0.2°C for a phone working in normal conditions (for a radiated power of about 600 mW). This conclusion is reached even though these studies used different frequencies and different dielectric and thermal properties for the various tissue types, different head and phone models, as well as different phone positioning with respect to the head. Clearly, lower values are obtained when GSM phones with lower powers are considered (250 mW at 900 MHz and 125 mW at 1,800 MHz).

When the SAR distribution is "adjusted" to reach the occupational basic restriction limit for the local SAR (i.e., $SAR_{10\,g}$ = 10.0 W/kg; ICNIRP, 1998), the maximum temperature increases in the head and in the brain, obtained from the different studies cited herein, are equal to 3.4°C in the head and 1.3°C in the brain. Since the threshold temperature increase for neuron damage is about 4.5°C (for more than 30 min) (Guyton, 1991), a safety factor of about 3.5 is obtained. This is considerably lower than the factor of 10 used to derive the whole-body basic restriction from the threshold for thermal effect.

4.2.2. Other Usage Positions of the Mobile Phones

If the phone is held in positions other than next to the ear during usage such as in front of the face, eyes are exposed to the RF source. The eye is an organ potentially susceptible to thermal damage since its internal structures (aqueous humor, lens and vitreous humor) lack blood supply and the lens has a limited repair capacity. Experimental studies on rabbits and other animals showed that a temperature increase of about 3–4°C can be considered a heating threshold in the lens for the induction of cataracts (Appleton et al., 1975; Guy et al., 1975; Kramar et al., 1975).

Thermal changes in the eye have been obtained for a monopole $3\lambda/16$ long on a perfectly conducting box of dimension $34 \times 50 \times 134$ mm^3 transmitting at 380 MHz (TETRA system) or at the two GSM frequencies (900 and 1,800 MHz). For the TETRA system, a more realistic helix was also considered. The phone model was positioned vertically in front of the left eye at different distances, and thermal increments have

Table 8. Maximum temperature increments in the eye for TETRA, GSM900 and GSM1800 handset for 1.0 W of radiated power (data from Wainwright, 2007)

Frequency (MHz)	Antenna type	Eye–antenna distance (mm)	T (°C)
380	Monopole	24	0.33
		84	0.077
	Helix	24	0.44
		84	0.09
900	Monopole	24	0.31
		84	0.017
1,800	Monopole	24	0.55
		84	0.043

been evaluated for 1-W radiated power. Table 8 shows the temperature increments obtained in the eye. The maximum temperature increase in the eye was always found in or very near the lens (Wainwright, 2007). The corresponding values when the handset is radiating a power close to the maximum allowed (e.g., 250 mW at 900 MHz) can be obtained linearly scaling the reported values for comparable exposure durations.

A relationship between the eye-averaged SAR (eye mass equal to 9.4 g) and temperature rise was given, so that for other exposure situations, the maximum temperature rise in the lens could be estimated by multiplying the eye-averaged SAR by a "heating factor" equal to 0.19 °C W^{-1} kg (Wainwright, 2007). When the local SAR was increased to the ICNIRP occupational exposure level (SAR, averaged over a 10 g region of arbitrary contiguous shape, equal to 10 W/kg), a maximum temperature increase of 1.4°C was obtained in the eye in steady state.

Maximum temperature increments in the lens have also been reported for a dipole antenna operating at 900 MHz, 1.5 GHz, and 1.9 GHz (Hirata, 2005). It was shown that for SAR values equal to the ICNIRP (1998) occupational basic restriction ($SAR_{10\,g}$–roughly corresponding to the SAR_{eye} = 10 W/kg), the maximum temperature increments are between 1.55 and 1.74°C for the different frequencies and dipole–eye distances considered. These temperature increments are below the reported threshold for the induction of cataracts, but the corresponding safety factor is about 2, smaller than the value of 10 used in the derivation of the whole-body SAR restriction in the safety guidelines (e.g., ICNIRP, 1998).

Somewhat higher temperature increments were obtained in the eye for a dipole antenna placed vertically at a distance of 5.0 cm from the center of the cornea of the right eye (Flyckt et al., 2007). In particular, with the dipole operating at three frequencies (900, 1,500, and 1,800 MHz) and radiating 1.0 W, maximum temperature rises between 0.22 and 0.27°C were obtained. The same authors noted that they also obtained higher SAR values compared to the previously published data.

Flyckt et al. (2007) used a head model that included the discrete vasculatures. To evaluate the influence of the vasculature on temperature results, they compared the temperature values to those obtained from the conventional BHE; they found no differences. However, the fully vascularized model gives slightly higher temperature rises in the anterior part of the eye, where the vessel density is lower than in the posterior part.

4.2.3 Variations of Tissue Thermal Parameters on Temperature

As mentioned previously, tissue dielectric parameters used for SAR computations are well established and are mostly based on a 4-term Cole and Cole formulation for adult humans (Gabriel, 1996). Considerable ambiguity is present in thermal properties, especially regarding blood perfusion rates. Indeed, values chosen by different research groups show variations that for some tissues can be as high as 80% (see Fig. 2). The variability in thermal parameters can have important consequences on computed temperature increases from SAR distribution. It appears that variations up to 50% in convective coefficient between the head and the environment have a relatively small effect, causing variations in peak temperature increase of about 10% (Wang and Fujiwara, 1999).

Clearly, a key parameter concerning temperature increases in the brain is blood perfusion. Moreover, blood perfusion in the brain varies among individuals and for different regions in the brain. It was demonstrated that by halving this parameter a modest increase in temperature rise in the brain could be observed, reaching 20% for the case of an adult head exposed to a monopole-over-metallic-box phone model (Wainwright, 2000). Moreover, studies have indicated that blood flow around where the peak SAR occurs is one of the most dominant factors in determining induced temperature increments (Hirata et al., 2006).

The influence of thermophysiological parameters of tissues on temperature calculations in heads of mobile phone users has been investigated by Samaras et al. (2007), using generic phones with a monopole on top of metallic box model at 900 and 1,800 MHz and two numerical head models (Visible Human and the smaller European Female). Results showed that the variations in the maximum calculated temperature rise in the brain, due to the tissue thermophysiological properties reported, range from 9 to 14%. However, if the maximum temperature rises in the ear are considered, the variation due to the themophysiological properties increases

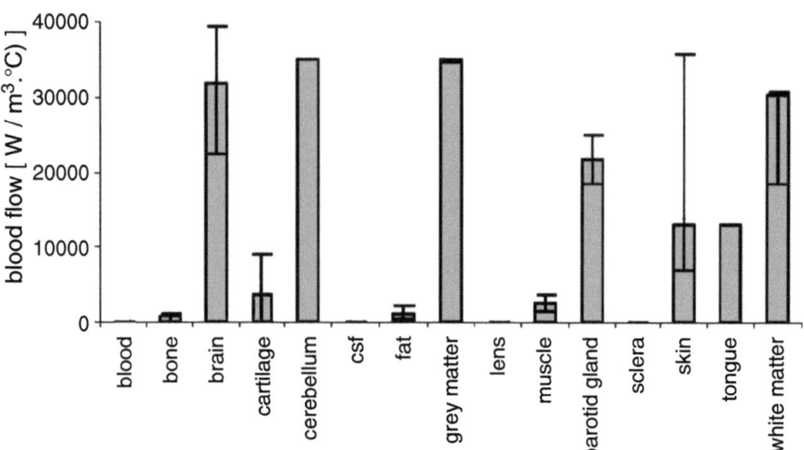

Figure 2. Mean, minimum, and maximum values used in the literature for blood perfusion (Samaras et al., 2007).

to 15–29%, because the cartilage and skin blood flow rates, which present a larger variation, play an important role in this case.

5. REGULATORY AND PUBLIC HEALTH ISSUES

The results summarized above clearly indicate that many of the computational studies has been aimed at the induced SAR and corresponding temperature change in biological systems exposed to RF electromagnetic fields. Moreover, a major effort has been devoted to the exposure of human and animal heads to mobile phones using anatomically correct models. These investigations allowed detailed studies of SAR distributions in different tissues and organs, effect of phone and antenna geometry on RF energy absorption, and level of heating induced in the brain and head tissues in general.

During the past decade, many computational studies in the literature have addressed regulatory issues, i.e., compliance of cellular phones on the market with existing exposure guidelines. A related issue pertains to the appropriateness of experimental phantoms used to test compliance of cellular phones. In fact, a variety of phantoms simulating the human head have been used to test compliance of mobile telecommunications equipment against safety standards and guidelines. Note that numerical compliance procedures have mostly been performed using complex anatomical phantoms based on MRI or photographic data. Experimental procedures have mainly relied on homogeneous phantoms filled with head-tissue simulating liquids since the entire volume must be accessible by the EM-field probes used for SAR measurements.

The most widely used homogeneous head phantom, namely SAM (see Fig. 3), has a shape derived from the size and dimensions of the 90th-percentile large adult

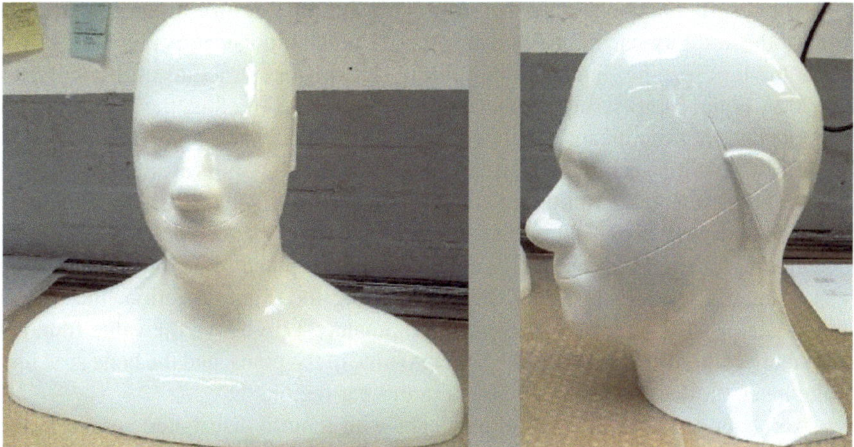

Figure 3. SAM phantom.

male, with the ear replaced by a lossless spacer of 4 mm, corresponding to the 10th percentile of ear thickness (IEC Int. Std. 62209-1, 2005). Appropriate parameters for the dielectric properties of the homogeneous liquid tissue-simulant filling the external plastic shell were chosen to give the same (or slightly higher) spatial peak absorption obtained in a worst-case layered model of the exposed head region. Some have doubted the appropriateness of such homogeneous lossless-ear phantom for giving conservative SAR estimates in the human head. Many papers have appeared in the literature, some reporting contrasting results. The main regulation-related aspects addressed in such computational studies will be briefly reviewed and discussed in the following sections.

5.1. Effect of the Pinna on Computed SAR Values

As stated above, in the SAM head phantom the ear was replaced by a lossless spacer of 4 mm. This choice was argued on the basis that although there is absorption in the pinna, the losses in the outer ear would be compensated by the replacement of the low-loss structure of the middle ear (bony structure with air cavities) through the lossy tissue-simulating material. However, modeling of the ear as a lossless dielectric spacer has led to great controversy. Because energy absorption in the pinna can be quite large due to the close proximity of this tissue to parts of the phone, thus leading to a possible underestimation of real exposure in the ear region.

The spatial-peak SAR in the ear region was evaluated in Burkhardt and Kuster (2000), based on a high-resolution phantom of a model with a collapsed ear. The aim was to find an appropriate modeling of the ear for experimental evaluations, which neither greatly overestimates nor underestimates the actual user exposure. To this end, a homogeneous phantom was created by cutting away the ear and smoothing the head surface in the ear region. The air-containing auditory canal was filled with tissue. Two different generic transmitters were used: a dipole and a generic phone, consisting of a simple box with a monopole antenna of realistic dimensions. Computed SAR values were directly normalized to the feed point current.

The results obtained in the heterogeneous head model for a dipole placed in front of the auditory canal, compared to those of the homogeneous phantom with a 4-mm lossless spacing, indicated that filling the low-loss structure of the inner ear with lossy liquid compensates for the losses in the pinna. Therefore, the homogeneous phantom does not significantly underestimate spatial-peak SAR in the head phantom. However, the volume containing the spatial-peak SAR value shifts away from the feed point to areas of greater wet tissue content. These results suggest that considerably higher spatial-peak SAR values are to be expected when the feed point of the dipole is shifted to locations where the tissue volume of the pinna in the proximity of the feed point is considerably larger. Indeed, shifting the dipole position, the largest spatial-peak SAR (1-g averaged) was approximately 3 dB above the value found at the center position. The reason is the larger volume of the pinna, and the bone structure of the inner ear does not entirely extend to this ear region.

To better represent the actual exposure with a more distributed source in the area of the ear, a generic phone held in the vertical position was also used. For the homogeneous phantom a lossless spacer of 4 and 6 mm was chosen. Results showed that a spacer of 4 mm represents the maximum exposure of the inhomogeneous phantom much better than a spacer of 6 mm. These findings were also verified for more realistic positions with respect to the head. As for the other position, the maximum underestimation of exposure for the dipole source with homogeneous modeling without the ear, using a spacing of 4 mm and appropriate dielectric parameters, was 20% for the 1-g spatially averaged SAR. Additional studies were performed at 1,800 MHz (Burkhardt and Kuster, 2000) since it is not obvious that the findings are also valid at higher frequencies due to the significantly reduced skin depth. The conclusions were similar to those found at 900 MHz.

The effect of replacing the lossy pinna with a lossless spacer was further investigated by Kanda et al. (2002). The investigations were performed for a box with a rectangular well simulating a compressed pinna during phone use. The bottom of the container and the well are made of 2-mm thick Lexan. The 2-mm deep well and the 2-mm bottom plate were chosen to simulate a 4-mm compressed ear, resulting in 6-mm total separation from the head liquid to the radiating device. A removable thin (2-mm) septum in line with the bottom of the box covered the well. The septum was perforated with a variety of openings to simulate the anatomical connection of the ear to the head. In all cases, the RF source was a balanced resonant dipole tuned for best match at 835 MHz. All results were normalized to 1-W radiated power. Evaluations were performed using a circular hole of 1.25-cm diameter in the septum; filled and unfilled ear canal showed that there is no difference in SAR values in the head with intact septum with or without liquid in the well. The circular opening presents a complex pattern of SAR distribution near the edge, but the SAR values in the head are still statistically similar to the values found with the intact septum.

A 4-mm wide and 3-cm long slit was also considered to model the ear's connection to the head (Kanda et al., 2002). The dipole antenna was considered both orthogonal and parallel to the slit, to also determine the effects due to the polarizability of a narrow slit. Peak 1-g averaged SAR was found to be not significantly different from the case of the circular opening. Even results obtained by considering an opening, resembling a "real ear" shape, showed no change in the overall 1 g average SAR in the head, both for vertical and horizontal dipole orientations, compared to the intact septum. In summary, these results showed that mere presence of a lossy ear, without an ohmic connection to the head, does not yield any measurable enhancement of SAR when compared to a lossless ear spacer. RF absorption enhancements exist at the edges of slots or holes, simulating the connection of the ear to the head. However, all edge effects on RF absorption in the simulated tissue were negligible when the SAR was averaged over 1 g.

While the results from one laboratory indicated that a plastic pinna did not produce significant underestimation of real exposure in the head, in terms of peak 1-g averaged SAR, on the contrary, results obtained by Gandhi and Kang (2002) showed just the opposite. The peak 1 g and 10 g SARs for two different average-size anatomically based models were compared with those of models with homogeneous brain-simulant dielectric properties and having 6-mm thick plastic ears. Relatively small handsets of dimensions typical of today's handsets and various nominal quarter-wavelength monopole antennas and helical antennas at both 800 and 1,900 MHz were studied. It was found that calculated SARs increased with reducing thickness of the pinna by a factor of 2.5 at 1,900 MHz and 1.4 at 835 MHz from 20-, 14-, 10-, to 6-mm thick layers simulating the pinna of the ear

The influence of a plastic ear was investigated using pinna thickness of 6 mm and smoothing the ear by filling in the crevices and the ear canal. A dielectric constant $\varepsilon_r = 2.56$ was assumed for the plastics in the shape of a smoothed ear. The calculated SARs, normalized to the radiated power, showed that the plastic ear model gave peak 1 g and 10 g SARs that are lower by factors of 2 or more than SARs in the original heterogeneous head models. This was attributed to a physical separation of 6 mm and the absence of the high SAR region in the plastic ear model, and that for an anatomic model, the lossy ear acted like a coupler conducting EM fields into the head since the SAR in the pinna region is substantial thus, resulting in higher SARs. Further studies were conducted using a 2-mm plastic shell filled with homogeneous liquid of appropriate dielectric properties ($\varepsilon_r = 41.5$, $\sigma = 0.9$ S/m at 835 MHz and $\varepsilon_r = 40.0$, $\sigma = 1.4$ S/m at 1,900 MHz) everywhere including the volume occupied by the smoothened pinna. In this case, the peak 10 g SAR for the "lossy pinna" homogeneous head phantom were found to be within ±15% of the SAR obtained for tissues in a realistic anatomic model of the head. Similar results were shown for peak 1 g SAR at 1,900 MHz but not at 835 MHz. These observations have led to the suggestion of using a thin-shelled head phantom including the "lossy pinna" vs. a lossless plastic spacer, for compliance testing of mobile phones.

The topic was further investigated using three-dimensional CAD files of mobile phones and the SAM head model (Gandhi and Kang, 2004). The calculated 1- and 10-g SARs in such model for some typical telephones were compared against the corresponding SARs calculated for some lossy ear anatomic models of the human head. Results confirmed that the peak 1- and 10-g SARs obtained for SAM with a plastic pinna are considerably lower than those for the anatomic models by a factor of up to two or more for some of the telephones having characteristics that are typical of today's devices. In particular, the peak SAR location for SAM shifted to the cheek region approximately 2.5 cm below the base of the radiating antenna, particularly at 835 MHz, while the anatomic models invariably give peak SAR locations close to the base of the antenna or to the top of the handsets. Since this is not the region of the highest electric and magnetic fields emanating from the antenna, according to the authors this was likely the reason why use of SAM resulted in greatly reduced 1- and 10-g SARs. In contrast, at 1,900 MHz, the peak 1- and 10-g SAR region for SAM is behind the plastic spacer. However, since this region of the lossy phantom is 5–10 mm further from the radiating antenna due to the plastic spacer, the calculated 1- and 10-g SARs are also a factor of two or more lower than those for anatomic models.

It was suggested that the gross underestimation of both 1- and 10-g SARs obtained for SAM is likely due to a separation on the order of 5–10 mm provided by the plastic shell from the highly radiating antenna region of the handset to the lossy tissue-simulant fluid. To remedy this high degree of underestimation of SAR for safety compliance testing, it was proposed that the plastic spacer of SAM be replaced by a lossy tissue-simulant fluid of 4-mm depth with an external shell of 2-mm thickness as for the rest of the SAM model. It was noted that, by using the modified SAM, an excellent agreement (within 20%) for the peak 10-g SAR could be obtained, comparable to the anatomical models, particularly at the higher frequency of 1,900 MHz. For the lower frequency of 835 MHz, the peak 1- and 10-g SARs were still fairly low for the modified SAM model. However, if a higher conductivity than the one suggested by international compliance testing procedures was used for the filler medium, higher SARs could be obtained, resulting in much better agreement with those for the anatomic models.

Subsequently, similar results, questioning the conservativeness of a lossless ear spacer for the evaluation of peak SAR in the head, were obtained by Christ et al. (2005), but with opposite conclusions. In this work, the authors compared the lossless ear homogeneous SAM model against three high-resolution anatomical head models. The head phantoms were exposed to the radiation of a generic mobile phone with different antenna types and a commercial mobile phone. The phones were placed in standardized testing positions and operated at 900 and 1,800 MHz. A comparison was made by neglecting the absorption in the pinna of all anatomical human phantoms. Under this assumption, for the standard positions evaluated, SAM always provided the highest 10-g averaged SAR. For the different phones and frequencies, different locations of the spatial peak SAR could be observed because the location of the cube containing the spatial peak SAR depends on the phone design but can also be different between phantoms for the same phone. Indeed, the authors observed that the locations and magnitudes depend on the local anatomy.

The situation was completely different when the authors considered the spatial peak SAR averaged over 10 g of pinna tissue only. Similar to the previously discussed results, the data showed that peak spatial SAR in the pinna of anatomical models could be up to two times higher than in the SAM phantom. However, by treating the pinna as an extremity, as in the 2005 IEEE standards with a correspondingly relaxed SAR limit twice that adopted in the body, the authors concluded that if the values for the head tissue are met, the values for the pinna treated as an extremity are intrinsically met as well, because the values do not exceed those assessed by SAM for the head tissue by more than a factor of 2.

A comprehensive investigation by Kainz et al. (2005) compared the SAM phantom to 14 anatomically correct head models to systematically evaluate whether or not SAM is conservative. A simple up-to-date phone model was used to determine the peak spatial SAR. To reduce uncertainties in phone positioning, which might have contributed to the conflicting results reported in the literature, a novel definition for the positioning of mobile phones next to anatomically correct head models was proposed (see Fig. 4). Moreover, since the net input power, feed-point impedance, and feed-point current all depend on the head model next to the mobile phone and the mobile phone position, the calculated SAR was normalized to either net input power or feed-point current. Using two phone positions and two frequencies, a total number of 112 different cases were computed. Results showed that only in 6 cases (5%) the 1-g averaged SAR calculated in the anatomical head model for head only tissue (excluding the pinna) was higher than the SAR in the SAM model. Considering only the compliance-relevant, worst-case configurations, SAM was conservative in all cases. For 10-g

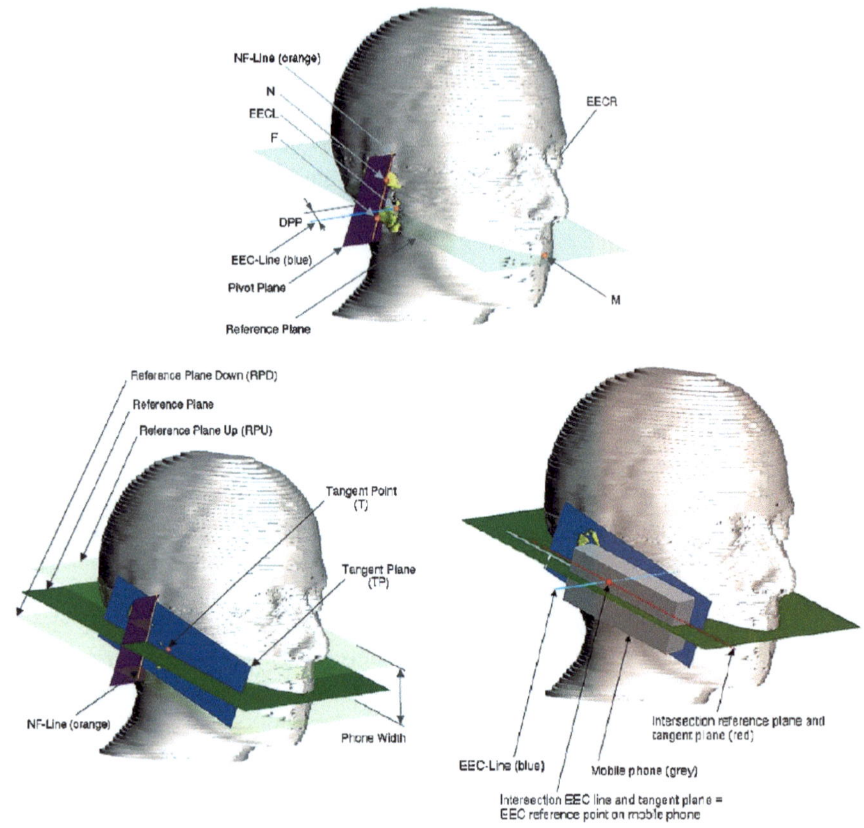

Figure 4. Positioning of the phone against anatomically correct head models (adopted with permission from Kainz et al., 2005).

averaged SAR in all tissue, the SAR was exceeded 27 times (24%) and in only 4 cases by more than 1 dB (26%). The maximum underestimation was 1.7 dB (48%), decreasing to only 0.6 dB (15%) for the worst-case configurations.

In summary, these data confirmed the strong dependence of SAR on the pinna shape, size, and deformation. Thus, for worst-case pinna geometries, SAM will not always result in conservative estimates in accordance with ICNIRP guidelines, which require inclusion of the pinna in SAR evaluations.

The SARs for the SAM phantom and anatomically correct models of the human head exposed to a mobile phone were also part of a study organized by IEEE SCC-34 (Beard et al., 2006). In this case, the SAM and two anatomically correct models were considered. Each participant in the comparison ran 12 simulations to fill an experiment matrix comprising the three head models, two frequencies (835 and 1,900 MHz), and two phone positions. Results were normalized to both net input power and feed-point current. However, statistical analysis of the results showed that the difference in normalization was not significant. The statistical analysis revealed that the average head only (excluding the pinna) SARs for the anatomic models were consistently lower than that for the SAM for all simulated conditions of frequency, position, SAR averaging volume (1 g or 10 g), and normalization. Thus, it was inferred that if the peak spatial SAR from compliance tests using SAM with a particular mobile phone meets the SAR limit for head tissue, then the exposure of a human head to that particular mobile phone would be below the limit. However, the analysis also showed that the average pinna only SAR for the anatomic models can exceed the SAR in the head for SAM. In particular, the highest pinna SAR seen was 1.44 times the maximum head SAR in SAM.

5.2. Dielectric Property of Head-Equivalent Homogeneous Tissue Phantoms

The experimental verification of compliance of mobile telecommunication devices with basic limits (IEEE, 2005; ICNIRP, 1998) is usually performed by using plastic shell head models filled with homogeneous liquid. Ideally, the head-tissue equivalent liquid should have dielectric parameters that give rise to SAR distributions mimicking the situation under actual user exposure.

To obtain the homogeneous tissue equivalent material, an attempt was made by Hombach et al. (1996) using power absorption at 900 MHz for four MRI-derived heterogeneous phantoms. The phone was modeled as a half-wavelength dipole placed at a distance of 15 mm from the head surface. The numerical results were compared with measurements in a multitissue phantom and two homogeneous phantoms of different shapes and sizes. This study proposed dielectric properties of homogeneous tissue equivalent liquid ($\varepsilon_r = 43.5$, $\sigma = 0.9$ S/m) for SAR compliance testing. However, this phantom material overestimated the worst-case SARs based on heterogeneous models. A similar conclusion was reached at 1,800 MHz by the same group for the same phantom with liquid material having dielectric parameters $\varepsilon_r = 41$, $\sigma = 1.65$ S/m (Meier et al., 1997).

An analytical planar model exposed to plane waves was used by Drossos et al. (2000) to derive worst-case tissue compositions for maximum spatial-peak SAR values. The tissue composition in the vicinity of the ear was approximated by a layered structure consisting of cartilage, skin, fat, muscle, skull, dura, CSF, and brain matter. Since effects from impedance matching or standing waves may result in enhanced spatial-peak absorption, the worst-case tissue composition was derived by varying the thicknesses of the layers in ranges that cover at least the 10th–90th percentiles of the user population, including adults and children. The entire frequency range covering the mobile communication bands between 300 and 3,000 MHz was considered. For each frequency, the tissue combinations and thicknesses that result in the highest spatial-peak SAR values averaged over 1 g and 10 g were evaluated. The homogeneous dielectric properties were derived by selecting the permittivity as the average of all tissues considered, while the conductivity was determined as that giving rise to the same (or higher) spatial-peak SAR value. This last condition produced different conductivity values for the 1- or 10-g averaged spatial peak SAR. The equivalent tissue properties are summarized in Table 9.

To verify whether the worst-case tissue composition for plane-wave excitation represents the worst-case for near-field sources, Drossos et al. (2000) studied the

Table 9. Plane wave derived equivalent homogeneous tissue permittivity properties

Frequency (MHz)	ε_r	σ (S/m) SAR 1 g	σ (S/m) SAR 10 g
300	45.3	0.87	0.70
900	41.5	0.97	0.85
1,800	40.0	1.40	1.40
2,450	39.2	1.70	1.80
3,000	38.5	2.10	2.40

exposure of anatomic MRI human head models either to a dipole source or to generic phones placed at various distances from the head. The study was performed at 900 and 1,800 MHz by comparing the spatial-peak SAR of nonhomogeneous phantoms to those obtained when all tissue properties were replaced by the generic tissue of Table 9. The spatial-peak SAR values obtained with the homogeneous modeling were always larger than those from nonhomogeneous modeling. Note that the dielectric permittivity (ε_r) and highest conductivity (σ) values given by Drossos et al. (2000) are presently included in the FCC (2001) and IEC-62209-1 (2005), recommendations for use in homogeneous phantom models.

5.3. Child and Adult Head Exposures

A major topic of interest is the exposure of children to mobile phone radiation; namely, whether exposure of a child's head produces SAR levels and distribution that are different to adults (Lin 2000, 2003). As already mentioned, SAR measurements for compliance testing of mobile terminals are currently performed using a phantom (SAM) based on the heads of adults. This approach was questioned in some studies showing increased power absorption in the heads of children compared to adults. This issue has become the subject of many publications, often reporting divergent results and conclusions.

The SAR distributions in mm-resolution models of the human adult head and scaled versions to represent the heads of 10- and 5-year-old children were reported by Gandhi et al. (1996). Cell phones were modeled as monopole antennas mounted on plastic-coated handsets of typical dimensions at 835 and 1,900 MHz. Because of the proximity of the hand to the telephone, the hand was also modeled as a region of 2/3 muscle-equivalent of 2-cm thick material wrapped around the handset on three sides, with the exception of the side facing the head. Computed SARs in the adult and two child models showed that although the peak 1-g SARs were similar at 1,900 MHz for all three models, the 1-g SARs were considerably higher (up to 50%) at 835 MHz for the smaller head sizes. This finding was reported as due to two concurrent effects: a larger extent of penetration of absorbed energy for the smaller models at both 835 and 1,900 MHz, and the thinner ears of the smaller models, which resulted in the antennas being closer to the region of highest SARs observed at the points of contact of the ear pressed against the scalp of the head, in general.

A follow-up paper from this group (Gandhi and Kang, 2002) expanded the previously reported study of energy deposition in models of adults and children to two distinctly different anatomically based models of the adult head, each of which was scaled up or down by 10% to obtain a larger and a smaller head models. Three different sizes of the handset and two different antennas (a monopole and a helix) were considered for two frequencies (835 and 1,900 MHz). Computed results, normalized to radiated power, showed that the peak 1 g SARs for both the head (pinna excluded) and the brain tissues increase monotonically with the reducing head size for both head models and all handset dimensions and antenna types. In particular, the peak 1 g SAR for head tissues for smaller models could be up to 60% higher at 1,900 MHz and 20% higher at 835 MHz compared to that for the larger head models. It was suggested that the shielding effect of the pinna was larger at the higher frequency of 1,900 MHz.

In contrast, a report by Schonborn et al. (1998) concluded that there were no differences in the absolute extent of SAR penetration between adult and child head models. Moreover, 1- and 10-g averaged SAR values for children and scaled head sizes differed only slightly from those of the adult. This investigation used head phantoms based on MRI scans of an adult and two children (3- and 7-year olds). In addition, scaled-down phantoms of the adult head, similar to the approach used in Gandhi et al. (1996), were also considered. Computations were performed for 900 and 1,800 MHz, using a dipole placed at a fixed distance of 15 mm from the head. While studies were also performed by varying the distance of the dipole from the head, it should be noted that the SAR values were normalized to the antenna feed-point current.

The apparent conflicts between the two sets of studies can be explained by noting that the increased extent of SAR penetration shown in Gandhi's papers referred to the relative penetration with respect to the head diameter, while Schonborn et al. based their conclusions on absolute penetration depth, which clearly do not depend on the size of the head. Furthermore, the phone was closer to the head for child head models in the Gandhi et al. study, thus causing an increase in average SAR values, while in Schonborn et al. the dipole was kept at a constant distance from the head. In addition, there are other obvious causes: namely, differences in the phantom models, normalizing SAR values to different quantities (antenna current vs. power) and, more importantly, different phone positioning (fixed distance vs. decreasing distance with decreasing pinna thickness).

The effect of normalization by antenna current vs. power was demonstrated in Wang and Fujiwara (2003). Two child-size models (3- and 7-year olds) were developed from a Japanese adult head model based on statistical data on external head shapes of Japanese children. In particular, for different parts of the head, different scaling factors were employed to derive scaled models that approximate the shape of real children. In addition to these 3- and 7-year-old head models, two more head models were developed with 90% and 82% scaling factors. The local peak SAR at 900 MHz was calculated under two exposure conditions, similar to those previously employed by the above-mentioned studies, respectively, but with the distance between the antenna and the head surface kept constant.

For the monopole antenna mounted on a rectangular metal box, SAR values were normalized to antenna output power. For the dipole antenna, SARs were normalized to antenna current.

For the monopole antenna, the statistical 3-year-old head model showed an increase of about 30% in the 1-g averaged spatial peak SAR and 20% in the 10-g averaged spatial peak SAR compared to the adult head model. The same observation was made in models with fixed scaling factors. Analysis indicated that, as the head size decreased, the antenna input impedance also decreased. Consequently, the antenna current increased to keep the output power constant. The increased antenna current produced stronger magnetic fields in the vicinity of the antenna, thus giving rise to the increased peak SAR in the children's heads. On the other hand, SAR distributions computed in various head models for dipoles did not demonstrate any significant differences between the adult and children's models for the 1- and 10-g averaged spatial peak SARs. The maximum differences between the adult and children were within 10%. Their analysis showed that in this case the variation on the antenna input impedance and, consequently, the antenna output powers were insignificant for different head models. These observations were supported by calculation of peak SARs for three different-sized homogeneous (brain-equivalent tissue) spheres having the same volumes as the adult, 7- and 3-year-old head models, respectively. The monopole mobile telephone model was used in this case. It was found that with decreasing sphere sizes, the resistive components of the antenna input impedance increased and, consequently, the peak SAR for a fixed output power decreased because of the decrease in antenna current.

There were other efforts devoted to identifying the parameters that may have an influence on determining significant differences in SARs induced by mobile phone fields in children's and adult's heads and brains. For example, Anderson (2003) computed the SARs using a three-layered (scalp, cranium, brain) spherical model of the head exposed to a dipole at 900 MHz. Martinez-Burdalo et al. (2004), evaluated SAR and field penetration into the head with a realistic and scaled three-dimensional human head model.

The study by Hadjem et al. (2005) is noteworthy in that SARs in adult, child-sized (CS), and child-like (CL) head models were compared. Specifically, two adult head models were used to develop two types of child head models. The CS head is obtained from a uniform reduction of the geometrical dimensions of the adult head, while the CL head is built by morphing deformation of an adult head to build 5- and 10-year-old children. Commercial dual-band mobile phone models were used as sources. The results supported the conclusion that since the brain is closer to the mobile phone in the case of the CS or CL heads, the power absorption in the child brain models is slightly higher than that of the adult. The difference between the heads of 5- and 10-year olds and between the CS head and the CL head is very small, except for brain tissues at 900 MHz.

Another attempt to clarify SARs in adult and child heads (Bit-Babik et al., 2005) employed two different adult head models and a set of child head models obtained either by linear scaling of the adult head or using more complex scaling procedures to attain anatomic correctness. Two monopole-over-box handset models were maintained at a fixed distance from the ear, thus giving rise to different distances from the head for each model. Average SAR was computed using both the standardized IEEE procedure and fixed volume cubes, which can introduce an error in SAR values depending on the deviation of tissue density from that of water. The study showed that 1- and 10-g average SAR values are not significantly different for adult- and child-head models, if the SAR averaging procedure standardized by IEEE is adopted. Adoption of the simpler fixed-cube averaging procedure yielded completely different results, with a marked increase in average SAR with decreasing head size.

The topic of child- vs. adult-head exposure at 835 and 1,900 MHz was also addressed as part of the IEEE SCC-34 comparisons using anatomically correct adult and 7-year-old child head models (Beard et al., 2006). The results were very different for the two cell-phones. For 1,900-MHz cell phones, the peak 1 g and 10 g SAR values in the head, pinna, and average tissue of the adult model were consistently higher than those for the child model, normalized either to the antenna current or to the power for the cheek and tilt positions . However, a majority of the SARs were higher in the child than the adult model, especially for the 835-MHz phone in tilt position when normalized to the antenna current.

The effect of anatomic details was studied using four MRI-derived head models, one female, one adult, and two children, age 3 and 7 years (Keshvari and Lang, 2005). The head models differed in size, external shape, and internal anatomy. Models were exposed to a half-wavelength dipole antenna at 900, 1,800, and 2,450 MHz frequencies, with the antenna placed vertically 2 cm away from the ear. The SARs were calculated for cases including and excluding pinna tissues and were normalized to the radiated antenna power. Results showed that when pinna was excluded, the SAR for adult male was the largest at all frequencies. However, the 3-year-old child and female model had higher SAR values if the pinna was present. The authors had concluded that in addition to the distance of separation between the antenna and the exposed tissue, tissue composition and anatomical differences between head models can contribute to differences in the RF energy absorption between anatomically correct MRI-based head models of adults and children.

To date, most dosimetric studies of children exposed to mobile phones have employed dielectric properties of biological tissues for adults. Nevertheless, the influence of variations in dielectric properties with age on SAR has received some attention (Keshvari et al., 2006; Wang et al., 2006).

Under the same exposure conditions and using the head models (without the pinna) that this group had used previously (see above), Keshvari et al. (2006) investigated the effect of a 5–20% increases in conductivity or conductivity and permittivity of head tissues on SARs for the ear and eye regions. Not surprisingly, the outcomes were different depending on whether it is for 900, 1,800, or 2,450 MHz, or whether the SAR values were averaged over a 1- or 10-g tissue mass. It is interesting to note that the 1-g SARs at 900 MHz were mostly higher for higher values of conductivity and permittivity in both adult- and child-sized head models; the SARs ranged from 5 to 15% at the highest level.

The investigation by Wang et al. (2006) used an empirical formula (based on rat tissues) to relate the complex permittivity of various tissues to the total body water, which is known to be age dependent. The spatial peak SARs for a 900-MHz mobile phone model (monopole antenna mounted on a rectangular metal box) was calculated for anatomically correct 3- and 7-year-old head models. The results showed that the 1- and 10-g averaged spatial peak SARs had nearly the same values in the child models for both child and adult equivalent dielectric properties. However, the empirical formula gave a 20% higher permittivity for the skull and underestimated by 20% the conductivity of brain tissues measured in rats. Adjusting for these differences produced 1- and 10-g averaged spatial peak SARs that differed by about 4% compared to the unadjusted data for the 3-year-old head. Results from other cases showed that the effect of the empirical, age-dependent dielectric properties on the spatial peak SAR was within 10%.

Clearly, the topic of child vs. adult head exposure is far from being settled. There are many variables and aspects of the problem that are often uncontrolled or poorly understood. Some of the differences and contrasting conclusions are attributable to different phone orientations, radio frequencies, and parameters used for SAR normalization (power or antenna current), or to procedures adopted for SAR averaging. The choice of different averaging procedures, starting with the same absorption profile, could lead to average SARs from being not significantly different to a marked increase with decreasing head sizes. One of the major causes for increased absorption in a child's head is the decreased distance of separation between the cell phone antenna and the head due to the smaller ear size. Moreover, how results are presented such as the extent of brain tissue involvement in children can vary depending on whether it is based on the conventional definition of penetration depth (i.e., e^{-2} depth) or on the quantity of brain tissue impacted relative to the head size. The paucity of data on age-dependent tissue conductivity and permittivity is a limitation also.

5.4. Effect of Averaging Procedure on Computed SAR Values

The choice of different averaging procedures for evaluating local SAR could lead to large differences in the computed values. The subject has already been mentioned in the section on the CEPHOS project (Nikita et al., 2000). The procedures used to perform averaging over a given mass also influence the quality of comparative dosimetry studies. This issue serves as a further complication in the various safety guidelines since the averaging masses for local SAR basic restriction are defined differently.

In the ICNIRP guideline, the basic restriction on local SAR is defined as 2.0 W/kg in the head and trunk, and 4.0 W/kg in the limbs for general population exposure (10 W/kg in the head and trunk, and 20 W/kg in the limbs, respectively, for occupational exposure). These SAR values are to be averaged over a mass defined as "any 10 g of contiguous tissue" (ICNIRP, 1998). However in the IEEE (1991; 1999) guideline, the SAR limit for localized exposures was set to 1.6 or 8.0 W/kg under controlled or uncontrolled environments, respectively, over 1 g of mass in the shape of a cube, which also is the case in the Rules of FCC (2001). In a harmonization attempt, the IEEE limit for localized exposures was increased to 2.0 W/kg for the general population and 10 W/kg for occupational exposure over 10 g of tissue (IEEE, 2005). However, in the arms and legs distal to the elbows and knees, and in the pinna, these limits are relaxed to 4.0 W/kg for the general population, and 20 W/kg for occupational exposure, respectively. The SAR values are to be evaluated with an averaging mass in the shape of a cube. Consequently, while the limiting SAR values are the same for the ICNIRP and IEEE guidelines, they are defined differently over masses of different shapes and for different parts of the human body (Lin, 2006).

Moreover, in neither the ICNIRP nor the IEEE guidelines, the quantity of air that could be included in the averaging mass, if any, is specified. However, when the local averaged SAR is to be evaluated at points of the human body which are located next to the body surface, a number of air voxels must be included in the volume unless the volume is taken deep into the body. A further complication arises from the weight of the chosen local mass (i.e., 1 g or 10 g), particularly when a cubic-shape is considered. Thus, a 2.15-cm side must be used for human tissues with a mass density about 1,000 kg/m^3 to obtain a cube having a mass of 10 g.

For numerical dosimetry studies using the FDTD method, the position of the *E*-field components in the FDTD cell (i.e., on the side of the cell – see Fig. 5) must be taken into account to evaluate SAR. Using the three *E*-field components as they are located in the FDTD lattice would mean having field components placed in different spatial positions (Fig. 5a). The three *E*-field components could be referred to the cell vertex, thus allowing for each component an average between the values in two adjacent cells (Fig. 5b), or to the cell center, in this case providing an average among four values for each field component (Fig. 5c). A comparison between the two schemes (vertex or center of the cell) for determining the field components showed no significant differences between them, whereas differences between 5 and 24 % were found when comparing the SAR values, computed with the other two schemes to those obtainable using the three *E*-field components as they are (Caputa et al., 1999).

The influence of FDTD resolution on mass-averaged SAR values was studied by Van de Kamer and Lagendijk (2002), using several FDTD resolutions (0.4, 1.0, and 2.0 mm) in a female adult head. The three-dimensional MRI dataset was scanned at a resolution of 1 mm^3, segmented into ten tissue types, and downscaled to a resolution of 2 mm^3. The cellular phone was modeled as a vertical dipole antenna operating at 915 MHz and placed at a distance of 2 cm from the head. For an effectively transmitted power of 0.25 W, the maximum SAR value averaged over a

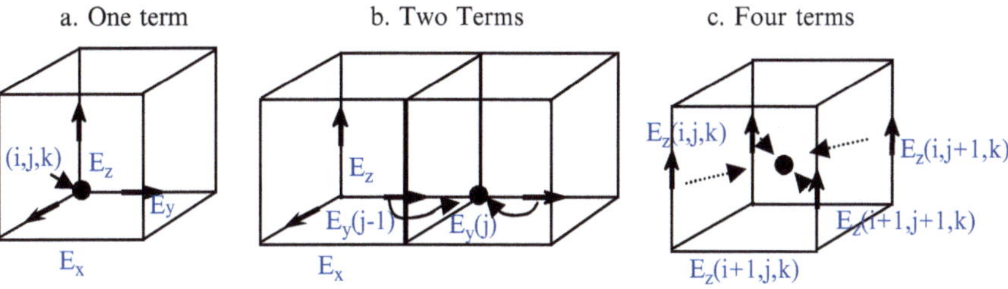

Figure 5. Summation schemes for the *E*-field components in the same spatial position to evaluate SAR (**a**) one term (**b**) two terms (**c**) four terms.

cube of 10 g of tissue was about 0.98 W/kg. This value varied by not more than 8% for the different resolutions, indicating that SAR computations at a resolution of 2 mm are sufficiently accurate in this case.

An important point in local averaged SAR evaluation is the percentage of air that is included in the averaging mass. In general, the more air is included in the mass, the more superficial the mass results. The corresponding SAR becomes higher, and, if the mass was originally shaped as a cube, its final shape may end up very different from it (see Fig. 6). For a spherical surface with a half-wavelength dipole positioned next to it, the computed $SAR_{10\,g}$ values for different positions of the averaging cube, differences up to 22% at 900 MHz and 30% at 1,800 MHz have been obtained (Stevens and Martens, 2000), with the highest SAR values found with the most external averaging cube.

Another approach to obtaining the cubical volume is by starting from a given location and expanding the volume in all directions until the desired value for the required mass is reached. In this case, a surface boundary of the averaging volume is not extended beyond the outer most surface of the body. If the averaging volume's surface extends beyond the exterior surface, it is discarded. A local SAR is assigned as equal to the average value obtained starting from a different location, but with the point under consideration included in the averaging volume (IEEE, 2002). Note that the local averaged SAR values obtained with this procedure is lower than those obtained from averaging cubes having a boundary surface completely outside the body and are assigned to values associated with the most superficial points of a biological body.

For averaging schemes involving different mass and shapes, a comparison of $SAR_{1\,g}$ values was performed using volumes shaped nearly as a cube and an anatomical head model based on parallelepiped cells ($1.974\times 1.974\times 3$ mm) and volumes made up with $5\times5\times3$, $5\times5\times4$, and $6\times6\times3$ cells (Gandhi et al., 1996). Only volumes of at least 80% of the cells occupied by tissues and no more than 20% of the cells filled with air were included. SAR data for different antennas and frequencies are presented in Table 10. It can be noted that there are wide variations in the SAR values for the different averaging shapes. In particular, up to twofold differences can be seen.

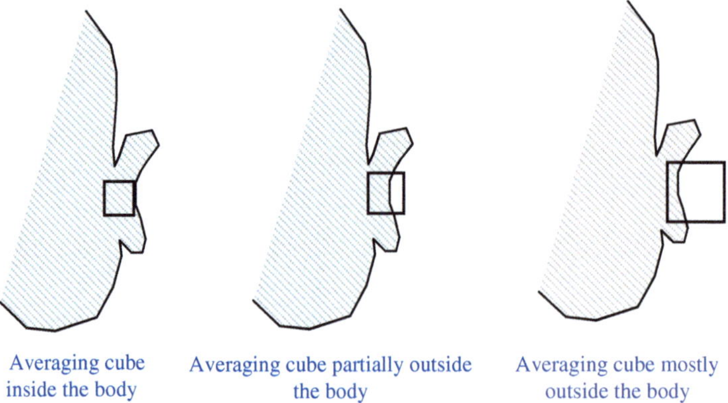

Figure 6. Different positions of the SAR averaging cube for the body surface with same tissue mass.

Table 10. SAR averaged over 1 g of mass obtained from different volume shapes

Antenna	*l*/4	3*l*/8	*l*/4 over box	*l*/2 dipole	*l*/4 over box	*l*/2 dipole
Frequency (MHz)	1,900	1,900	900	900	1,900	1,900
Radiated power (mW)	125	125	600	600	125	125
Averaging volume (cells)	SAR (W/kg)					
5 × 5 × 3	0.52 (1.01 g; 98.7%)	0.32 (1.01 g; 98.7%)	2.07 (1.00 g; 92.0%)	2.10 (1.00 g; 92%)	0.53 (1.00 g; 98.7%)	0.55 (1.00 g; 98.7%)
5 × 5 × 4	1.11 (1.03 g; 81.0%)	0.69 (1.6 g; 86.0%)	2.49 (1.07 g; 83.0%)	2.71 (1.07 g; 83.0%)	0.87 (1.03 g; 80.0%)	0.81 (1.04 g; 81.0%)
6 × 6 × 3	1.03 (1.10 g; 82.4%)	0.69 (1.11 g; 86.1%)	2.45 (1.18 g; 86.1%)	2.47 (1.13 g; 81.5%)	0.83 (1.13 g; 81.5%)	0.76 (1.16 g; 83.3%)

The actual mass and percentage of air included in the volume are given in parentheses (from Gandhi et al., 1996)

Moreover, similar SAR values from different schemes are found for different masses and with different percentage of air included in the averaging volume.

The difference in SAR values obtained from different averaging volumes was also reported by Wainwright (2007) by comparing cubic and arbitrary shaped volumes. For an example, the $SAR_{10\,g}$ averaged over a cubic shape was equal to 1.49 W/kg, while the $SAR_{10\,g}$ averaged over a generic contiguous shape was 2.71 W/kg for a monopole antenna operating at 900 MHz (radiated power 1.0 W). At 1,800 MHz, for the same antenna and radiated power, the reported values are 2.58 W/kg and 4.00 W/kg, respectively. Thus, a ratio of about 1.5 is obtained between the $SAR_{10\,g}$ for a contiguous tissue and a cubic-shaped tissue. A similar ratio was obtained by considering a dipole antenna, operating at 915 MHz, placed at a distance of 2 cm from the head and radiating 250 mW (Van de Kamer and Lagendijk, 2002).

6. CONCLUDING REMARKS

A complexity in the interaction of RF electromagnetic field with biological bodies is that the same exposure or incident field does not necessarily provide the same field inside biological bodies of different size, shape, or constitution. Therefore, an important task in assessing the health and safety of RF exposure from wireless communication devices and systems is the determination of induced fields in biological tissues and the widely accepted dosimetric quantity, SAR. The best-known biological effects resulting from either partial-body or whole-body exposures is that associated with SAR-induced temperature rises in humans and laboratory animals. We have provided in this chapter a comprehensive summary of the SAR and temperature aspects of human exposures to the cellular mobile telephone system. The descriptions include SAR distributions and peak temperature elevations, their derivation and computation, and implications for guidelines designed to limit human exposure in the wireless communication frequency band. It is hoped that they will serve as a common ground for a better understanding of human exposure to mobile phone systems.

Indeed, fields induced inside a biological body exposed to RF electromagnetic energy have been the subject of a large number of studies beginning in the 1960s. More recently, the availability of high-performance computational resources and advances in numerical algorithms have combined to provide a powerful and precise methodology for studying SAR distributions and investigating temperature elevations from the interaction of wireless communication fields with the human body. The development is facilitated by three-dimensional biological models with about 1-mm spatial resolutions. In particular, major emphasis has been devoted to dosimetry and temperature studies concerning human exposure to cellular mobile telephone fields because of the popularity and rapid introduction of new technologies.

The topics of numerical dosimetry and temperature elevation have been discussed in considerable detail. Some descriptions of experimental dosimetry have been included to illustrate the complementary need for numerical investigations in experimental studies designed for testing compliance of mobile phones with exposure guidelines. The results from recent dosimetric research have summarized include the influence of different metallic implants worn by mobile phone users and the environment in which exposure occurs, such as inside a vehicle. Many of the results will provide the much needed information to the general public, research scientists, or cell phone manufacturers and operators, because of their importance in mobile phone compliance testing. Other topics discussed the specific concerns of mobile phone use by children. Among the topics of technical interest described are the influence of the pinna on computed SAR values and temperature increments, the effect of averaging procedures on numerical SAR, and the variation of results due to the uncertainties associated with the dielectric parameters used to characterize human tissues. However, some issues still remain concerning the evaluation of specific energy absorption in the head of a cell phone user. These include, among the others, the assessment of the variation of SAR with different geometrical and electrical characteristics of the human head, the most reliable or accurate approach to perform the local SAR averaging.

The march toward higher operating frequencies in wireless technologies and the recent development of applications using wideband signals and transmission schemes (ICNIRP, 2008) will push toward the development of increasingly more accurate models of the interaction. Accurate numerical dosimetry and temperature computation require the use of detailed anatomical models of the human head or torso in the case of body-worn devices. Moreover, detailed modeling will be required to take into account the antenna structure, phone case, and internal components of the mobile terminals or body-worn devices. There are increasing concerns and interest on specific categories of exposed subjects such as children and pregnant women, with the latter also requiring dosimetry and temperature investigations of the fetus in its various development stages. Finally, the protection of workers exposed to RF radiating sources has begun to attract attention at the international level. Assessment of the risk and protection afforded workers from exposure to different sources will become a necessity. Numerical techniques for SAR and temperature will be essential for such assessments. They could also provide the scientific information to help refine exposure guidelines and establish compliance with exposure limits promulgated by relevant guidelines.

REFERENCES

Ackerman, M.J., 1998, The visible human project. *Proc IEEE* 86: 504.

Ali, M., Douglas, M.G., Sayem, A.T.M., Faraone, A., and Chou, C.K., 2007, Threshold power of canonical antennas for inducing SAR at compliance limits in the 300–3000 MHz frequency range. *IEEE Trans Electromagn Compat* 49(1): 143–152.

Anderson, V., 2003, Comparisons of peak SAR levels in concentric sphere head models of children and adults for irradiation by a dipole at 900 MHz. *Phys Med Biol* 48: 3263–3275.

Appleton, B., Hirsch, S.E., and Brown, P.V.K., 1975, Investigation of single exposure microwave ocular effects at 3000 MHz. *Ann N Y Acad Sci* 247: 125–134.

ARIB STD-T56, 2002, Specific absorption rate (SAR) estimation for cellular phone. Association of Radio Industries and Businesses, Japan

Bach-Andersen, J., et al., co-ordinated by G. d'Inzeo, 1994, Proposal for Numerical Canonical Models in Mobile Communication. COST244 WG3 Numerical Program.

Balzano, Q., Garay, O., and Steel, F.R., 1978, Energy deposition in simulated human operators of 800 MHz portable transmitters. *IEEE Trans Veh Technol* 27: 174–181.

Beard, B.B., Kainz, W., Onishi, T., Iyama, T., Watanabe, S., Fujiwara, O., Wang, J., Bit-Babik, G., Faraone, A., Wiart, J., Christ, A., Kuster, N., Lee, A.-K., Kroeze, H., Siegbahn, N., Keshvari, J., Abrishamkar, H., Simon, W., Manteuffel, D., and Nikoloski, N., 2006, Comparisons of computed mobile phone induced SAR in the SAM phantom to that in anatomically correct models of the human head. *IEEE Trans Electromagn Compat* 48: 397–407.

Berenger, J.P., 1994, A perfectly matched layer for the absorption of electromagnetic waves. *J Comput Phys* 114: 185–200.

Bernardi, P., Cavagnaro, M., and Pisa, S., 1996, Evaluation of the SAR distribution in the human head for cellular phones used in a partially closed environment. *IEEE Trans Electromagn Compat* 38(3): 357–366.

Bernardi, P., Cavagnaro, M., and Pisa, S., 1998, SAR distribution and temperature increase in an anatomical model of the human eye exposed to the field radiated by the user antenna in a wireless LAN. *IEEE Trans Microw Theory Tech* 46(12): 2074–2082.

Bernardi, P., Cavagnaro, M., Pisa, S., and Piuzzi, E., 2000, Specific absorption rate and temperature increases in the head of a cellular phone user. *IEEE Trans Microw Theory Tech* 48(7): 1118–1126.

Bernardi, P., Cavagnaro, M., Pisa, S. and Piuzzi, E., 2001a, A graded-mesh FDTD code for the study of human exposure to cellular phones equipped with helical antennas. *Appl Comput Electromag Soc J* 16: 90–96.

Bernardi, P., Cavagnaro, M., Pisa, S., and Piuzzi, E., 2001b, Power absorption and temperature elevations induced in the human head by a dual-band monopole-helix antenna phone. *IEEE Trans Microw Theory Tech* 49(12): 2539–2546.

Bernardi, P., Cavagnaro, M., Pisa, S., and Piuzzi, E., 2003, Specific absorption rate and temperature elevation in a subject exposed in the far-field of radio-frequency sources operating in the 10–900-MHz range. *IEEE Trans Biomed Eng* 50: 295–304.

Bernardi, P., Cavagnaro, M., Pisa, S. Piuzzi, E., and Lin, J., 2005, SAR distribution produced by cellular phones radiating close to the human body. *Bioelectromagnetics 2005*. Dublin, Ireland, Abstract Book, pp. 131–133.

Bernsten, S., and Hornsleth, S.N., 1994, Retarded lime absorbing boundary conditions. *IEEE Trans Antennas Propagat* 42: 1059–1064.

Bit-Babik, G., Guy, A.W., Chou, C.-K., Faraone, A., Kanda, M., Gessner, A., Wang, J., and Fujiwara, O., 2005, Simulation of exposure and SAR estimation for adult and child heads exposed to radiofrequency energy from portable communication devices. *Radiat Res* 163: 580–590.

Burkhardt, M., and Kuster, N., 2000, Appropriate modeling of the ear for compliance testing of handheld MTE with SAR safety limits at 900/1800 MHz. *IEEE Trans Microw Theory Tech* 48: 1927–1934.

Caputa, K., Okoniewski, M., and Stuchly, M.A., 1999, An algorithm for computations of the power deposition in human tissue. *IEEE Antennas Propag Mag* 41(4): 102–107.

Caputa, K., Stuchly, M.A., Skopec, M., Bassen, H.I., Ruggera, P., and Kanda, M., 2000, Evaluation of electromagnetic interference from a cellular telephone with hearing aid. *IEEE Trans Microw Theory Tech* 48: 2148–2154.

Cavagnaro, M., and Pisa, S., 1996, Simulation of cellular phone antennas by using inductive lumped elements in the 3D-FDTD algorithm. *Microw Opt Technol Lett* 13: 324–327.

CENELEC, 2001, Basic standard for the measurement of specific absorption rate related to human exposure to electromagnetic fields from mobile phones (300 MHz–3 GHz). *European Standard EN50361*. European Committee for Electrotechnical Standardization. Central secretariat, rue de Stassart 35, B-1050 Brussels, Belgium.

Cerri, G., Russo, P., Schiavoni, A., Tribellini, G., and Bielli, P., 1998, A new MoM-FDTD hybrid technique for the analysis of scattering problems. *Electron Letts* 34: 438.

Chatterjee, I., Gu, Y.G., and Gandhi, O.P., 1985, Quantification of electromagnetic absorption in humans from body-mounted communication transceivers. *IEEE Trans Veh Technol* 34: 56–62.

Chavannes, N., Tay, R., Nikoloski, N., and Kuster, N., 2003, Suitability of FDTD-based TCAD tools for RF design of mobile phones. *IEEE Antennas Propag Mag* 45: 52–66.

Chen, J.Y., and Gandhi, O.P., 1989, Electromagnetic deposition in an anatomically based model of man for leakage fields of a parallel-plate dielectric heater. *IEEE Trans Microw Theory Tech* 37(1): 174–180.

Chen, H.Y., and Wang, H.H., 1994, Current and SAR induced in a human head model by the electromagnetic fields irradiated from a cellular phone. *IEEE Trans Microw Theory Tech* 42(12): 2249–2254.

Chen, J.Y., Gandhi, O.P., and Conover, D.L., 1991, SAR and induced current distributions for operator exposure to RF dielectric sealers. *IEEE Trans Electromagn Compat* 33(3): 252–261.

Christ, A., Chavannes, N., Nikoloski, N., Gerber, H.-U., Pokovic, K., and Kuster, N., 2005, A numerical and experimental comparison of human head phantoms for compliance testing of mobile telephone equipment. *Bioelectromagnetics* 26: 125–137.

Christ, A., Samaras, T., Klingenbock, A., and Kuster, 2006, Characterization of the electromagnetic near-field absorption in layered biological tissue in the frequency range from 30 MHz to 6000 MHz. *Phys Med Biol* 51: 4951–4965.

Chuang, H.R., 1994, Human operator coupling effects on radiation characteristics of a portable communication dipole antenna. *IEEE Trans Antennas Propag* 42: 556–560.

Colburn, J.S., and Rahmat-Samii, Y., 1998, Human proximity effects on circular polarized handset antennas in personal satellite communications. *IEEE Trans Antennas Propag* 46: 813–820.

Cooper, J., and Hombach, V., 1996, Increase in specific absorption rate in human heads arising from implantations. *Electron Letts* 32(24): 2217–2219.

D'Inzeo, G., 1994, Proposal for numerical canonical models in mobile communications. In: Simunic, D., Ed., *Biomedical Effects of Electromagnetic Fields – Reference Models in Mobile Communications*. Proceedings of COST 244 Meeting, 1–7, Rome, Italy.

Davis, C.C., Beard, B.B., Tillman, A., Rzasa, J., Merideth, E., and Balzano, Q., 2006, International intercomparison of specific absorption rates in a flat absorbing phantom in the near-field of dipole antennas. *IEEE Trans Electromagn Compat* 48(3): 579–588.

Dimbylow, P.J., 1993, FDTD calculation of the SAR for a dipole closely coupled to the head at 900 MHz and 1.9 GHz. *Phys Med Biol* 38(3): 361–368.

Dimbylow, P.J., 1998, FDTD calculation of the whole-body averaged SAR in an anatomically realistic voxel model of the human body from 1 MHz to 1 GHz. *Phys Med Biol* 42: 479–490.

Dimbylow, P.J., and Gandhi, O.P., 1991, Finite-difference time-domain calculations of SAR in a realistic heterogeneous model of the head for plane wave exposure from 600 MHz to 3 GHz. *Phys Med Biol* 36(8): 1075–1089.

Dimbylow, P.J., and Mann, S.M., 1994, SAR calculations in an anatomically realistic model of the head for mobile communication transceivers at 900 MHz and 1.8 GHz. *Phys Med Biol* 39(10): 1537–1553.

Dimbylow, P., Khalid, M., and Mann, S., 2003, Assessment of specific energy absorption rate (SAR) in the head from a TETRA handset. *Phys Med Biol* 48: 3911–3926.

Dominguez, H., Raizer, A., Carpes, W.P., Jr., 2002, Electromagnetic fields radiated by a cellular phone in close proximity to metallic walls. *IEEE Trans Magn* 38(2 Part 1): 793–796.

Drossos, A., Santomaa, V., and Kuster, N., 2000, The dependence of electromagnetic energy absorption upon human head tissue composition in the frequency range 300–3000 MHz. *IEEE Trans Microw Theory Tech* 48: 1988–1995.

Edwards, R.M., and Whittow, W.G., 2005, Applications of a genetic algorithm for identification of maxima in specific absorption rates in the human eye close to perfectly conducting spectacles. *IEE Proc Sci Meas Technol* 152(3): 89–96.

EN 50361, 2001, Basic standard for the measurement of specific absorption rate related to exposure to electromagnetic fields from mobile phones (300 MHz–3 GHz).

Fayos-Fernandez, J., Arranz-Faz, C., Martinez-Gonzalez, A.M., and Sanchez-Hernandez, D., 2006, Effect of pierced metallic objects on SAR distributions at 900 MHz. *Bioelectromagnetics* 27: 337–353.

FCC (Federal Communication Commission), 2001, Evaluating compliance with FCC guidelines for human exposure to radio frequency electromagnetic fields. *Supplement C to OET Bulletin 65 (Edition 97-01)*, Washington, DC 20554.

Flyckt, V.M.M., Raaymakers, B.W., Kroeze, H., and Lagendijk, J.J.W., 2007, Calculation of SAR and temperature rise in a high-resolution vascularized model of the human eye and orbit when exposed to a dipole antenna at 900, 1500 and 1800 MHz. *Phys Med Biol* 52: 2691–2701.

Fujimoto, M., Hirata, A., Wang, J., Fujiwara, O., and Shiozawa, T., 2006, FDTD-derived correlation of maximum temperature increase and peak SAR in child and adult head models due to dipole antenna. *IEEE Trans Electromagn Compat* 48: 240–247.

Gabriel, C., 1996, Compilation of the dielectric properties of body tissues at RF and microwave frequencies. Brooks AFB, TX, Tech. Rep., AL/OE-TR-1996-0037.

Gabriel, S., Lau, R.W., and Gabriel, C., 1996, The dielectric properties of biological tissues: III. Parametric models for the dielectric spectrum of tissues. *Phys Med Biol* 41: 2271–2293.

Gandhi, O.P., and Chen, J.Y., 1995, Electromagnetic absorption in the human head from experimental 6-GHz handheld transceivers. *IEEE Trans Electromagn Compat* 37: 547–558.

Gandhi, O.P., and Kang, G., 2002, Some present problems and a proposed experimental phantom for SAR compliance testing of cellular telephones at 835 and 1900 MHz. *Phys Med Biol* 47: 1501–1518.

Gandhi, O.P., and Kang, G., 2004, Inaccuracies of a plastic "pinna" SAM for SAR testing of cellular telephones against IEEE and ICNIRP safety guidelines. *IEEE Trans Microw Theory Tech* 52: 2004–2012.

Gandhi, O.P., Gu, Y., Chen, J.Y., Bassen, H.I., 1992, Specific absorption rates and induced current distributions in an anatomically based human model for plane wave exposures. *Health Phys* 63(3): 281–290.

Gandhi, O.P., Lazzi, G., and Furse, C.M., 1996, Electromagnetic absorption in the human head and neck for mobile telephones at 835 and 1900 MHz. *IEEE Trans Microw Theory Tech* 44: 1884–1897.

Gandhi, O.P., Lazzi, G., Tinniswood, A., and Yu, Q.S., 1999. Comparison of numerical and experimental methods for determination of SAR and radiation patterns of handheld wireless telephones. *Bioelectromagnetics* 20: 93–101.

Gandhi, O.P., Li, Q.X., and Kang, G., 2001. Temperature rise for the human head for cellular telephones and for peak SARs prescribed in safety guidelines. *IEEE Trans Microw Theory Tech* 49(9): 1607–1613.

Gedney, S.D., 1996, An anisotropic perfectly matched layer-absorbing medium for the truncation of FDTD lattices. *IEEE Trans Antennas Propag* 44: 1630–1639.

Gjonaj, E., Bartsch, M., Clemens, M., Schupp, S., and Weiland, T., 2002, High-resolution human anatomy models for advanced electromagnetic field computations. *IEEE Trans Magn* 38: 357–360.

Guy, A.W., Lin, J.C., Kramar, P.O., and Emery, A.F., 1975, Effect of 2450 MHz radiation on the rabbit eye. *IEEE Trans Microw Theory Tech* 23: 492–498.

Guyton, A.C., 1991, *Textbook of Medical Physiology*. WB Saunders, Philadelphia, PA.

Hadjem, A., Lautru, D., Dale, C., Wong, M.F., Hanna, V.F., and Wiart, J., 2005, Study of specific absorption rate (SAR) induced in two child head models and in adult heads using mobile phones. *IEEE Trans Microw Theory Tech* 53: 4–11.

Hirata, A., 2005, Temperature increase in human eyes due to near-field and far-field exposures at 900 MHz, 1.5 GHz, and 1.9 GHz. *IEEE Trans Electromagn Compat* 47(1): 68–76.

Hirata, A., and Shiozawa, T., 2003, Correlation of maximum temperature increase and peak SAR on the human head due to handset antennas. *IEEE Trans Microw Theory Tech* 51(7): 1834–1841.

Hirata, A., Matsuyama, S., and Shiozawa, T., 2000, Temperature rises in the human eye exposed to EM waves in the frequency range 0.6–6 GHz. *IEEE Trans Microw Theory Tech* 42: 386–393.

Hirata, A., Fujimoto, M., Asano, T., Wang, J., Fujiwara, O., and Shiozawa, T., 2006a, Correlation between maximum temperature increase and peak SAR with different average schemes and masses. *IEEE Trans Electromagn Compat* 48: 569–578.

Hirata, A., Fujiwara, O., and Shiozawa, T., 2006b, Correlation between peak spatial-average SAR and temperature increase due to antennas attached to human trunk. *IEEE Trans Biomed Eng* 53: 1658–1664.

Hombach, V., Meier, K., Burkhardt, M., Kuhn, E., and Kuster, N., 1996, The dependence of EM energy absorption upon human head modeling at 900 MHz. *IEEE Trans Microw Theory Tech* 44: 1865–1873.

ICNIRP, 1998, Guidelines for limiting exposure to time-varying electric, magnetic, and electromagnetic fields (up to 300 GHz). *Health Phys* 74: 494–522.

ICNIRP, 2008, ICNIRP statement on EMF-emitting new technologies. *Health Phys* 94(4): 376–392.

IEC 62209-2, 2007, Human exposure to radio frequency fields from hand-held and body-mounted wireless communication devices – human models, instrumentation, and procedures – Part 2: Procedure to determine the specific absorption rate (SAR) for mobile wireless communication devices used in close proximity to the human body (frequency range of 30 MHz to 6 GHz), Draft Version IEC Technical Committee 106 Geneva, Switzerland.

IEC Int. Std. 62209-1, 2005, Human exposure to radio frequency fields from hand-held and body-mounted wireless communication devices – Human models, instrumentations, and procedures – Part 1:

Procedure to determine the specific absorption rate (SAR) for hand-held devices used in close proximity to the ear (frequency range of 300 MHz to 3 GHz).

IEEE, 1991, Standard for safety levels with respect to human exposure to radio frequency electromagnetic fields, 3 kHz to 300 GHz. *IEEE Standard C95.1*. Institute of Electrical and Electronics Engineers, New York, NY.

IEEE, 1999, IEEE Standard for safety levels with respect to human exposure to radio frequency electromagnetic fields, 3 kHz to 300 GHz. *IEEE Std C95.1*, 1999 Ed. Institute of Electrical and Electronics Engineers, New York, NY.

IEEE, 2002, IEEE recommended practice for measurements and computations of radio frequency electromagnetic fields with respect to human exposure to such Fields, 100 kHz–300 GHz. *IEEE Std C95.3-2002*.

IEEE, 2003, IEEE recommended practice for determining the peak spatial-average specific absorption rate (SAR) in the human head from wireless communications devices: measurement techniques. *IEEE Standard 1528 SCC34*. IEEE, New York, NY.

IEEE, 2005, IEEE standard for safety levels with respect to human exposure to radio frequency electromagnetic fields, 3 kHz to 300 GHz. *IEEE Std C95.1-2005*. IEEE, New York, NY.

Jensen, M.A., and Rahmat-Samii, Y., 1994, Performance analysis of antennas for hand-held transceiver using FDTD. *IEEE Trans Antennas Propag* 42: 1106–1113.

Jensen, M.A., and Rahmat-Samii, Y., 1995, EM interaction of handset antennas and a human in personal communications. *Proc IEEE* 83(1): 7–17.

Kainz, W., Christ, A., Kellom, T., Seidman, S., Nikoloski, N., Beard, B., and Kuster, N., 2005, Dosimetric comparison of the specific anthropomorphic mannequin (SAM) to 14 anatomical head models using a novel definition for the mobile phone positioning. *Phys Med Biol* 50: 14, 3423–3445.

Kanda, M., Balzano, Q., Russo, P., Faraone, A., and Bit-Babik, G., 2002, Effects of ear-connection modeling on the electromagnetic-energy absorption in a human-head phantom exposed to a dipole antenna field at 835 MHz. *IEEE Trans Electromagn Compat* 44: 4–10.

Kang, G., and Gandhi, O.P., 2002, SARs for pocket-mounted mobile telephones at 835 and 1900 MHz. *Phys Med Biol* 47(23): 4301–4313.

Keshvari, J., and Lang, S., 2005, Comparison of radio frequency energy absorption in ear and eye region of children and adults at 900, 1800 and 2450 MHz. *Phys Med Biol* 50: 4355–4369.

Keshvari, J., Keshvari, R., and Lang, S., 2006, The effect of increase in dielectric values on specific absorption rate (SAR) on eye and head tissues following 900, 1800 and 2450 MHZ radio frequency (RF) exposure. *Phys Med Biol* 51: 1463–1477.

Koulouridis, S., and Nikita, K.S., 2004, Study of the coupling between human heads and cellular phone helical antennas. *IEEE Trans Electromagn Compat* 46: 62–70.

Kramar, P.O., Emery, A.F., Guy, A.W., and Lin, J.C., 1975, The ocular effects of microwaves on hypothermic rabbits: A study of microwave cataractogenic mechanisms. Ann N Y Acad Sci 247: 155–165.

Kunz, K.S., and Luebbers, R.J., 1993, *The Finite-Difference Time-Domain Method for Electromagnetics*. CRC, Boca Raton, FL.

Kuster, N., 1992, Multiple multipole method applied to an exposure safety study. *Appl Comput Electromag Soc J* 7: 43–60.

Kuster, N., and Balzano, Q., 1992, Energy absorption mechanism by biological bodies in the near field of dipole antennas above 300 MHz. *IEEE Trans Veh Technol* 1: 17–23.

Lazzi, G., and Gandhi, O.M., 1997, Realistically tilted and truncated anatomically based models of the human head for dosimetry of mobile telephones. *IEEE Trans Electromagn Compat* 39: 55–61.

Lazzi, G., and Gandhi, O.P., 1998, On modelling and personal dosimetry of cellular telephone helical antennas with the FDTD code. *IEEE Trans Antennas Propag* 46(4): 525–530.

Li, L.W., Leong, M.S., Kooi, P.S., and Yeo, T.S., 2000, Specific absorption rates in human head due to handset antennas: A comparative study using FDTD method. *J Electromag Waves Appl* 14: 987–1000.

Lin, J.C., 1986, Computer methods for field intensity predictions, In Polk, C., and Postow, E., Eds., *CRC Handbook of Biological Effects of Electromagnetic Fields*. CRC, Boca Raton, FL, pp. 273–313.

Lin, J.C., 2000a, Mechanisms of field coupling into biological systems at ELF and RF frequencies. In: Lin, J.C., Ed., *Advances in Electromagnetic Fields in Living Systems*, Vol. 3. Kluwer, New York, NY, pp. 1–38.

Lin, J.C., 2000b, Specific absorption rates (SARs) induced in head tissues by microwave radiation from cell phones. *IEEE Antennas Propag Mag* 42(5): 138–140.

Lin, J.C., 2003, Risk to children from cellular telephone radiation (Health Effects Column). *IEEE Microw Mag* 4(1): 20–26.

Lin, J.C., 2006, The new IEEE standard for human exposure to radio frequency radiation and the current ICNIRP guidelines. *URSI Radio Sci Bull* 317: 61-63.

Lin, J.C., 2007, Dosimetric comparison between different possible quantities for limiting exposure in the RF band: Rationale for the basic one and implications for guidelines. *Health Phys* 92(6): 547–553.

Lin, J.C., and Bernardi, P., 2007, Computer methods for predicting field intensity and temperature change. In: Barnes, F., and Greenebaum, B., Eds., *Bioengineering and Biophysical Aspects of Electromagnetic Fields*, Chapter 10. CRC, Boca Raton, FL, pp. 293–380.

Mangoud, M.A., Abd-Alhameed, R.A., and Excell, P.S., 2000, Simulation of human interaction with mobile telephones using hybrid techniques over coupled domains. *IEEE Trans Microw Theory Tech* 48: 2014–2021.

Martens, L., De Moerloose, J., De Zutter, D., De Poorter, J., and De Wagter, C., 1995, Calculation of the electromagnetic fields induced in the head of an operator of a cordless telephone. *Radio Sci* 30: 283–290.

Martinez-Burdalo, M., Martin, A., Anguiano, M., and Villar, R., 2004, Comparison of FDTD-calculated specific absorption rate in adults and children when using a mobile phone at 900 and 1800 MHz. *Phys Med Biol* 49: 345–354.

Mason, P.A., Hurt, W.D., Walters, T.J., D'Andrea, J.A., Gajsek, P., Ryan, K.L., Nelson D.A., Smith, K.I., and Ziriax, J.M., 2000, Effects of frequency, permittivity, and voxel size on predicted specific absorption rate values in biological tissue during electromagnetic-field exposure. *IEEE Trans Microw Theory Tech* 48: 2050–2058.

Mazzurana, M., Sandrini, L., Vaccari, A., Malacarne, C., Cristoforetti, L., and Pontalti, R., 2003, A semi-automatic method for developing an anthropomorphic numerical model of dielectric anatomy by MRI. *Phys Med Biol* 48: 3157–3170.

Meier, K., Hombach, V., Kastle, R., Tay, R.Y.S., and Kuster, N., 1997, The dependence of electromagnetic energy absorption upon human-head modeling at 1800 MHz. *IEEE Trans Microw Theory Tech* 45: 2058–2062.

Mur, G., 1981, Absorbing boundary conditions for the finite-difference approximation of the time-domain electromagnetic-field equations. *IEEE Trans Electromagn Compat* EMC-23(4): 377–382.

Nagaoka, T., Watanabe, S., Sakurai, K., Kunieda, E., Watanabe, S., Taki, M., and Yamanaka, Y., 2004, Development of realistic high-resolution whole-body voxel models of Japanese adult males and females of average height and weight, and application of models to radio-frequency electromagnetic-field dosimetry. *Phys Med Biol* 49: 1–15.

Nikita, K.S., Cavagnaro, M., Bernardi, P., Uzunoglu, N.K., Pisa, S., Piuzzi, E., Sahalos, J.N., Krikelas, G.I., Vaul, J.A., Excell, P.S., Cerri, G., Chiarandini, S., De Leo, R., and Russo, P., 2000, A study of uncertainties in modeling antenna performance and power absorption in the head of a cellular phone user. *IEEE Trans Microw Theory Tech* 48(12): 2676–2685.

Okoniewski, M., and Stuchly, M.A., 1996, A study of the handset antenna and human body interaction. *IEEE Trans MTT* 44(10): 1855–1864.

Onishi, T., Iyama, T., and Uebayashi, S., 2005, The estimation of the maximum SAR with respect to various types of wireless device usage. *Procedings of EMC Zurich.* pp. 151–154.

Ozisik, N., 1985, *Heat Transfer: A Basic Approach.* Mc Graw Hill, New York, NY.

Pennes, H.H., 1948. Analysis of tissue and arterial blood temperatures in resting forearm. *J Appl Physiol* 1: 93–122.

Pisa, S., Cavagnaro, M., Piuzzi, E., Bernardi, P., and Lin, J.C., 2003, Power density and temperature distributions produced by interstitial arrays of sleeved-slot antennas for hyperthermic cancer therapy. *IEEE Trans Microwave Theory Tech* 51: 2418–2426.

Pisa, S., Cavagnaro, M., Lopresto, V., Piuzzi, E., Lovisolo, G.A., and Bernardi, P., 2005a, A procedure to develop realistic numerical models of cellular phones for an accurate evaluation of SAR distribution in the human head. *IEEE Trans Microw Theory Tech* 53(4): 1256–1265.

Pisa, S., Cavagnaro, M., Piuzzi, E., and Lopresto, V., 2005b, Numerical–experimental validation of A GM-FDTD code for the study of cellular phones. *Microw Opt Technol Letts* 47(4): 396–400.

Pisa, S., Cavagnaro, M., Lopresto, V., Piuzzi, E., Lovisolo, G.A., and Bernardi, P., 2006, A numerical model of a commercial dual-band cellular phone equipped with a planar antenna. In: *Proceedings of EMC Europe 2006 Barcelona*, Barcelona, Spain, September 2006, pp. 586–591.

Riu, P.J., and Foster, K.R., 1999, Heating of tissue by near-field exposure to a dipole: A model analysis. *IEEE Trans Biomed Eng* 46(8): 911–917.

Rowley, J.T., and Waterhouse, R.B., 1999, Performance of shorted patch antennas for mobile communication handsets at 1800 MHz. *IEEE Trans. Microwave Theory Tech* 47: 815–822.

Rowley, J.T., Waterhouse, R.B., and Joyner, K.H., 2002, Modeling of normal-mode helical antennas at 900 MHz and 1.8 GHz for mobile communication handsets using the FDTD technique. *IEEE Trans Antennas Propag* 50: 812–820.

Samaras, T., Kalampaliki, E., and Sahalos, J.N., 2007, Influence of thermophysiological parameters on the calculations of temperature rise in the head of mobile phone users. *IEEE Trans Electromagn Compat* 49: 936–939.

Schiavoni, A., Bertotto, P., Richiardi, G., and Bielli, P., 2000, SAR generated by commercial cellular phones-phone modeling, head modeling, and measurements. *IEEE Trans Microw Theory Tech* 48(11, Part 2): 2064–2071.

Schonborn, F., Burkhardt, M., and Kuster, N., 1998, Differences in energy absorption between heads of adults and children in the near field of sources. *Health Phys* 74: 160–168.

Stevens, N., and Martens, L., 2000, Comparison of averaging procedures for SAR distributions at 900 and 1800 MHz. *IEEE Trans Microw Theory Tech* 48(11): 2180–2184.

Stuchly, M.A., Kraszewski, A., Stuchly, S.S., Hartsgrove, G.W., and Spiegel, R.J., 1987, RF energy deposition in a heterogeneous model of man: Near field exposure. *IEEE Trans Biomed Eng* 34: 944–950.

Taflove, A., and Hagness, S.C., 2000, *Computational Electrodynamics: The Finite-Difference Time-Domain Method.* Artech House, Boston, MA.

Tinniswood, A.D., Furse, C.M., and Gandhi, O.P., 1998a, Computations of SAR distributions for two anatomically based models of the human head using CAD files of commercial telephones and the parallelized FDTD code. *IEEE Trans Antennas Propag* 46: 829–833.

Tinniswood, A.D., Furse, C.M., and Gandhi, O.P., 1998b, Power deposition in the head and neck of an anatomically-based human body model for plane wave exposures. *Phys Med Biol* 43: 2361–2378.

Toftgard, J., Hornsleth, S.N., and Bach Andersen, J., 1993. Effects on portable antennas of the presence of a person. *IEEE Trans Antennas Propag* 41(6): 739–746.

Troulis, S.E., Scanlon, W.G., and Evans, N.E., 2003, Effect of a hands-free wire on specific absorption rate for a waist-mounted 1.8 GHz cellular telephone handset. *Phys Med Biol* 48(12): 1675–1684.

US FCC (Federal Communications Commission), 2001. Evaluating compliance with FCC guidelines for human exposure to radiofrequency electromagnetic fields. *Supplement C (Edition 01–01) to OET Bulletin 65 (Edition 97–01)*, Washington, DC 20554.

Van de Kamer J.B., and Lagendijk, J.J.W., 2002, Computation of high-resolution SAR distributions in a head due to a radiating dipole antenna representing a hand-held mobile phone. *Phys Med Biol* 47: 1827–1835.

Van Leeuwen, G.M.J., Lagendijk, J.J.W., Van Leersum, B.J.A.M., Zwamborn, A.P.M., Hornsleth, S.N., and Kotte, A.N.T.J., 1999, Calculation of change in brain temperatures due to exposure to a mobile phone. *Phys Med Biol* 44: 2367–2379.

Virtanen, H., Huttunen, J., Toropainen, A., and Lappalainen, R., 2005, Interaction of mobile phones with superficial passive metallic implants. *Phys Med Biol* 50: 2689–2700.

Virtanen, H., Keshvari, J., and Lappalainen, R., 2006, Review interaction of radio frequency electromagnetic fields and passive metallic implants: A brief review. *Bioelectromagnetics* 27: 431–439.

Virtanen, H., Keshvari, J., and Lappalainen, R., 2007, The effect of authentic metallic implants on the SAR distribution of the head exposed to 900, 1800 and 2450 MHz dipole near field. *Phys Med Biol* 52(5): 1221–1236.

Wainwright, P., 2000, Thermal effects of radiation from cellular telephones. *Phys Med Biol* 45: 2363–2372.

Wainwright, P.R., 2007, Computational modelling of temperature rises in the eye in the near field of radiofrequency sources at 380, 900 and 1800 MHz. *Phys Med Biol* 52: 3335–3350.

Wang, J., and Fujiwara, O., 1999, FDTD computation of temperature rise in the human head for portable telephones. *IEEE Trans Microw Theory Tech* 47: 1528–1534.

Wang, J., and Fujiwara, O., 2003, Comparison and evaluation of electromagnetic absorption characteristics in realistic human head models of adult and children for 900-MHz mobile telephones. *IEEE Trans Microw Theory Tech* 51: 966–971.

Wang, J., Fujiwara, O., Watanabe, S., and Yamanaka, Y., 2004, Computation with a parallel FDTD system of human-body effect on electromagnetic absorption for portable telephones. *IEEE Trans Microw Theory Tech* 52: 53–58.

Wang, J., Fujiwara, O., and Watanabe, S., 2006, Approximation of aging effect on dielectric tissue properties for SAR assessment of mobile telephones. *IEEE Trans Electromagn Compat* 48: 408–413.

Watanabe, S.I., Taki, H., Nojima, T., and Fujiwara, O., 1996, Characteristics of the SAR distributions in a head exposed to electromagnetic fields radiated by a hand-held portable radio. *IEEE Trans Microw Theory Tech* 44(10): 1874–1883.

Whittow, W.G., and Edwards, R.M., 2004, A study of changes to specific absorption rates in the human eye close to perfectly conducting spectacles within the radio frequency range 1.5 to 3.0 GHz. *IEEE Trans Antennas Propag* 52(12): 3207–3212.

Yee, K.S., 1966, Numerical solutions of initial boundary value problems involving Maxwell's equations in isotropic media. *IEEE Trans Antennas Propag* 14: 302–307.

Zubal, I.G., Harrel, C.R., Smith, E.O., Rattner, Z., Gindi, G., and Hoffer, P.B., 1994, Computerized three-dimensional segmented human anatomy. *Med Phys* 21: 299–302.

Index

MIX
Papier aus verantwortungsvollen Quellen
Paper from responsible sources
FSC® C105338

If you have any concerns about our products,
you can contact us on
ProductSafety@springernature.com

In case Publisher is established outside the EU,
the EU authorized representative is:
Springer Nature Customer Service Center GmbH
Europaplatz 3, 69115 Heidelberg, Germany

Printed by Libri Plureos GmbH
in Hamburg, Germany